Solid Mechanics and Its Applications

Volume 252

Aims and Scope of the Series

The fundamental questions arising in mechanics are: *Why?*, *How?*, and *How much?* The aim of this series is to provide lucid accounts written by authoritative researchers giving vision and insight in answering these questions on the subject of mechanics as it relates to solids.

The scope of the series covers the entire spectrum of solid mechanics. Thus it includes the foundation of mechanics; variational formulations; computational mechanics; statics, kinematics and dynamics of rigid and elastic bodies: vibrations of solids and structures; dynamical systems and chaos; the theories of elasticity, plasticity and viscoelasticity; composite materials; rods, beams, shells and membranes; structural control and stability; soils, rocks and geomechanics; fracture; tribology; experimental mechanics; biomechanics and machine design.

The median level of presentation is to the first year graduate student. Some texts are monographs defining the current state of the field; others are accessible to final year undergraduates; but essentially the emphasis is on readability and clarity.

More information about this series at http://www.springer.com/series/6557

Ebrahim Esmailzadeh · Davood Younesian
Hassan Askari

Analytical Methods in Nonlinear Oscillations

Approaches and Applications

Springer

Ebrahim Esmailzadeh
Faculty of Engineering and Applied Science
University of Ontario Institute of Technology
Oshawa, ON
Canada

Hassan Askari
Department of Mechanical and Mechatronics Engineering
University of Waterloo
Waterloo, ON
Canada

Davood Younesian
School of Railway Engineering
Iran University of Science and Technology
Tehran
Iran

ISSN 0925-0042 ISSN 2214-7764 (electronic)
Solid Mechanics and Its Applications
ISBN 978-94-024-1540-7 ISBN 978-94-024-1542-1 (eBook)
https://doi.org/10.1007/978-94-024-1542-1

Library of Congress Control Number: 2018941529

Printed on acid-free paper

This Springer imprint is published by the registered company Springer Nature B.V.
The registered company address is: Van Godewijckstraat 30, 3311 GX Dordrecht, The Netherlands

Dedicated to my wife, Rouhangiz and my parents Mohammad and Fakhri

Ebrahim Esmailzadeh

Dedicated to my wife, Seyedeh Shirin and my parents Mohammad Hassan and Azam Sadat

Davood Younesian

Dedicated to my wife, Rosa and my parents Eiraj and Asieh

Hassan Askari

Preface

This book is intended to bring together a scientific platform that covers very recent analytical methods in nonlinear systems, which mainly developed during the past few decades. The scope of this book is to serve researchers and graduate students, who are pursuing research studies on nonlinear oscillations, analytical dynamics, and applied mathematics. It may also be considered as a broad reference source for those seeking advanced knowledge in the area of nonlinear differential equations and systems. Moreover, the detailed analyses and results presented here are an essential source for modern courses in the vibration of micro- and nanosystems.

Over five decades of learning and teaching courses and research on nonlinear vibration, advanced control and dynamics of systems are behind the development of the present book. The fascinating world of nonlinear vibration was first introduced to me at the University of London in 1966, when late Professor E. J. Le Fevre said, *Ebrahim*, take *Nonlinear Vibration and Nonlinear Control as elective courses in your final-year of undergraduate program.* I was very happy to take these courses, since they opened a new horizon to the real-world systems and processes. There were not that many books, mostly English translation of Russian ones. "Nicolai Minorsky, *Nonlinear Oscillations*, 1962", "James J. Stoker, *Nonlinear Vibrations*, 1950", "Stephen P. Timoshenko, *Vibration Problems in Engineering*, 1955", "Norman W. McLachlan, *Theory and Application of Mathieu Functions*, 1947", and "Jacob P. Den Hartog, *Mechanical Vibrations*, 1960", to name a few. Most of my learning was utilized in my PhD and subsequently, I taught "*Nonlinear Dynamic Systems*" course for the first time at Arya-Mehr University of Technology in 1973. My appointment at Massachusetts Institute of Technology in 1976 to teach Dynamics guided by late Professor Stephen H. Crandall and Vibrations supervised by late Professor Den Hartog enriched my knowledge. Acquaintance with them encouraged me to pursue teaching and research in areas of nonlinear vibrations and control. Many of my students are now engaged with teaching courses and doing research in nonlinear dynamic systems, among them are my co-authors, Davood Younesian (Ph.D. 2005) and Hassan Askari (M.A.Sc. 2014).

The material presented in every chapter is intended to be in-depth and self-contained to ease the understanding of the subject. The contents of all chapters are cohesively divided into two separate main parts, first analytical approaches and then followed by their applications. Initially, the main concept of all analytical methods as well as their highlights and recent developments are brought to the reader's attention. Subsequently, the practical aspects and engineering applications of the analytical methods are covered in the second part, together with a wide range of solved problems.

Chapter 1 presents an introduction to nonlinear dynamical systems, categorized into different types and input sources. A detailed historical review of nonlinear oscillations with highlights of the main contributions since early 1990 is presented. Parametric oscillation and several well-known types of linear and nonlinear oscillators are discussed. The concept of resonance occurring in nonlinear systems is examined. Real-world modeling of nonlinear systems in different disciplines, namely, global warming, chemical reactions, sociology, structural vibrations, epidemic diseases, and ecology, are developed.

Several classical methods for solving nonlinear oscillatory systems and different types of singular points and perturbation methods are discussed in Chap. 2. The concept of parametric excitation and the well-known Hill and Mathieu equations are presented in detail. For better understanding of those methods, more than 15 different examples were analyzed and their results are fully discussed.

Most semi-analytical techniques, based on the energy balance method and the concept of the energy balance method with its modified versions, are discussed in Chap. 3. The Hamiltonian approach is comprehensively analyzed and the rational energy balance method, as a potent technique for solving nonlinear oscillators, is presented. These techniques were utilized for solving large number of nonlinear differential equations with applications in nanosystems, structural vibration, and fluid sloshing.

Chapter 4 is focused on the residual-based methods to analyze nonlinear oscillations. Detailed discussions on the max-min analytical methods with their various types and modifications, known as Chinese ancient mathematics, are presented. The frequency–amplitude formulation and its modified versions for solving conservative nonlinear oscillators are explained in detail.

The fundamental theory and applications of variational principle with several higher order approximations are investigated in Chap. 5. Direct relationship between the Hamiltonian and variational approaches is mathematically discussed in detail. Potential applications of these techniques in studying rational and irrational oscillators, nonlinear oscillators with either discontinuity or non-integer fractional characteristic equation, and the rational elastic characteristic equation were examined. Detailed analyses of Duffing equation, nonlinear Schrodinger equation, Lane–Emden equation, Thomas–Fermi equation, heat conduction equation, and Duffing–Harmonic equation were carried out.

The integral-based approach and three well-known semi-analytical techniques, namely, the Adomian decomposition method, the homotopy analysis method, and the variational iteration method, are presented in Chap. 6. Historical development and application of different types of well-known nonlinear differential equations, including the Volterra integro-differential equation, the nonlinear Schrödinger equation, and the van der Pol and Korteweg-de Vries equation are fully discussed.

Existence of nonlinearities in small-size structures, being the first attempt to show the occurrence of nonlinear behavior in nano- and microstructures, is described in Chap. 7. A mathematical model of micro- and nanosystems, based on the Duffing equation, is derived and modeling examples of nanoresonators for mass detection, protein microtubules vibration, and nonlinear vibration of carbon nanotube resonators are analyzed. Further analyses on systems having their governing equations of motion comprised of either self-excitation terms or parametrically excited terms are performed. Nonlinearities that appeared in the coupled nonlinear micro- and nanoscaled systems are fully discussed and possibilities of quadratic nonlinearities, reported in nanoscale systems, are examined. This chapter serves as a platform for analyzing and categorizing the nonlinear behavior of low-dimensional systems and nanostructures.

Attempt is made to present performance sensitivity analyses, as well as the error evaluation study for different solution methods in every chapter. Several miscellaneous problems are given at the end of each chapter to gain a better insight into theory and solution methods. They are selected from recently published articles in refereed journals and conference proceedings within the realm of applied mathematics, nonlinear vibration, and nanosystems. These practice problems could be considered as course assignments and research topics in the areas of nonlinear vibrations, applied mathematics, and structural dynamics.

It has been gratifying to work with the staff of Springer through the development of this book. We are indebted to the Springer team Ms. Nathalie Jacobs, Ms. Cynthia Feenstra, Mr. Albert Paap, and our toughest editors, an enthusiastic, professional, and supportive team, especially, Ms. Almitra Ghosh, Ms. Nishanthi Venkatesan, and Ms. Divya Prabha Karthikesan, who made all efforts to make the book up to the highest standards, concise, and easily readable. We also want to thank all those specialists whose names we do not know, but nevertheless appreciate their efforts very much.

Last but not least, we wish to thank our families: without their patience, encouragement, and unconditional support, this book might never have been completed.

Oshawa, ON, Canada — Ebrahim Esmailzadeh
Tehran, Iran — Davood Younesian
Waterloo, ON, Canada — Hassan Askari

Contents

Chapter 1
Introduction

Abstract This chapter outlines an introduction to nonlinear oscillations with a historical overview of this field considering three different time periods since 1900. Real-world modelling of nonlinear systems in different disciplines, namely, the global warming, chemical reactions, sociology, structural vibrations, epidemic diseases and ecology, are discussed. Several well-known types of linear and non-linear oscillators are briefly discussed. Parametric oscillation is introduced and the background, terminology, applications and examples of these oscillators are presented. The phenomenon of resonance occurring in nonlinear systems is briefly discussed in the last section. The main outline of the book is also presented in this chapter.

Keywords Nonlinear oscillation · Historical background · Real-world applications · Resonance phenomenon · Parametric oscillations

Nonlinear oscillation is a hot topic and involves a number of systems with applications in automotive, sensing, fluid-solid interaction, aerospace, micro- and nano-scale, and bioengineering. In view of that, it is vital to have knowledge about nonlinear phenomena and be familiar with theoretical and practical approaches to understand them. In this chapter, we first introduce the concept of nonlinearity and also provide a brief review on nonlinear oscillations history. Section 2 provides an overview of the organization of the book. In Sect. 3, conservative and non-conservative oscillatory systems are illustrated. Furthermore, the unique and rich behavior of parametrical excited oscillations is fully presented. Finally, the concept behind the primary and combinatory resonances is fully discussed.

E. Esmailzadeh et al., *Analytical Methods in Nonlinear Oscillations*, Solid Mechanics and Its Applications 252, https://doi.org/10.1007/978-94-024-1542-1_1

1.1 Why Nonlinear Oscillations?

Linear oscillations theory has been successfully applied by many researchers for analyzing of oscillatory devices. In addition, it shows that it can be effectively employed in industrial applications. Hence, it can be a potential question for anyone as why so many researchers from different disciplines have become intrigued by the theory of nonlinear oscillations. In fact, there could be many reasons for the fascination of the nonlinear oscillation theory, for example, developing new devices in micro- and nano-scales; analyzing real world cases with consideration of nonlinearity to have a better insight into the oscillatory devices; having uncertainties in the model parameters; and also improving the design of nonlinear systems.

1.2 Brief Review of Nonlinear Oscillations History

Nonlinear behavior emerges in plenty of real world phenomena. As a result, researchers from different fields explore nonlinear systems. The beauty of nonlinear dynamics may show why many researchers become interested in the analysis of nonlinear systems [1].

Analysis of dynamical systems began in mid-1600s, when Newton introduced differential equations. In the road of development of the nonlinear dynamics, Poincare has had a crucial role by employing qualitative analysis for dynamical systems in the last days of 1800s. His work was in fact a breakthrough in the area of nonlinear systems and results in the first glimpse from chaos. The topic of nonlinear oscillations is certainly one of the cornerstones of dynamics. In the early years of 1900s, nonlinear oscillations and their applications in physics and engineering were the popular topic among researchers. During these years, pioneering scientists, namely van der Pol, Duffing, Cartwright, Levinson, Littlewood, Bogoliubov, Krylov, Levenson, Minorsky, Vitt, Andronov, Birkhoff and Kolomogonov have made a breakthrough in the area of nonlinear oscillatory systems. Table 1.1 briefly describes their contributions in the above-mentioned area.

As depicted in Table 1.1, Russian scientists were of major contributors in the area of nonlinear oscillatory systems during the early years of 1900s [13]. Development of high speed computers after 1950s was indeed a milestone in the field of nonlinear oscillations. Following the pioneering works in the area of nonlinear oscillatory systems and using new invented computers engender many theoretical and practical intuition about nonlinear systems [1].

From the 1950 to 1955, Hayashi published a few papers and studied subharmonic, forced oscillations and stability of nonlinear systems [14, 15–16]. During 1950s and 1960s, a few researchers focused on nonlinear oscillations in plasma physics [17, 18]. During these years, Crandall was one of the most active scientists in the field of nonlinear vibrations who made significant contribution in random vibrations and also implementation of perturbation theory [19]. Scientists who have

Table 1.1 Important contribution in the field of nonlinear oscillatory systems (early 1900s)

Renowned Scientists	Reported contributions in nonlinear oscillations
B. van der Pol	Introduction of relaxation-oscillations [2]
G. Duffing	Observation of cubic nonlinearity [3]
M. Cartwright and J. E. Littlewood	Relaxation-oscillations and the topological approach for solving of nonlinear problems [4]
N. N. Bogoliubov and N. M. Krylov	One of the first educational tools in the field of nonlinear mechanics and also developed the asymptotic methods in nonlinear mechanics [5]
A. A. Andronov and A. A. Vitt	Self-excited oscillations and one of the first educational tools in the field of nonlinear oscillations [6]
A. Kolmogorov	Nonlinear diffusion equation [7]
G. D. Birkhoff	Dynamical systems with two-degree-of-freedom [8]
N. Levinson	Transformation theory utilized for nonlinear equations [9]
N. Minorsky	Parametric excitation [10]
M. E. Levenson	Analyzed the Duffing equation [11]
J. A. Shohat	Studied the van der Pol equation [12]

made major contributions in nonlinear oscillations with special focus on the structural mechanics during the second half of the 20th century include T. Yamamoto, Y. Ishida, A. H. Nayfeh, D. T. Mook, R. Rand, F. C. Moon, E. H. Dowell, among others. It is worth to mention that Lorenz's discovery about chaos was a breakthrough in the field of nonlinear dynamics that had significant influence on many research works in the area of nonlinear oscillatory systems.

Table 1.2 introduces some main contributions of the above-mentioned scientists in the area of nonlinear oscillatory systems.

Table 1.2 Important contributions in the area of nonlinear oscillatory systems (1950–2000)

Scientist	Major contributions in nonlinear oscillations
A. H. Nayfeh	Enriching field of nonlinear vibrations with writing more than 5 highly cited books and 400 papers during 1970–2000 [20–24]
T. Yamamoto	He published more than 130 papers and a number of books in the field of nonlinear vibration with special focus on rotor dynamics [25]
Y. Ishida	The main contribution of Ishida is nonlinear vibration, vibrations suppression and rotor dynamics [25]
R. Rand	High influence in the field of nonlinear oscillations with publishing more than 150 papers about vibrations of beam, van der Pol oscillators, parametric excitation and chaos [26–28]
F. C. Moon	He is one of the pioneers in the field of chaotic vibrations and explored chaos in structural systems [29, 30]
E. H. Dowell	One of the highly influential researchers in the field of nonlinear dynamics with special focus on fluid-structure interaction [31]
M. P. Païdoussis	Païdoussis is one of the most influential scientists in the field of dynamical modelling of pipes and shells [32]
R. A. Ibrahim	He is one of the pioneers in the field of fluid sloshing and nonlinear oscillations [33]

Table 1.3 Major contributions in the area of nonlinear oscillatory systems (Last two decades)

Scientist	Contribution in nonlinear oscillations
M. Amabili	An influential researcher in the area of nonlinear oscillatory systems with focus on shell structures and fluid solid interaction [34]
L. Q. Chen	An important figure in nonlinear vibration energy harvesting, and vibration of axially moving continua [35]
L. A. Bergman	Development of vibration-based energy harvesting and nonlinear structural dynamics [37]
B. Balachandran	Analysis of nonlinear phenomena in beams, milling process and different types of structures [24, 44]
R. E. Mickens	Development of new types of nonlinear differential equations arising in nonlinear oscillatory systems [39]
A. F. Vakakis	Significant contribution in the field nonlinear normal modes and its applications in vibrations [37, 45]
P. D. Spanos	Analyzing of nonlinear and random vibrations is the main focus of his research [43]
E. Esmailzadeh	Analyzing of nonlinear vibrations of beam- and plate- type structures [46–48]

The above-mentioned scientists developed the major concepts behind the theory of nonlinear oscillations. Following their principal works, a number of researchers extended the field even beyond macro structures and focused on nonlinear oscillations in micro- and nano-systems. Currently, several research works were done on the fluid-structures interaction and nonlinear modelling of their nonlinear oscillatory systems. Chaotic vibration analysis of structures is another growing research topics in the analysis of nonlinear oscillatory systems.

During the last two decades, the number of research projects in the field of nonlinear oscillatory systems have been drastically increased. Researchers namely, Amabili [34], Chen [35], Kerschen [36], Bergman [37], Awrejcewicz [38], Mickens [39], Cveticanin [40], Inman [41], Balachandran [24], Esmailzadeh [42], and Spanos [43] played important roles in developing innovative works in the field of nonlinear oscillations. Table 1.3 illustrates part of their recent contributions in the area of nonlinear oscillatory systems.

1.3 Overview of the Book

This book will provide introductory materials regarding different theories applied to nonlinear oscillatory systems. A number of analytical methods and nonlinear systems will be illustrated for the purpose of providing a better understanding of the vibration behavior of nonlinear systems. The organization of this book has been presented in three main parts.

The first part encompasses Chaps. 1 and 2, which illustrates the major classical methods in the field of nonlinear oscillations. Few examples are presented on the work that based on the recently published articles in well-known international journals specifically in the field of nonlinear oscillations.

The focus of the second part of the book will be on recently proposed analytical methods for analyzing nonlinear systems. Energy balance method and its modified version are fully discussed in Chap. 3. More than 12 examples are provided in order to give a better understanding of the illustrated methods.

Chapter 4 presents the residual methods and their applications in analyzing different nonlinear dynamical systems. Several examples were considered and detailed analyses for each one of them have been presented.

Chapter 5 discusses the semi-inverse and the variational based approaches, and analyzed few examples based on recently published articles.

The integral based methods including the variational iteration, Homotopy analysis, and the Adomian decomposition methods are fully described in Chap. 6. Furthermore, attempts were made to investigate as how these methods can be applied in solving the well-known nonlinear differential equations. Also, the concept and methodology of the self-excited oscillatory systems have fully been discussed in Chap. 6.

Finally, the existence of nonlinearities in the micro- and nano-devices have been fully discussed in Chap. 7. Throughout the entire book, more than 80 examples and 90 practice problems are presented, which have been developed from an extensive search of topics recently published as journal articles, and they can be used to assist the readers for better understanding of nonlinear phenomena. Moreover, this book can be a useful research source and graduate textbook for professors and graduate students who are engaged in research on vibration of nonlinear systems and advanced mathematics.

1.4 Nonlinear Dynamical Systems

Generally, a nonlinear dynamical system is mathematically described by an ordinary differential equation with the following form:

$$\ddot{x}+f(x,\dot{x},t)=0 \tag{1.4.1}$$

where $f(x,\dot{x},t)$ is nonlinear function and dot $(.)$ indicates the derivative with respect to time. The above equation can be generalized to model dynamical systems with higher order degrees of freedom. It should be noted that many dynamical systems are described by partial differential equations, but only few of them can be converted into an ordinary differential form.

Considering the function f in Eq. (1.4.1) and its dependency on time, it can be said that there are two main categories for the nonlinear differential equations:

(a) The autonomous type, where there is no explicit relationship between f and time.

(b) The non-autonomous type, known as forced oscillation, in which time appears in the equation.

Nonlinear differential equations can be seen in different applications, such as, system vibration, bio-systems, nano-systems, fluid mechanics, material science, chemical reactions, political science, astronomy, economy, and social science. The main focus of this book is the nonlinear behavior of dynamical systems, but in this section, different nonlinear systems with their mathematical models will be presented to show the diversity of nonlinear differential equations in different fields of science and engineering.

1.4.1 Oscillation of a Pendulum

Oscillation of a pendulum about a vertical equilibrium axis is one of the classical examples of the nonlinear oscillations, and has been explored so far by many researchers. It is often used as an introductory example for illustrating a few major concepts in nonlinear oscillations. The nonlinear differential equation of a simple pendulum oscillation in the field of gravity is formulated as:

$$\ddot{x} + (g/l)\sin(x) = 0 \tag{1.4.2}$$

where $\sin(x)$ is the nonlinear term of equation. Parameters g and l represent the respective gravitational acceleration and the pendulum length, which are kept constant. As depicted from the above equation, there is not an explicit relationship between f and time, and so, it is an autonomous equation. In this book, different methods are presented, which will be capable of finding the solutions of Eq. (1.4.2).

1.4.2 Predator-Prey Dynamics

A mathematical model for predator-prey dynamics is described by the following set of equations. This set of equations has been investigated by many researchers to either find analytical solutions or perform stability analysis. The principal equation of prey-predator model is described by the following expressions:

$$\dot{V} = bV - aVP \tag{1.4.3}$$

and

$$\dot{P} = caVP - dP \tag{1.4.4}$$

The first equation shows the prey, while the second is related to the predator. Parameters a, b, c and d are the constant predation rate, density-independent exponential growth of the prey population V in the absence of predators, a constant conversion rate c of eaten prey into new predator abundance, and constant per capita mortality rate of predators, respectively. The predator-prey dynamical model has been modified by many researchers in the field of ecology for better under standing of population dynamics. There are two nonlinear terms in the above set of equations, namely, aVP and $caVP$. In addition, this is also an autonomous type equation. In Chap. 6, few potent approaches are presented, which are capable of analyzing the above set of equations.

1.4.3 Brusselator System in Chemical Reaction

The Brusselator system is a well-known theoretical model for a type of autocatalytic reaction. This model has been developed to mathematically describe the following reaction mechanism:

$$\begin{aligned} & A \rightarrow X \\ & 2X + Y \rightarrow 3X \\ & B + X \rightarrow Y + C \\ & X \rightarrow D \end{aligned} \tag{1.4.5}$$

This system is mathematically described by the following differential equations:

$$\begin{aligned} \dot{X} &= 1 - (b - 1)X + aX^2Y \\ \dot{Y} &= bX - aX^2Y \end{aligned} \tag{1.4.6}$$

The above set of nonlinear differential equations is represented in its dimensionless form. In addition, the above equation is also related to an autonomous system.

1.4.4 Mathematical Model of Crime in a Society

Nonlinear differential equations can also be used for modeling of social activities. A very interesting model in this field shows the effect of a police force in controlling of crime in a society as presented below [49]:

$$\begin{aligned}
\frac{\mathrm{d}S}{\mathrm{d}t} &= (1-\rho(P))A - \frac{\beta SC}{N} + \theta \nu R - \mathrm{d}S \\
\frac{\mathrm{d}C}{\mathrm{d}t} &= \rho(P)A + \frac{\beta SC}{N} - \frac{\gamma CP}{N} + (1-\theta)\nu R - (\alpha + d)C \\
\frac{\mathrm{d}R}{\mathrm{d}t} &= \frac{\gamma CP}{N} - \nu R - \mathrm{d}R \\
\frac{\mathrm{d}P}{\mathrm{d}t} &= \phi C - \phi_0 (P - P_0)
\end{aligned} \tag{1.4.7}$$

The above model, which was formulated by Misra [49] is an autonomous nonlinear differential equation. Parameter S shows people who have not committed any crime so far but are prone to criminality. C represents people who are involved in various criminal activities. R shows people who essentially are the criminals and residing in jails, and P is the number of individuals serving the police force at time t. Other parameters involved in the above equation can be studied in Ref. [49]. For further review of this example, few analytical methods are proposed in Chap. 6, which can be used to find a solution for the above equation.

1.4.5 *Mathematical Model for Global Warming*

Interestingly enough, nonlinear differential equation can be exploited for the modelling of CO_2 emission, considering the effect of human population and forest biomass. Misra and Verma [50], formulated a set of nonlinear differential equations that shows dynamics of atmospheric carbon dioxide CO_2 gas. Equation (1.4.8) shows their developed model, which is an autonomous type of nonlinear ordinary differential equation [50]. The stability analysis and finding a solution for this equation can be considered as a critical topic for more exploration.

$$\begin{aligned}
\frac{\mathrm{d}X}{\mathrm{d}t} &= Q_0 + \lambda N - \alpha X - \lambda_1 XF \\
\frac{\mathrm{d}N}{\mathrm{d}t} &= sN(1 - \frac{N}{L}) - \theta XN + \pi \phi NF \\
\frac{\mathrm{d}F}{\mathrm{d}t} &= uF(1 - \frac{F}{M}) - \phi NF + \pi_1 \lambda_1 XF
\end{aligned} \tag{1.4.8}$$

Parameters X, N and F show the respective amount of CO_2, human population, and the forest biomass. Interested readers can find the definitions of other parameters of Eq. (1.4.8) in Ref. [50].

To show the presence of nonlinear differential equation in bioscience, we will examine the next example that is used for the modelling of epidemic diseases, which illustrates the application of nonlinear differential equations in biosciences.

1.4.6 Mathematical Model of Epidemic Disease

The focus of this example will be on the epidemic diseases. SIR model is a standard mathematical formulation that is used to describe many epidemiological diseases. The following nonlinear differential equation illustrates the well-known SIR model:

$$\begin{aligned}\frac{\mathrm{d}S}{\mathrm{d}t} &= -\beta SI \\ \frac{\mathrm{d}I}{\mathrm{d}t} &= \beta SI - \gamma I \\ \frac{\mathrm{d}R}{\mathrm{d}t} &= \gamma I\end{aligned} \tag{1.4.9}$$

where the population of susceptible people is assumed as S, the number of people infected being I, and the number of people who have recovered and developed immunity to the infection being R. The coefficients β and γ represent the infection and the recovery rate, respectively. There are two nonlinear terms present in the above set of equations, which are SI in S and I equations. Equation (1.4.9) shows an autonomous nonlinear differential equation.

We have seen few nonlinear differential equations arising in different fields to represent a wide range of applications of nonlinearities. The remaining parts of Chap. 1 will focus only on the nonlinear differential equations derived from several mechanical systems.

1.5 Conservative Oscillatory Systems

This part of the Chapter is related to the nonlinear conservative oscillatory systems with focus on the mechanical systems. In order to fully understand the meaning of the conservative oscillations, we need to be familiar with the conservative forces and the potential energy. Mathematically, a conservative force should meet at least one of the following equivalent conditions:

(a) The work done by the force is zero when a moving particle through a trajectory that starts and terminates at the same place. Accordingly, we can have the following mathematical definition:

$$W \equiv \oint_c \vec{F} \cdot \mathrm{d}r = 0 \tag{1.5.1}$$

(b) The curl of a force is a zero vector, so, we can have the following mathematical formulation:

$$\nabla \times F = 0 \tag{1.5.2}$$

(c) For the last point, the force can be written as a negative gradient of potential:

$$\vec{F} = -\nabla U \tag{1.5.3}$$

The conservative force can also be defined in a better way. Suppose that we have a differential equation given in the form of

$$\ddot{x} + f(x) = 0 \tag{1.5.4}$$

If $U(-x) = U(x)$ and $f(-x) = -f(x)$, then Eq. (1.5.4) would represent a conservative nonlinear oscillatory systems. Note that $U(x)$ corresponds to the potential energy pertinent to $f(x)$. It is indeed one of the necessary condition but not the sufficient one. The best way to define a conservative oscillation is to consider the potential and kinetic energies. If during any particular motion, one could observe the following behavior, then one can say that the oscillatory system is conservative. This definition can be formulated as

$$U + K = \text{Constant} \tag{1.5.5}$$

where K and U represent the kinetic and potential energies of the oscillatory system, respectively. For better understanding of the conservative oscillatory systems, few examples are presented in the next parts.

1.5.1 Duffing Equation

The undamped Duffing equation is mathematically formulated by the following equation:

$$\ddot{x} + \alpha x + \beta x^3 = 0 \tag{1.5.6}$$

with the assumption that both α and β are positive coefficients. The restoring force of the above equation is:

$$f(x) = \alpha x + \beta x^3 \tag{1.5.7}$$

The potential energy function of this equation can be written in the form of

$$U(x) = \frac{\alpha}{2}x^2 + \frac{\beta}{4}x^4 \tag{1.5.8}$$

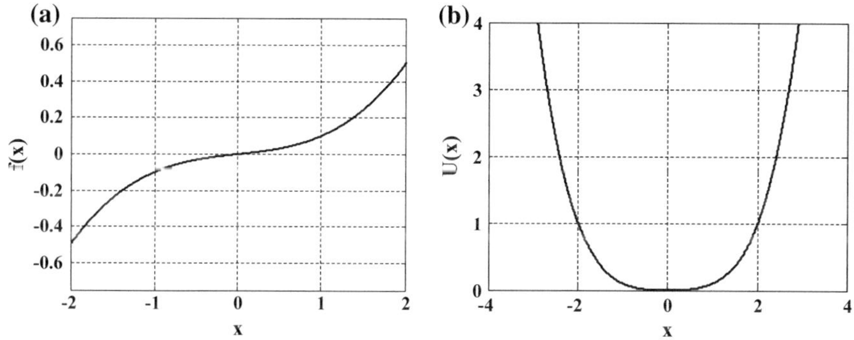

Fig. 1.1 **a** Restoring force, and **b** potential energy level for Duffing equation, $\alpha = 0.1$, and $\beta = 0.2$

Figure 1.1 shows the restoring force and the potential energy of the Duffing equation.

As depicted from Fig. 1.1, the potential energy level remains the same for both $U(-x)$ and $U(x)$. Also, it is observed that $f(-x) = -f(x)$. Readily, it can be concluded that the given Eq. (1.5.6) is a conservative type oscillator. We will review a generalized form of Duffing equation in Chap. 3.

1.5.2 Oscillator with Fractional Power

In this example, we will see a nonlinear differential equation with a fractional power. Suppose that the oscillation of a mechanical system are mathematically modelled by the following equation:

$$\ddot{x} + \alpha x^{1/3} = 0 \tag{1.5.9}$$

One could easily obtain the restoring force of the system as:

$$f(x) = \alpha x^{1/3} \tag{1.5.10}$$

and, the potential energy term is obtained by the following equation:

$$U(x) = \frac{3}{4}\alpha x^{4/3} \tag{1.5.11}$$

Figure 1.2 shows the restoring force and the potential energy correspond to Eq. (1.5.9). It shows that the restoring force is an odd function while the potential energy is an even function.

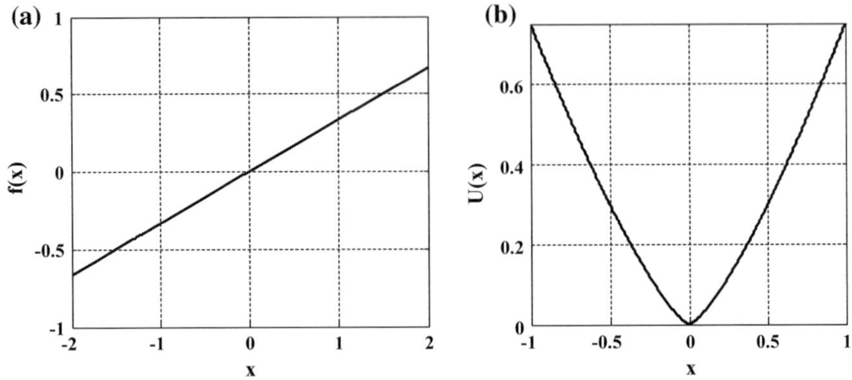

Fig. 1.2 **a** Restoring force, and **b** potential energy level for Eq. (1.5.9) $\alpha = 1$

1.5.3 *Relativistic Oscillator*

Relativistic oscillator is mathematically modelled by the following equation:

$$\ddot{x} + \frac{\alpha x}{\sqrt{1 + x^2}} = 0 \tag{1.5.12}$$

Accordingly, the restoring force of Eq. (1.5.12) is given as:

$$f(x) = \frac{\alpha x}{\sqrt{1 + x^2}} \tag{1.5.13}$$

and its potential energy is formulated in the following form

$$U(x) = \alpha\sqrt{1 + x^2} \tag{1.5.14}$$

Figure 1.3 represents the restoring force and the potential energy of the relativistic equation. As expected, one could observe that the restoring force is an odd function, while the potential energy behaves as an even function.

1.5.4 *Oscillator with Discontinuity*

Consider an oscillator with a nonlinear term that contains a discontinuity in the form of

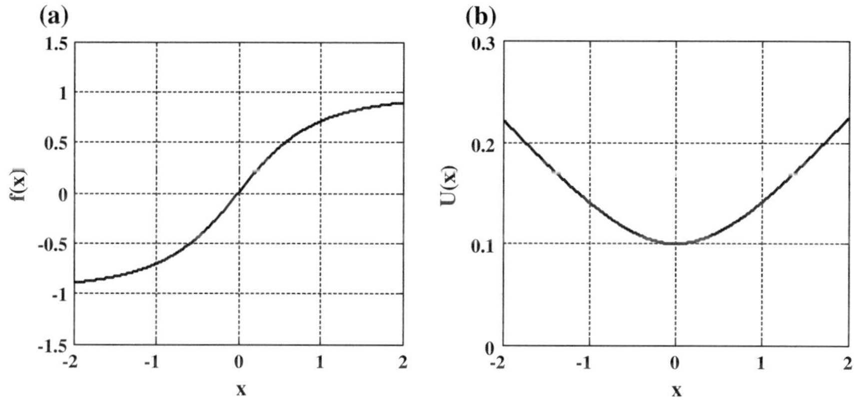

Fig. 1.3 **a** Restoring force, and **b** potential energy level for Eq. (1.5.12) $\alpha = 1$

$$\ddot{x} + x^2 sign(x) = 0 \tag{1.5.15}$$

The restoring force of Eq. (1.5.15) is described by Eq. (1.5.16):

$$f(x) = x^2 sign(x) \tag{1.5.16}$$

The potential energy deduced from Eq. (1.5.16) is written as:

$$U(x) = \frac{1}{3}x^3 sign(x) \tag{1.5.17}$$

Figure 1.4 shows the restoring force and the potential energy of the nonlinear oscillator with discontinuity. It can be seen that the restoring force is an odd function, while the potential energy behaves as an even function.

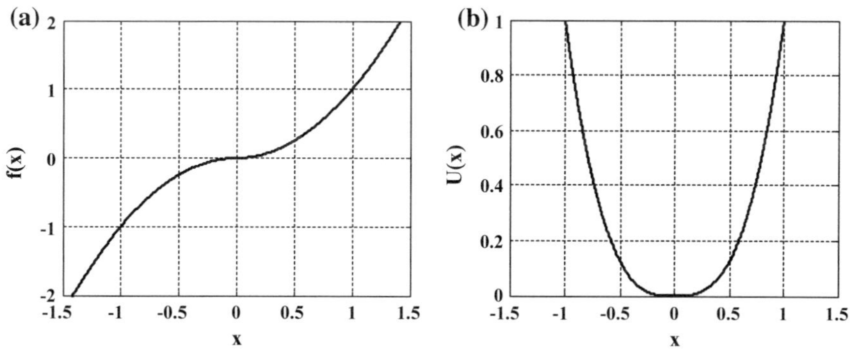

Fig. 1.4 **a** Restoring force, **b** potential energy level for Eq. (1.5.15)

We will focus our attention on the concept of the non-conservative oscillatory systems in the next section.

1.6 Non-conservative Oscillatory Systems

There could be few reasons for having a non-conservative oscillatory system. When a non-conservative force is applied to a body, there would be a work done during any oscillation cycle, and this being the major difference between the conservative and non-conservative forces. Frictional force is the most common type of a non-conservative force. The damping force in an oscillatory system is a non-conservative one, and is a function of velocity.

In an oscillatory system, one could have different types of damping, e.g., dry friction, hysteresis damping, material damping and linear and nonlinear damping. These forces can be the main reason for having a non-conservative oscillation. Generally, free vibration of a mechanical system with the damping term can be modelled by the typical form of differential equation as:

$$\ddot{x} + f(\dot{x}) + f(x) = 0 \tag{1.6.1}$$

in which $f(x)$ and $f(\dot{x})$ could be either a linear or nonlinear functions which represent the elastic restoring force and the damping force, respectively. Considering the dry friction, known as Coulomb damping, one develops the following differential equation:

$$\ddot{x} + \mu N\, sign(\dot{x}) = 0 \tag{1.6.2}$$

where the constant parameters μ and N are the friction coefficient and the normal force, respectively.

1.7 Parametrically Excited Vibration

The concept of parametric excitation is introduced in this section and few examples are given to illustrate the importance of this type of vibration. Whenever one parameter undergoes a periodic excitation in a dynamic system, there would be time-varying coefficients in the corresponding differential equations. Generally, the differential equation of a nonlinear oscillator with parametric excitation could have the following general form of

$$\ddot{x} + \alpha(t)\, L(x, \dot{x}) + \beta(t) N(x, \dot{x}) = f(t) \tag{1.7.1}$$

where $L(x, \dot{x}), N(x, \dot{x}), \alpha(t), \beta(t)$ and $f(t)$ correspond to the linear term, nonlinear term, time-varying coefficient for the linear term, time-varying coefficient for the nonlinear term, and the excitation term, respectively. There is an important difference between the oscillators with external excitation and oscillators with parametric excitation. When an external excitation is applied to a system at rest, it will displace the system from its static equilibrium position. This is unlike the case with the parametric excitation. In this case the system must initially be displaced from its equilibrium position before the parametric excitation could start the oscillation of the system.

It is hence, impossible to reach a large amplitude of vibration in the oscillators with an external excitation, unless the frequency of excitation is close to one of the natural frequencies of the system. On the other hand, a small perturbation in the oscillators with parametric excitation can result in large oscillations when the excitation frequency is adjacent to twice of one of the natural frequencies of the system [20]. One simple form of Eq. (1.7.1) in the absence of the nonlinear term $N(x, \dot{x})$ and the excitation term $f(t)$ is called Hill's equation as shown below:

$$\ddot{x} + \alpha(t)\,L(x) = 0 \tag{1.7.2}$$

Equation (1.7.2) was first developed by G. W. Hill for the motion of the lunar perigee [51]. It is in fact a linear form of parametric excitation.

Parametric excitation emerged in many different mechanical and electrical systems. The first observation of parametric excitation was reported by Faraday [52]. In a pioneering article, Faraday reported that the free surface of a fluid layer is unstable to the standing surface waves when subjected to the vertical periodic motion. Faraday showed that the waves vibrate at the frequency of one half of the excitation frequency [53]. Few years later, Lord Rayleigh confirmed Faraday's results with his own experimental work. In one of his articles, he has reported of having three different types of vibration with their corresponding differential equations. The last differential equation of his article is in fact a type of parametric oscillators. Rayleigh described the system with his following illustration:

> If the force of restitution, or "spring," of a body susceptible of vibration be subjected to an imposed periodic variation, the differential equation becomes
>
> $$\frac{d^2\theta}{dt^2} + k\frac{d\theta}{dt} + (n^2 - 2\alpha\,\sin(2pt))\theta = 0 \tag{1.7.3}$$
>
> A similar equation would apply approximately in the case of a periodic variation in the effective mass of the body.

His assumption was that the values of parameters k and α are small. Rayleigh's work is indeed one of the earliest endeavors for the mathematical modelling of parametric excitations [54]. Following the works of Rayleigh, actual parametric excitations were observed in many different mechanical and electrical devices. One of the principal works in the area of parametric oscillation was developed in the form of Mathieu equation. This equation was first introduced by the French mathematician, known as Mathieu [55].

The well-known Mathieu equation is described by the following differential equation:

$$\frac{d^2x}{dt^2} + (a + 2h^2 \cos(2t))x = 0 \qquad (1.7.4)$$

where a and h are constant coefficients. Mathieu equation has been studied by many researchers both theoretically and experimentally. Whittaker is one of the pioneers who developed the analytical solution for Mathieu equation. In his work, he focused on studying an equation with the following terms:

$$\frac{d^2y}{dz^2} + (a + 16q \cos(2z))y = 0 \qquad (1.7.5)$$

He is indeed one of the first mathematician who introduced new parameters of σ and μ. The parameter σ, instead of a, was initially defined by him and eventually, he obtained the solution based on the new defined parameters.

Young [56], Burges [57], Jeffreys [58], Dougall [59], Ince [60], Goldstein [61], Langer [62], and Minorsky [63] are well-known mathematicians who have played an influential role on better understanding the Mathieu equation, and also parametric excitation during 1900–1950.

One of the leading researchers in the analysis of Mathieu equation and parametric excitation is R. H. Rand from Cornell University [64, 65–66]. He has studied different phenomena in parametric excitation, specifically in Mathieu equation and reported in his many publications. In Sect. 1.7.1, one of his recently developed nonlinear differential equation pertinent to parametric excitation will be briefly introduced.

Table 1.4 presents the recent advances and applications in the study of Mathieu equation.

In most cases, many researchers have focused their attention mainly on the stability analysis of parametric oscillators. In the next section, a general example is briefly described based on Ref. [66].

1.7.1 Fractional Mathieu Equation

Recently, Rand et al. [66] studied a form of Mathieu equation with the following expression:

$$\frac{d^2x}{dt^2} + (\delta + \varepsilon \cos t)x + CD^{\alpha}x = 0 \qquad (1.7.6)$$

Table 1.4 Recent advances in the analysis of parametric oscillators

Mathieu equation forms	Applications
$M\ddot{x}+Kx-(F_s+F_d\cos pt)Qx(t)+\eta x^3=0$	Laminated composite cylindrical shells subjected to periodic axial loads [67]
$\ddot{x}+2C_r\dot{x}+(a_r+2q_r(\cos 2t+d_r\cos 4t))x=f_t$	Submerged floating pipeline [68]
$\ddot{x}+\omega^2(1-P_{O_1}-P_{t_1}\cos(Pt))x=0$	FGM conical shells under periodic lateral pressure [69]
$M\ddot{x}+C\dot{x}+(k_1+k_2\cos(2\Omega t)+k_3\sin(2\Omega t))x$ $=F_1\cos\Omega t+F_1\cos\Omega t+F_g$	Cracked rotors with time varying stiffness [70]
$\ddot{x}+\omega^2(1-K_v\cos(2T))x=0$	Liquid filled vessel under seismic excitations [71]
$\ddot{x}+2\mu\dot{x}+\Omega_0^2(1-\varepsilon\cos\omega_v t)x+\beta x^3$ $+kx(\dot{x}^2+x\ddot{x})=\bar{\delta}\cos(\omega_H t+\psi)$	Slender beam under horizontal-vertical excitations [72]
$\frac{d^2p_\eta}{d\eta^2}+(\alpha-2q\cos(2\eta))p_\eta=0$ $\frac{d^2p_\xi}{d\xi^2}+(\alpha-2q\cosh(2\xi))p_\xi=0$	Submerged elliptical cylindrical shell [73]
$\ddot{x}+2(k_0+\mu(1+\cos 2\omega t)x^2)\dot{x}$ $+(\omega_L^2-\xi\cos 2\omega t)x=0$	Bimaterial magneto-elastic cantilever beam with thermal loading [74]
$\ddot{x}+(a+b\cos\theta t)x+\beta x^3=0$	Beams with periodically variable length; stability of floating offshore structures; vibrations of beam type structures under harmonic support motion [75]
$\ddot{x}+(a+2q\cos(t))(x+\alpha_1x^2+\alpha_2x^3)=0$	Paul trap mass spectrometers [76]
$\ddot{x}+(\delta+\varepsilon\cos(t))x=\varepsilon(Ax^3+Bx^2\dot{x}$ $+Cx\dot{x}^2+D\dot{x}^3)$	Cable being towed by a submarine [77]
$\frac{\partial^2\phi}{\partial x^2}+\frac{\Omega^2}{c^2}\left(1-\frac{K_B^2c^2}{\Omega^2}\varepsilon\cos(kx)\right)\phi=0$	Neutrino oscillations by density ripple in dense plasmas [78]
$\frac{d^2x}{d\theta^2}+\left(\frac{D-1}{2}-\frac{(D-1)(D-3)}{4}\cot^2(\theta)\right)x$ $-\omega^2(1-\cos(\theta))x+Ex=0$	Hooke's law correlation in two-electron systems [79]
$\ddot{x}-(a+bx^2)\dot{x}+\omega^2(1+h\cos\gamma t)x=0$	Dynamics of dust grain charge in dusty plasmas [80]
$\ddot{x}+\sum_{n=1}^{N}(a_n-2\varepsilon q_n\cos 2nt)x^n=0$ $n=1,2,3$	Parametrically excited pendulum [81]
$\ddot{x}+Q_1^2x-\varepsilon x\cos 2t+\varepsilon[\beta_1-\cos 2t]y=0$ $\ddot{y}+Q_2^2y-\varepsilon y\cos 2t+\varepsilon[\beta_2-\cos 2t]x=0$	Accelerator dynamics [82]

where the index α represents the fractional derivative term. It is evident that Eq. (1.7.6) would convert to a classical damped Mathieu equation when $\alpha = 1$. They obtained the transition curves for the above example by employing the method of Harmonic Balance. The first-order transition curve of the above equation is given as [66]:

$$\delta = \frac{1}{4} - \frac{C}{2^{\alpha}}\cos\frac{\alpha\pi}{2} \pm \frac{\sqrt{2^{2\alpha}\varepsilon^2 - 4C^2\sin^2\frac{\alpha\pi}{2}}}{2^{\alpha+1}} \tag{1.7.7}$$

Let there will be no fractional term and accordingly, Eq. (1.7.7) is converted to the following from [66]:

$$\ddot{x} + (\delta + \varepsilon\cos t)x = 0 \tag{1.7.8}$$

The transition curves for Eq. (1.7.8) can be constructed from the expression below [66]:

$$\delta = \frac{1}{4} \pm \frac{1}{2}\varepsilon - \frac{1}{8}\varepsilon^2 \tag{1.7.9}$$

when $\alpha = 1$ in Eq. (1.7.7), one will have the classical damped Mathieu equation as [66]:

$$\ddot{x} + C\dot{x} + (\delta + \varepsilon\cos t)x = 0 \tag{1.7.10}$$

The transition curves of Eq. (1.7.8) would then become [66]:

$$\delta = \frac{1}{4} \pm \frac{1}{2}\sqrt{\varepsilon^2 - c^2} \tag{1.7.11}$$

When constructing the transition curves by using Eq. (1.7.8), one could explore the stable and the unstable regions in the objective oscillatory systems. Figure 1.5 represents the transition curves of Eq. (1.7.8). The areas shown by the gray color represent the unstable oscillatory regions.

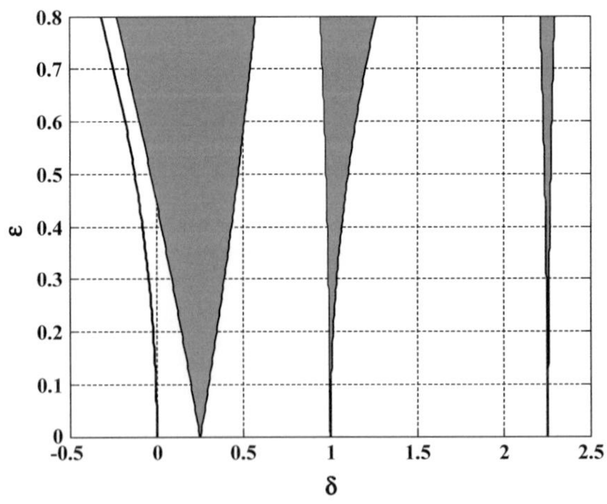

Fig. 1.5 Transition curves for the Mathieu equation

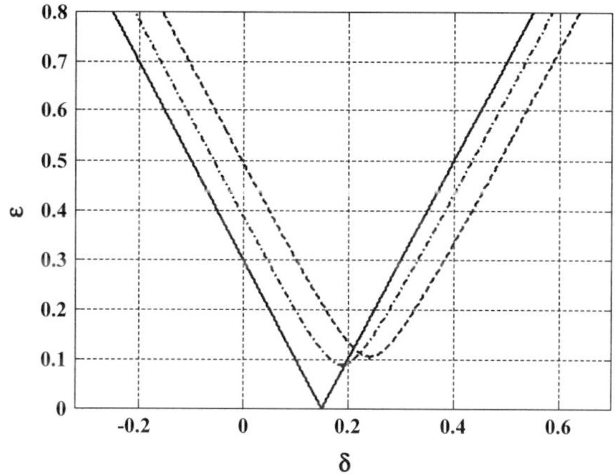

Fig. 1.6 Transition curves for the Mathieu equation. (Fraction derivative effect, $C = 0.1$) Solid line $\alpha = 0$; dot-dashed line $\alpha = 0.4$; dashed line $\alpha = 0.9$

Figure 1.6 shows the effect of the fractional derivative on the transition curves and the stability regions deduced from the Mathieu equation. As presented by this figure, when increasing the value of α from zero to one, it would result in increasing the middle value of parameter δ.

Figure 1.7 illustrates the influence of the damping coefficient on the transition curves.

Kovacic et al. [83] studied the Mathieu-Duffing equation represented in the following form:

$$\ddot{x} + (\delta + 2\varepsilon \cos 2t)x + cx^3 = 0 \tag{1.7.12}$$

Based on their work, the transition curves of Eq. (1.7.12) in the presence of strong softening nonlinearity is

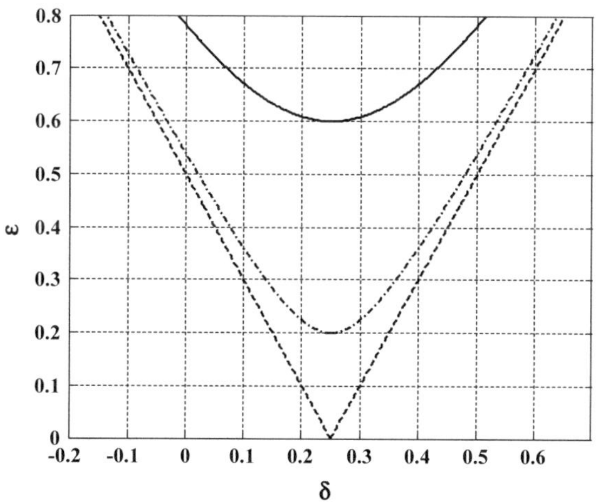

Fig. 1.7 Transition curves for the Mathieu equation. Solid line $c = 0.6$; dashed-dot line $c = 0.2$; dashed line $c = 0$

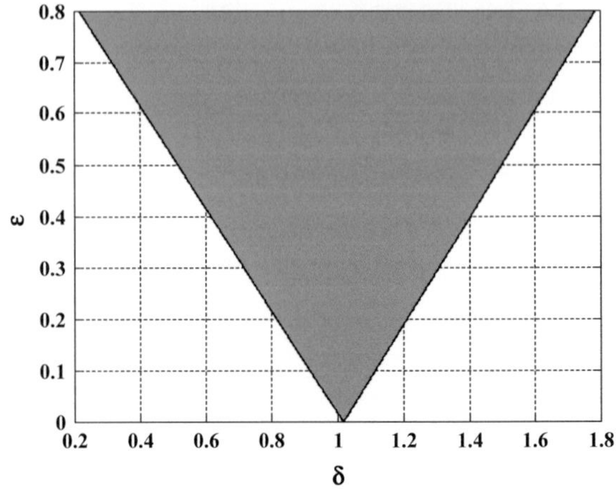

Fig. 1.8 Transition curves for Mathieu-Duffing equation, softening nonlinearity ($\varepsilon_1 = 0.1$)

$$\delta = 1 + \frac{\varepsilon_1}{6} - \varepsilon \text{ and } \delta = 1 + \frac{\varepsilon_1}{6} + \varepsilon\left(1 - \frac{\varepsilon_1}{8\left(1 - \frac{\varepsilon_1}{3}\right)}\right) \tag{1.7.13}$$

However, for the case of hardening nonlinearity, they obtained the transition curve formula as [83]:

$$\delta = 1 - \frac{2\varepsilon_1}{3} - \varepsilon \text{ and } \delta = 1 - \frac{2\varepsilon_1}{3} + \varepsilon\left(1 + \frac{\varepsilon_1(3 - \varepsilon_1)}{4(3 + \varepsilon_1 - \varepsilon_1^2)}\right) \tag{1.7.14}$$

Interested readers can find a detailed solution procedure in Refs. [83, 84]. Figure 1.8 shows the transition curves of the considered nonlinear system for the case of softening nonlinearity.

The area with the gray color in Fig. 1.8 shows the unstable region. The gray area shown in Fig. 1.9 represents the unstable region when the system has the hardening nonlinearity.

In the next section, different types of resonant cases in nonlinear oscillatory systems are discussed.

1.8 Resonance in Nonlinear Systems

Generally, resonance occurs when an external force drives a dynamic system at a specific frequency to oscillate with greater amplitude. In a nonlinear system, one could look at this phenomenon from another angle. Understanding different types of resonance in nonlinear systems is crucial since they may cause severe failure in the

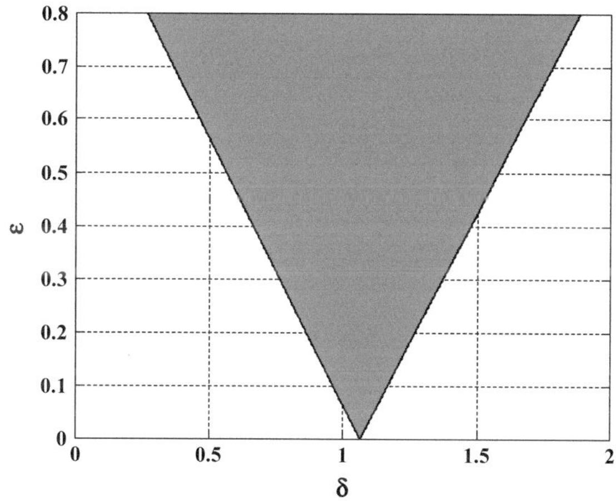

Fig. 1.9 Transition curve for Mathieu-Duffing equation, hardening nonlinearity ($\varepsilon_1 = 0.1$)

considered system. This phenomenon is very important in different fields of science and engineering.

Resonance can be observed in musical instruments, tall buildings, bridges, micro- and nano- devices, electrical systems, space structures, aircrafts, and also road vehicles. In order to illustrate different types of resonance in an oscillator, one could consider the following nonlinear differential equations:

$$\ddot{x} + f(x, \dot{x}) = F\cos(\Omega t) \tag{1.8.1}$$

The initial assumptions could be in the form of

$$f(x, \dot{x}) = 2\varepsilon\mu\dot{x} + \alpha x + \varepsilon\beta x^3 \tag{1.8.2}$$

Accordingly, one would arrive at the following equation

$$\ddot{x} + 2\varepsilon\mu\dot{x} + \omega_0^2 x + \varepsilon\beta x^3 = F\cos(\Omega t) \tag{1.8.3}$$

It is required to investigate three types of possible resonant cases in Eq. (1.8.3). The first case of resonance being the primary or the main resonance in which, the frequency of excitation is set to be equal to the linear natural frequency.

The second type of resonance, which is only possible in nonlinear systems, is referred to the super-harmonic resonance. Owing to have a cubic nonlinearity in Eq. (1.8.3), one could define the resonant frequency, $\omega_0 \approx 3\Omega$ as the super-harmonic resonance of Eq. (1.8.3).

The third type of possible resonance case for Eq. (1.8.3) is the sub-harmonic resonance. Due to having cubic nonlinearity in Eq. (1.8.3), one could define the resonant frequency, $\omega_0 \approx \frac{1}{3}\Omega$ as the sub-harmonic resonance of Eq. (1.8.3).

Table 1.5 Different types of resonances in Eq. (1.8.1)

Resonance case	Mathematical solution
Primary resonance $\Omega = \omega_0 + \varepsilon\sigma$	$\sigma = \frac{3}{8}\frac{\beta}{\omega}a^2 \pm \left(\frac{F^2}{4\omega_0^2 a^2} - \mu^2\right)^{1/2}$
Super-harmonic resonance $\Omega = \frac{1}{3}\omega_0 + \varepsilon\sigma$	$\sigma = 3\frac{\beta\Lambda^2}{\omega} + \frac{3}{8}\frac{\beta}{\omega}a^2 \pm \left(\frac{\beta^2\Lambda^6}{\omega_0^2 a^2} - \mu^2\right)^{1/2}$ $\Lambda = \frac{F}{2(\omega_0^2 - \Omega^2)}$
Sub-harmonic resonance $\Omega = 3\omega_0 + \varepsilon\sigma$	$\left[9\mu^2 + \left(\sigma - \frac{9\alpha\Lambda^2}{\omega_0} - \frac{9\alpha\Lambda^2}{8\omega_0}a^2\right)\right]a^2 = \frac{81\alpha^2\Lambda^2}{\omega_0}a^2$

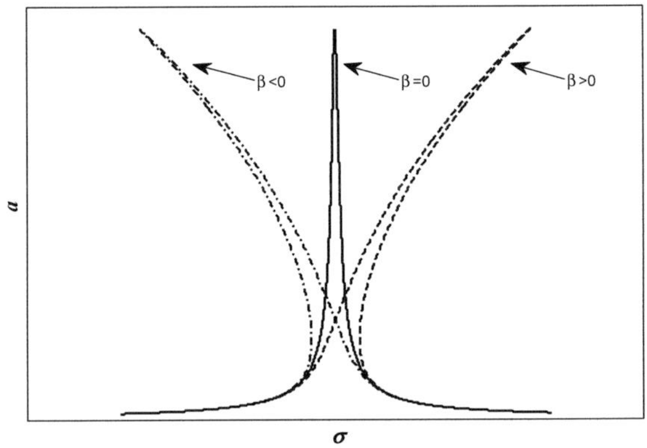

Fig. 1.10 Frequency response curves for different values of β

Table 1.5 shows the mathematical solutions of each cases of resonance related to Eq. (1.8.3).

Figure 1.10 illustrates a comparison between the resonance case of linear and the nonlinear vibrations based on Eq. (1.8.3). The dashed line represents the primary resonance for $\beta > 0$ and presents the hardening nonlinearity. Whenever $\beta = 0$, there is no nonlinearity in the system, and the solid line in Fig. 1.10 indicates the resonance in this case.

The dashed-dot line in Fig. 1.10 represents the primary resonance of the system when $\beta < 0$ and the softening nonlinearity is observed for that range of β. The illustration relevant to the jump phenomenon in nonlinear oscillators can be seen in Fig. 1.11. One could observe that there are multivalued amplitudes due to the bending of the frequency-response curve and consequently, the jump phenomenon would occur.

As can be seen from Fig. 1.11, by slowly decreasing Ω along the path AFB, the amplitude of oscillations gradually increases. Then a slight change in Ω results in an abrupt jump from point B to point C. By decreasing the frequency from point C, the amplitude of oscillation decreases toward point D. Now, let us assume that the excitation frequency is far below ω_0. Hence, as we increase Ω along the path DCE, there is no jump from point C to B this time, and therefore, the amplitude of

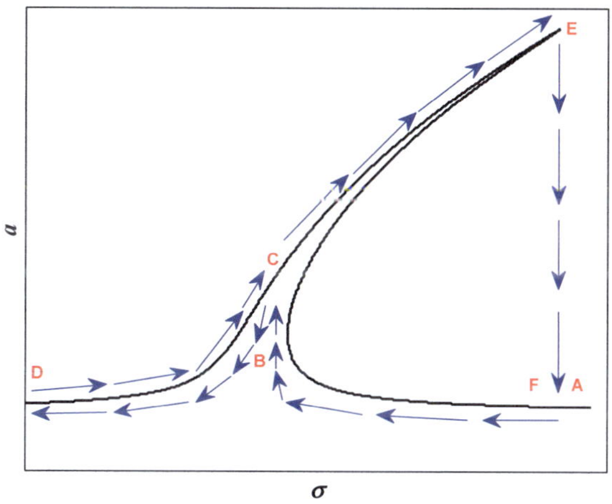

Fig. 1.11 Jump phenomenon in frequency response curve

Table 1.6 Frequency changing process to detect jump phenomenon

Decreasing frequency Ω	A to B then upward jump to C and move towards D
Increasing frequency Ω	D to C then to E then a downward jump to F and A

oscillations increases by rising Ω to reach the maximum point E. At point E, a slight increment in Ω results in a downward jump from point E to F. As depicted from Fig. 1.11, an increment in Ω from point E toward F decreases the amplitude of oscillation.

Table 1.6 briefly describes the two jump phenomenon observed in Fig. 1.11.

Now suppose Eq. (1.8.3) undergoes multi-frequency excitations. Accordingly, one will arrive at Eq. (1.8.4) as:

$$\ddot{x} + 2\varepsilon\mu\dot{x} + \omega_0^2 x + \varepsilon\beta x^3 = \sum_{n=1}^{N} F_n \cos(\Omega_n t + \theta_n) \tag{1.8.4}$$

Having multi-frequency excitations in a nonlinear system may cause many different interesting phenomena. For example, when $N = 2$, we have to look at new cases as:

$$\omega_0 = \frac{1}{2}(\Omega_1 \pm \Omega_2), \omega_0 = 2\Omega_1 \pm \Omega_2 \text{ and } \omega_0 = 2\Omega_2 \pm \Omega_1 \tag{1.8.5}$$

The above cases must be studied because the free oscillation term does not vanish when we consider cubic nonlinearity and double frequency of excitation. It must be noted that the steady-state response will not have the periodic form, unless $\Omega_n = \Psi_n \Omega_1$ in which Ψ_n must be a rational fraction [20]. Another interesting phenomenon that can occur in the considered system is the simultaneous resonances.

As an example, one could have both cases of $\Omega_1 \approx \frac{1}{3}\omega_0$ and $\Omega_2 \approx 3\omega_0$.

Another interesting phenomenon that can be seen is the internal resonance, which happens in nonlinear systems having multi-degrees of freedom.

Let us assume that we would have a nonlinear system with cubic nonlinearities shown below [20]:

$$\ddot{x}_1 + \omega_1^2 x_1 + 2\mu_1 \dot{x}_1 + \alpha_1 x_1^3 + \alpha_2 x_1^2 x_2 + \alpha_3 x_2^2 x_1 + \alpha_4 x_2^3 = 0 \tag{1.8.6}$$

$$\ddot{x}_2 + \omega_2^2 x_2 + 2\mu_2 \dot{x}_2 + \alpha_5 x_2^3 + \alpha_6 x_2^2 x_1 + \alpha_7 x_1^2 x_2 + \alpha_8 x_2^3 = 0 \tag{1.8.7}$$

It is found that the internal resonance will occur when $\omega_2 \approx 3\omega_1$ and $\omega_1 \approx 3\omega_2$ in the above system.

References

1. Strogatz SH (2014) Nonlinear dynamics and chaos: with applications to physics, biology, chemistry, and engineering. Westview press, Boulder
2. van der Pol B (1920) Theory of the amplitude of free and forced triode vibration. Radio Rev 1:701–710
3. Kovacic I, Brennan MJ (2011) Background: on Georg Duffing and the Duffing equation. Wiley, New Jersey, pp 1–23
4. Cartwright ML (1952) Non-linear vibrations: a chapter in mathematical history. Presidential Address to the Mathematical Association, January 3, 1952. Math Gaz 36(316):81–88
5. Krylov NM, Bogoliuboff N, Lefschetz S (1949) Introduction to non-linear mechanics. Princeton University Press, Princeton
6. Reissig R, Andronov AA, Vitt AA, Khaikin SE (1967) Theory of oscillators (International Series of Monographs in Physics, Vol. 4) XXXII + 815 S. m. 598 Fig. Pergamon Press, Oxford/London/Edinburgh/New York/Toronto/Paris/Frankfurt 1966. Preis geb. £ 10 net. ZAMM—J Appl Math Mech Zeitschrift für Angew Math und Mech 47(7):480–481
7. Publications of A. N. Kolmogorov (1989) pp 945–964
8. Birkhoff GD (1917) Dynamical systems with two degrees of freedom. Trans Am Math Soc 18 (2):189–199
9. Levinson N (1944) Transformation theory of non-linear differential equations of the second order. Ann Math 45(4):723–737
10. Minorsky N (1945) On parametric excitation. J Franklin Inst 240(1):25–46
11. Levenson ME (1949) Harmonic and subharmonic response for the duffing equation ẍ + αx + βx3 = F cosωt (α > 0). J Appl Phys 20(11):1045
12. Shohat J (1944) On Van der Pol's and related non-linear differential equations. J Appl Phys 15(7):568
13. LaSalle JP, Lefschetz S (1961) Recent Soviet contributions to ordinary differential equations and nonlinear mechanics. J Math Anal Appl 2(3):467–499
14. Hayashi C (1953) Subharmonic oscillations in nonlinear systems. J Appl Phys 24(5):521
15. Hayashi C (1953) Forced oscillations with nonlinear restoring force. J Appl Phys 24(2):198
16. Hayashi C (1953) Stability investigation of the nonlinear periodic oscillations. J Appl Phys 24 (3):344

17. Dolph CL (1962) A unified theory of the nonlinear oscillations of a cold plasma. J Math Anal Appl 5(1):94–118
18. Wang HSC (1963) Nonlinear stationary waves in relativistic plasmas. Phys Fluids 6(8):1115–1123
19. Crandall SH (1963) Perturbation techniques for random vibration of nonlinear systems. J Acoust Soc Am 35(11):1700
20. Nayfeh AH, Mook DT (1995) Nonlinear oscillations. Wiley-VCH Verlag GmbH, New Jersey
21. Nayfeh AH (2011) Introduction to perturbation techniques. Wiley, New Jersey
22. Nayfeh AH, Pai PF (2008) Linear and nonlinear structural mechanics. Wiley, New Jersey
23. Nayfeh AH (2011) The method of normal forms. Wiley, New Jersey
24. Nayfeh AH, Balachandran B (2008) Applied nonlinear dynamics: analytical, computational and experimental methods. Wiley, New Jersey
25. Ishida Y, Yamamoto T (2013) Linear and nonlinear rotor dynamics: a modern treatment with applications. Wiley, New Jersey
26. Storti DW, Rand RH (1982) Dynamics of two strongly coupled van der pol oscillators. Int J Non Linear Mech 17(3):143–152
27. Li GX, Rand RH, Moon FC (1990) Bifurcations and chaos in a forced zero-stiffness impact oscillator. Int J Non Linear Mech 25(4):417–432
28. Rand RH, Holmes PJ (1980) Bifurcation of periodic motions in two weakly coupled van der Pol oscillators. Int J Non Linear Mech 15(4–5):387–399
29. Moon FC (1987) Chaotic vibrations: an introduction for applied scientists and engineers. Wiley, New Jersey
30. Moon FC (2008) Chaotic and fractal dynamics: introduction for applied scientists and engineers. Wiley, New Jersey
31. Dowell EH, Il′gamov MA (1988) Studies in nonlinear aeroelasticity. Springer, Berlin
32. Paidoussis MP (1998) Fluid-structure interactions: slender structures and axial flow, vol 1. Elsevier Science, New York
33. Ibrahim RA (2005) Liquid Sloshing dynamics: theory and applications. Cambridge University Press, Cambridge
34. Amabili M (2008) Nonlinear vibrations and stability of shells and plates. Cambridge University Press, Cambridge
35. Liu Y, Chen LQ (2013) Chaos in attitude dynamics of spacecraft. Springer, Berlin, Heidelberg
36. Kerschen G, Peeters M, Golinval JC, Vakakis AF (2009) Nonlinear normal modes, Part I: a useful framework for the structural dynamicist. Mech Syst Signal Process 23(1):170–194
37. Vakakis AF, Gendelman OV, Bergman LA, McFarland DM, Kerschen G, Lee YS (2008) Nonlinear targeted energy transfer in mechanical and structural systems. Springer, Netherlands
38. Awrejcewicz J (1991) Bifurcation and Chaos in coupled oscillators. World Scientific, Singapore
39. Mickens RE (2010) Truly nonlinear oscillations: harmonic balance, parameter expansions, iteration, and averaging methods. World Scientific, Singapore
40. Cveticanin L (2014) Strongly nonlinear oscillators: analytical solutions. Springer, Berlin
41. Erturk A, Inman DJ (2011) Piezoelectric energy harvesting. Wiley, New Jersey
42. Diba F, Esmailzadeh E, Younesian D (2014) Nonlinear vibration analysis of isotropic plate with inclined part-through surface crack. Nonlinear Dyn 78(4):2377–2397
43. Roberts JB, Spanos PD (1986) Stochastic averaging: an approximate method of solving random vibration problems. Int J Nonlinear Mech 21(2):111–134
44. Balachandran B (2001) Nonlinear dynamics of milling processes. Philos Trans R Soc A Math Phys Eng Sci 359(1781):793–819
45. Kerschen G, Worden K, Vakakis AF, Golinval J-C (2006) Past, present and future of nonlinear system identification in structural dynamics. Mech Syst Signal Process 20(3):505–592

46. Younesian D, Sadri M, Esmailzadeh E (2014) Primary and secondary resonance analyses of clamped-clamped micro-beams. Nonlinear Dyn 76(4):1867–1884
47. Ansari M, Esmailzadeh E, Jalili N (2011) Exact frequency analysis of a rotating cantilever beam with tip mass subjected to torsional-bending vibrations. J Vib Acoust 133(4):41003–41009
48. Ansari M, Esmailzadeh E, Jalili N (2009) Coupled vibration and parameter sensitivity analysis of rocking-mass vibrating gyroscopes. J Sound Vib 327(3):564–583
49. Misra AK (2014) Modeling the effect of police deterrence on the prevalence of crime in the society. Appl Math Comput 237:531–545
50. Misra AK, Verma M (2013) A mathematical model to study the dynamics of carbon dioxide gas in the atmosphere. Appl Math Comput 219(16):8595–8609
51. Hill GW (1886) On the part of the motion of the lunar perigee which is a function of the mean motions of the sun and moon. Acta Math 8(1):1–36
52. Faraday M (1831) On a peculiar class of acoustical figures; and on certain forms assumed by groups of particles upon vibrating elastic surfaces. Philos Trans R Soc London 121:299–340
53. Douady S (1990) Experimental study of the Faraday instability. J Fluid Mech 221(1):383
54. Rayleigh L (1883) XXXIII On maintained vibrations. Philos Mag Ser 5 15(94):229–235
55. Mathieu E (1868) Mémoire sur le mouvement vibratoire d'une membrane de forme elliptique. J mathématiques pures appliquées 2e série 13:137–203
56. Young AW (1913) On the quasi-periodic solutions of Mathieu's differential equation. Proc Edinb Math Soc 32:81–90
57. Burgess AG (1914) Determinants connected with the periodic solutions of Mathieu's equations. Proc Edinb Math Soc 33:122–138
58. Jeffreys H (1925) On certain solutions of Mathieu's equation. Proc Lond Math Soc S2 (1):437–448
59. Dougall J (1925) The solutions of Mathieu's differential equation: representation by contour integrals, and asymptotic expansions. Proc Edinb Math Soc 44:57–71
60. Ince EL (1926) The second solution of the Mathieu equation. Math Proc Cambridge Philos Soc 23(1):47–49
61. Goldstein S (1928) The second solution of Mathieu's differential equation. Math Proc Cambridge Philos Soc 24(2):223–230
62. Langer RE (1934) The solutions of the Mathieu equation with a complex variable and at least one parameter large. Trans Am Math Soc 36(3):637–695
63. Minorsky N (1945) On parametric excitation. J Franklin Inst 240(1):25–46
64. Anne Month L, Rand RH (1982) Bifurcation of 4:1 subharmonics in the nonlinear mathieu equation. Mech Res Commun 9(4):233–240
65. Holmes CA, Rand RH (1981) Coupled oscillators as a model for nonlinear parametric excitation. Mech Res Commun 8(5):263–268
66. Rand RH, Sah SM, Suchorsky MK (2010) Fractional Mathieu equation. Commun Nonlinear Sci Numer Simul 15(11):3254–3262
67. Darabi M, Ganesan R (2016) Non-linear dynamic instability analysis of laminated composite cylindrical shells subjected to periodic axial loads. Compos Struct 147:168–184
68. Yang H, Wang Z, Xiao F (2016) Parametric resonance of submerged floating pipelines with bi-frequency parametric and vortex-induced oscillations excitations. Taylor and Francis, Abingdon, pp 1–9
69. Sofiyev AH (2016) Parametric vibration of FGM conical shells under periodic lateral pressure within the shear deformation theory. Compos Part B Eng 89:282–294
70. AL-Shudeifat MA (2015) Stability analysis and backward whirl investigation of cracked rotors with time-varying stiffness. J Sound Vib 348:365–380
71. Kolukula SS, Sajish SD, Chellapandi P (2015) Experimental investigation of slosh parametric instability in liquid filled vessel under seismic excitations. Ann Nucl Energy 76:218–225
72. Chiba M, Shimazaki N, Ichinohe K (2014) Dynamic stability of a slender beam under horizontal–vertical excitations. J Sound Vib 333(5):1442–1472

73. Li TY, Xiong L, Zhu X, Xiong YP, Zhang GJ (2014) The prediction of the elastic critical load of submerged elliptical cylindrical shell based on the vibro-acoustic model. Thin-Walled Struct 84:255–262
74. Wu G-Y (2013) Non-linear vibration of bimaterial magneto-elastic cantilever beam with thermal loading. Int J Non Linear Mech 55:10–18
75. Younesian D, Esmailzadeh E, Sedaghati R (2013) Existence of periodic solutions for the generalized form of Mathieu equation. Nonlinear Dyn 39(4):335–348
76. Abraham GT, Chatterjee A (2003) Approximate asymptotics for a nonlinear Mathieu equation using harmonic balance based averaging. Nonlinear Dyn 31(4):347–365
77. Ng L, Rand R (2002) Bifurcations in a Mathieu equation with cubic nonlinearities: Part II. Commun Nonlinear Sci Numer Simul 7(3):107–121
78. Shukla PK (2011) Amplification of neutrino oscillations by a density ripple in dense plasmas. J Plasma Phys 7(3):289–291
79. Loos PF (2010) Hooke's law correlation in two-electron systems. Phys Rev A 81(3):032510
80. Momeni M, Kourakis I, Moslehi-Fard M (2007) A Van der Pol-Mathieu equation for the dynamics of dust grain charge in dusty plasmas. J Phys A: Math Theor 40(24):F473
81. El-Dib YO (2001) Nonlinear Mathieu equation and coupled resonance mechanism. Chaos, Solitons Fractals 12(4):705–720
82. Mahmoud GM (1997) Stability regions for coupled Hill's equations. Phys A Stat Mech Appl 242(1–2):239–249
83. Kovacic I, Cveticanin L (2009) The effects of strong cubic nonlinearity on the existence of periodic solutions of the Mathieu-Dufng equation. J Appl Mech Trans ASME 76(5):1–3
84. Cveticanin L, Kovacic I (2007) Parametrically excited vibrations of an oscillator with strong cubic negative nonlinearity. J Sound Vib 304(1–2):201–212

Chapter 2
Classical Methods

Abstract Several classical approaches for solving nonlinear oscillatory systems and different types of singular points are introduced in this chapter. Different perturbation techniques, such as straightforward expansion method (SEM), Lindstedt–Poincaré perturbation method (LPPM), multiple time-scales method (MTSM), and the Bogoliubov–Krylov averaging method (BKAM) are presented in the second section. The focus of the next section is on the parametric excitations and the well-known Hill and Mathieu equations. For better understanding of those methods, more than 15 different examples were analyzed and their results are fully discussed. The last section consists of 14 practice problems, which could be useful for researchers and instructors working in the field of nonlinear oscillations.

Keywords Straightforward expansion method · Lindstedt–Poincaré perturbation method · Multiple time-scales method · Bogoliubov–Krylov averaging method · Hill and Mathieu equations

This chapter presents classical approaches for solving nonlinear oscillatory systems. The concept of nonlinear differential equations is briefly discussed in the first section. Singular points are important tools for studying stability, and the methods to find them are given in this section. Subsequently, different classical techniques for solving several nonlinear problems are presented. Perturbation-based methods including the straightforward expansion, Lindstedt–Poincaré method, multiple time-scales method, and the method of strained parameters are fully discussed in the second section. In addition, the method of averaging is explained here. In the last part of this chapter, a brief illustration is provided about parametric excitation and the well-known Hill's equation.

E. Esmailzadeh et al., *Analytical Methods in Nonlinear Oscillations*, Solid Mechanics and Its Applications 252, https://doi.org/10.1007/978-94-024-1542-1_2

2.1 Nonlinear Differential Equations

Nonlinear differential equations have the following general form, which were previously described in Sect. 1.4:

$$\ddot{x} + f(x, \dot{x}, t) = 0 \tag{2.1.1}$$

One known example being the following equation, which is called the nonlinear conservative Duffing equation:

$$\ddot{x} + x + x^3 = 0 \tag{2.1.2}$$

Another form of the nonlinear nonconservative Duffing equation is given as

$$\ddot{x} + \dot{x} + x + x^3 = 0 \tag{2.1.3}$$

Different methods to find the singular points from differential equations are presented in the next section.

2.1.1 Singular Points

Mathematically speaking, a singular point is a point in the solution space in which all the derivatives with respect to time would be equal to zero. This means

$$\frac{\mathrm{d}}{\mathrm{d}t} = 0, \ \frac{\mathrm{d}}{\mathrm{d}t^2} = 0, \ \frac{\mathrm{d}}{\mathrm{d}t^3} = 0, \tag{2.1.4}$$

.

Table 2.1 shows the above definition:

It must be noted that singular points are actually those points where the system remains at dynamic equilibrium.

The oscillation of a simple pendulum about its vertical equilibrium position is mathematically formulated by the following incomplete second-order nonlinear ordinary differential equation:

$$\ddot{\theta} + \frac{g}{l}\sin\theta = 0 \tag{2.1.5}$$

Table 2.1 Singular points and their definitions

$\ddot{x} + f(x) = 0$	$\xrightarrow{\ddot{x}=0}$	By solving equation $f(x) = 0$, singular points are determined
$\ddot{x} + f(x, \dot{x}) = 0$	$\underset{\ddot{x}=0}{\xrightarrow{\dot{x}=0}}$	By solving equation $f(x, 0) = 0$, singular points are determined

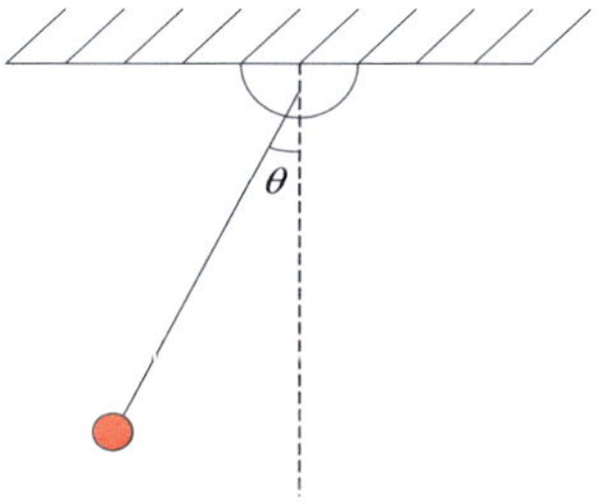

Fig. 2.1 Oscillating of a simple pendulum

To obtain the singular points for the above differential equation, one should set $\ddot{\theta} = 0$, and, hence $\sin\theta = 0$. Accordingly, the singular points of the pendulum are given as $\theta = n\pi, n = 0, 1, 2, \infty\ldots$ (Fig. 2.1).

2.1.2 Linearization Around Singular Points

Suppose one could have a set of nonlinear differential equations as

$$\begin{cases} \dot{x}_1 = f_1(x_1, x_2) \\ \dot{x}_2 = f_2(x_1, x_2) \end{cases} \tag{2.1.6}$$

In order to determine the singular points of the above set of equations, one should solve the following set of equations:

$$\begin{cases} f_1(x_1, x_2) = 0 \\ f_2(x_1, x_2) = 0 \end{cases} \tag{2.1.7}$$

The roots of Eq. (2.1.7) are known as the singular points of Eq. (2.1.6). The goal is to linearize Eq. (2.1.6) around its singular points, and could be assumed as x_{10} and x_{20}. Applying Taylor's expansion series for (f_1, f_2) around the points (x_{10}, x_{20}) results in having

$$f_1 = f_1(x_{10}, x_{20}) + \left.\frac{\partial f_1}{\partial x_1}\right|_{\substack{x_1 = x_{10} \\ x_2 = x_{20}}} (x_1 - x_{10}) + \left.\frac{\partial f_1}{\partial x_2}\right|_{\substack{x_1 = x_{10} \\ x_2 = x_{20}}} (x_2 - x_{20}) + \cdots \tag{2.1.8}$$

and

$$f_2 = f_2(x_{10}, x_{20}) + \left.\frac{\partial f_2}{\partial x_2}\right|_{\substack{x_1 = x_{10} \\ x_2 = x_{20}}} (x_2 - x_{20}) + \left.\frac{\partial f_2}{\partial x_1}\right|_{\substack{x_1 = x_{10} \\ x_2 = x_{20}}} (x_1 - x_{10}) + \cdots \tag{2.1.9}$$

By neglecting the higher order terms in the above equation, the following set of equations will be obtained, which is the linearized form of Eq. (2.1.6):

$$\begin{Bmatrix} \dot{x}_1 \\ \dot{x}_2 \end{Bmatrix} = \begin{bmatrix} \frac{\partial f_1}{\partial x_1} & \frac{\partial f_1}{\partial x_2} \\ \frac{\partial f_2}{\partial x_1} & \frac{\partial f_2}{\partial x_2} \end{bmatrix} \begin{Bmatrix} x_1 - x_{10} \\ x_2 - x_{20} \end{Bmatrix} \tag{2.1.10}$$

When the linearized form of the nonlinear differential equation is obtained, then one could easily analyze the stability of their corresponding singular points.

2.1.2.1 Example on Damped System

Consider a second-order system that has the following nonlinear differential equation:

$$\ddot{x} + 0.5\dot{x} + 2x + x^2 = 0 \tag{2.1.11}$$

By setting $\ddot{x} = \dot{x} = 0$, one could easily obtain the singular points of the system as

$$x^2 + 2x = 0 \Rightarrow x_1^* = 0, \text{ and } x_2^* = -2 \tag{2.1.12}$$

In order to linearize Eq. (2.1.11), one should first convert it into state space form as

$$\begin{aligned} \dot{x}_1 &= x_2 = f_1(x_1, x_2) \\ \dot{x}_2 &= -0.5x_2 - 2x_1 - x_1^2 = f_2(x_1, x_2) \end{aligned} \tag{2.1.13}$$

Then linearize the above equations around $x_1^* = 0$ as

$$\begin{Bmatrix} \dot{x}_1 \\ \dot{x}_2 \end{Bmatrix} = \begin{bmatrix} 0 & 1 \\ -2 & -0.5 \end{bmatrix}_{x^*=0} \begin{Bmatrix} x_1 \\ x_2 \end{Bmatrix} \tag{2.1.14}$$

Similarly, the following set of equations could be written for $x_2^* = -2$:

$$\begin{Bmatrix} \dot{x}_1 \\ \dot{x}_2 \end{Bmatrix} = \begin{bmatrix} 0 & 1 \\ -2 & -0.5 \end{bmatrix}_{x^*=-2} \begin{Bmatrix} x_1 \\ x_2 \end{Bmatrix} \tag{2.1.15}$$

The following set of equations by applying the change of variables is:

$$\begin{cases} u_1 = x_1 + 2 \\ u_2 = x_2 \end{cases}$$

$$\begin{Bmatrix} \dot{u}_1 \\ \dot{u}_2 \end{Bmatrix} = \begin{bmatrix} 0 & 1 \\ 2 & -0.5 \end{bmatrix} \begin{Bmatrix} u_1 \\ u_2 \end{Bmatrix} \tag{2.1.16}$$

2.1.3 *Classification of Singular Points*

Suppose we have an inhomogeneous nonlinear differential equation in the form of

$$\ddot{x}+f(x)=0 \tag{2.1.17}$$

Integrating both sides of Eq. (2.1.17) will lead to

$$\int \dot{x}\mathrm{d}\dot{x}+\int f(x)\mathrm{d}x=H \tag{2.1.18}$$

The above equation is indeed the total energy of the system when x represents its displacement. In other words, the above integral form is the corresponding energy equation for the conservative system, where H represents the initial level of the energy.

Integrating both sides of Eq. (2.1.18), one would have

$$\frac{1}{2}\dot{x}^2+F(x)=H \tag{2.1.19}$$

Mathematically speaking, $F(x)$ could have different forms of energy. The next section illustrates different possible forms of $F(x)$.

2.1.3.1 $F(x)$ with Monotonic Form

The first possible form of energy is monotonicity for function $F(x)$. The following example shows $F(x)$ with monotonic form:

$$\ddot{x}+\mathrm{e}^x=0\rightarrow f(x)=\mathrm{e}^x\rightarrow F(x)=\mathrm{e}^x \tag{2.1.20}$$

Figure 2.2 schematically illustrates a monotonic function of $F(x)$ and its corresponding phase-plane trajectories.

2.1.3.2 $F(x)$ with Concave Form

Suppose that $F(x)$ is a concave function. As a simple example would be

$$\ddot{x}-x=0\rightarrow f(x)=-x\rightarrow F(x)=-\frac{x^2}{2} \tag{2.1.21}$$

Fig. 2.2 **a** Schematic of a monotonic $F(x)$.
b Corresponding phase-plane trajectories

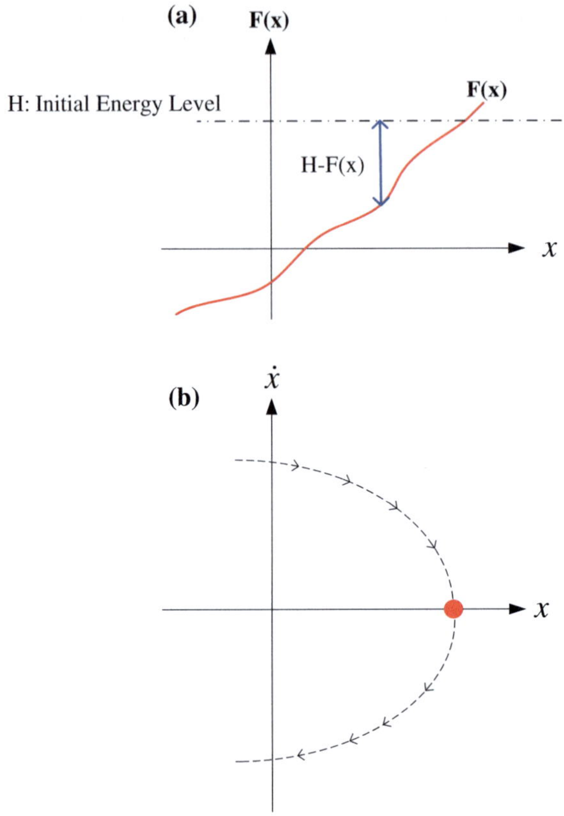

$$\ddot{x} - x + 1 = 0 \rightarrow f(x) = -x + 1 \rightarrow F(x) = -\frac{x^2}{2} + x \tag{2.1.22}$$

Figure 2.3 shows a concave form function and the point S refers to a singular point that is of a known saddle type.

Fig. 2.3 A schematic representation of a concave function

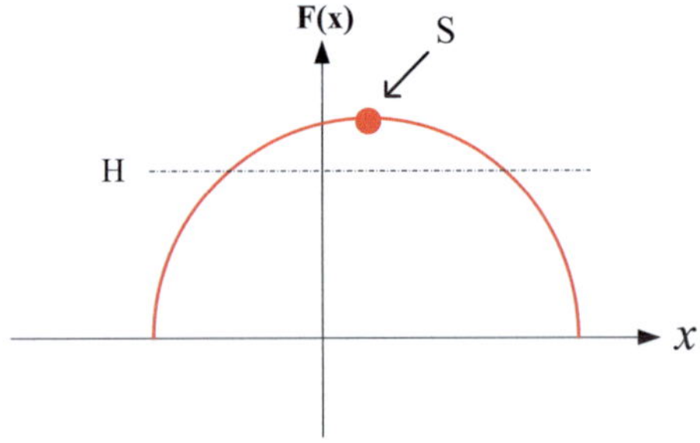

2.1.3.3 $F(x)$ with Convex Form

Assume that $F(x)$ is a convex function as illustrated in the following example:

$$\ddot{x} + x = 0 \rightarrow f(x) = x \rightarrow F(x) = \frac{x^2}{2} \tag{2.1.23}$$

To obtain the phase-plane trajectories of the above equation

$$\dot{x} = \pm\sqrt{2(h - F(x))} \tag{2.1.24}$$

Thus, it can be seen that the phase-plane trajectory is a closed curve. Hence, for this special case, there will be a periodic solution as

$$\frac{\mathrm{d}x}{\mathrm{d}t} = +\sqrt{2(h - F(x))} \quad \text{(Upper half part)} \tag{2.1.25}$$

Accordingly, one could have

$$\mathrm{d}t = +\frac{\mathrm{d}x}{\sqrt{2(h - F(x))}} \quad \text{(Upper half part)} \tag{2.1.26}$$

Finally, the following function is given to evaluate the period of oscillation of the system:

$$T = 2\int_{x_1}^{x_2} \frac{\mathrm{d}x}{\sqrt{2(h - F(x))}} \tag{2.1.27}$$

Example Find the period of oscillation for the following differential equation:

$$\ddot{x} + x = 0 \quad x(0) = a, \quad \dot{x}(0) = 0 \tag{2.1.28}$$

where $F(x) = \frac{x^2}{2}$ and $h = \frac{a^2}{2}$.

Accordingly, the period of oscillation for the motion described in Eq. (2.1.28) is:

$$T = 2\int_{-a}^{a} \frac{\mathrm{d}x}{\sqrt{2(a^2/2 - x^2/2)}} = 2\pi \tag{2.1.29}$$

2.1.3.4 Turning Point

The governing differential equation of motion for a nonlinear dynamic system is given as

$$\ddot{x} + x^2 = 0 \quad f(x) = x^2, \quad F(x) = \frac{x^3}{3} \tag{2.1.30}$$

Point D shown in Fig. 2.4a is a degenerative singular point with the following characteristics:

$$f''(x) = f'(x) = f(x) = 0 \tag{2.1.31}$$

Accordingly, the phase-plane trajectories touch the horizontal axis tangentially at point D as shown in Fig. 2.4b.

Explanations about singular points and their types, based on Sect. 2.1.2, are presented in Table 2.2.

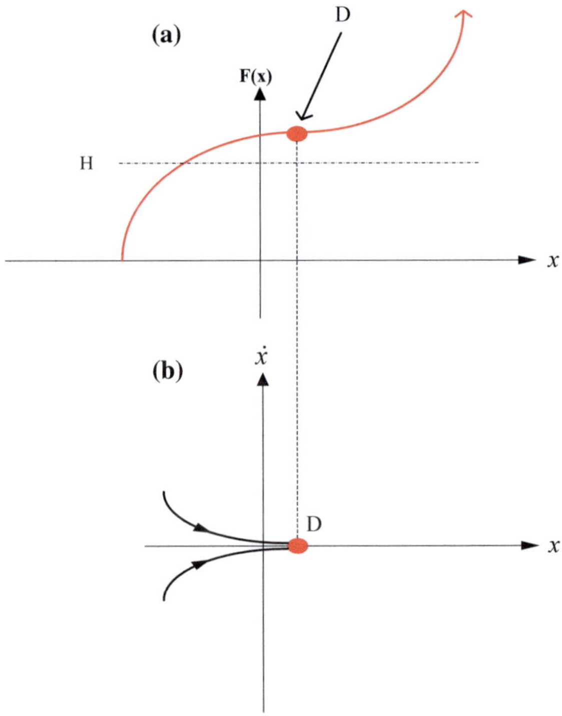

Fig. 2.4 **a** Function with a turning point. **b** Corresponding phase-plane trajectories

Table 2.2 Highlights and summary about singular points

$F(x)$	Monotonic	No singular point	Type
$F'(x) = 0$	$F''(x) < 0$	$F(x)$ has a maximum value	Saddle point
$F'(x) = 0$	$F''(x) > 0$	$F(x)$ has a minimum value	Center point
$F'(x) = 0$	$F''(x) = 0$	$F(x)$ has a turning point	Degenerative saddle point

2.2 Perturbation Methods

Perturbation methods are highly reliable and applicable in solving nonlinear differential equation, especially when the coefficient of nonlinear term is of smaller order than its other terms. Also, one could assume terms of small order based on the physics of the considered equation. Indeed, these methods rely on a dimensionless parameter in the problem that is relatively small [1, 2].

Generally, there are two possible cases when a term of nonlinear differential equation is highly smaller than the other terms:

(a) The coefficient pertinent to the above term is related to the highest derivative in the considered nonlinear differential equation (nth derivation);
(b) The coefficient of the above term is not related to the highest derivative in the considered nonlinear differential equation ($(n-1)$th derivation).

Accordingly, the first case illustrates the singular perturbation shown in the following equations:

$$\varepsilon\ddot{x} + x = 0 \tag{2.2.1}$$

and

$$\varepsilon\ddot{x} + \dot{x} + x = 0 \tag{2.2.2}$$

The second case is called the regular perturbation as shown in the nonlinear differential equation

$$\ddot{x} + \dot{x} + \varepsilon x^3 = 0 \tag{2.2.3}$$

where $\varepsilon \ll 1$. From the mechanics point of view, the first approach works when the terms related to the mass and acceleration have a very small coefficient, while the second approach is appropriate for the nonlinear differential equations with very small coefficient for the stiffness term.

2.2.1 Straightforward Expansion Method (SEM)

This method could be well explained by the following two examples:

Example 1: The following nonlinear differential equation is derived for a dynamic system [1–4]:

$$\dot{x}+x+\varepsilon x^2=0 \text{ with the initial condition of } x(0)=a \tag{2.2.4}$$

Based on the straightforward expansion method (SEM), one could apply the Taylor expansion to perturb its solution around the small parameter ε as

$$x(t)=x_0(t)+\varepsilon x_1(t)+\varepsilon^2x_2(t)+\cdots \tag{2.2.5}$$

where $x_0(t)$, $x_1(t)$ and $x_2(t)$ are the unknown functions that need to be determined.

Initially, Eq. (2.2.5) is substituted in Eq. (2.2.4), and the following equation is obtained:

$$\begin{aligned}&(\dot{x}_0(t)+\varepsilon\dot{x}_1(t)+\varepsilon^2\dot{x}_2(t)+\cdots)+(x_0(t)+\varepsilon x_1(t)+\varepsilon^2x_2(t))\\&\varepsilon(x_0(t)+\varepsilon x_1(t)+\varepsilon^2x_2(t)+\cdots)^2=0\end{aligned} \tag{2.2.6}$$

By collecting the terms with $\varepsilon^0, \varepsilon^1, \varepsilon^2$ from Eq. (2.2.6), one could have

$$(\dot{x}_0+x_0)\varepsilon^0+(\dot{x}_1+x_1+x_0^2)\varepsilon+(\dot{x}_2+x_2+2x_0x_1)\varepsilon^2=0 \tag{2.2.7}$$

Since the sequence of $\{\varepsilon^0, \varepsilon^1, \varepsilon^2, \ldots\}$ is a linear independent sequence, then all the coefficients of ε^i must be set to zero. Hence, the following set of equations, based on Eq. (2.2.7), is obtained:

$$\begin{cases}\varepsilon^0: -> \dot{x}_0+x=0\\ \varepsilon^1: -> \dot{x}_1+x+x_0^2=0\\ \varepsilon^2: -> \dot{x}_2+x=-2x_0x_1\end{cases} \tag{2.2.8}$$

The solution for the first equation is as follows:

$$x_0=Ae^{-t} \tag{2.2.9}$$

Substituting Eq. (2.2.9) into the second row of Eq. (2.2.8) and applying the iteration procedure to the higher order of the above equations would result in equation below

$$\dot{x}_1 + x = -A^2 e^{-2t} \tag{2.2.10}$$

Considering the initial condition as $x(0) = a$, the following form of equation is obtained:

$$x_0(0) + \varepsilon x_1(0) + \varepsilon^2 x_2(0) + \cdots = a \to x_0(0) = a, x_1(0) = 0, x_2(0) = 0, \ \ldots \tag{2.2.11}$$

$$x_0 = Ae^{-t} \to x_0(0) = a \to A = a \to x_0 = ae^{-t} \tag{2.2.12}$$

$$\dot{x}_1 + x_1 = -a^2 e^{-2t} \to x_1(t) = Be^{-t} + a^2 e^{-2t} \text{ and } x_1(0) = 0 \Rightarrow B = -a^2 \tag{2.2.13}$$

and

$$x_1(t) = a^2(e^{-2t} - e^{-t}) \tag{2.2.14}$$

therefore

$$\dot{x}_2 + x_2 = -2x_0 x_1 \Rightarrow \dot{x}_2 + x_2 = -2a^3(e^{-3t} - e^{-2t}) \tag{2.2.15}$$

and then

$$x_2(t) = a^3(e^{-t} - 2e^{-2t} + e^{-3t}) \tag{2.2.16}$$

Therefore, the approximate solution of Eq. (2.2.4) is obtained as

$$x(t) = ae^{-t} + \varepsilon a^2(e^{-2t} - e^{-t}) + \varepsilon^2 a^3(e^{-t} - 2e^{-2t} + e^{-3t}) + O(\varepsilon^3) \tag{2.2.17}$$

Example 2: The following nonlinear differential equation is given for a dynamic system:

$$\ddot{x} + x + \varepsilon x^2 = 0 \quad x(0) = a, \quad \dot{x}(0) = 0 \tag{2.2.18}$$

The following expansion for the solution of the above equation similar to the first example is used:

$$x(t) = x_0(t) + \varepsilon x_1(t) + \varepsilon^2 x_2(t) + \varepsilon^3 x_3(t) + \cdots \tag{2.2.19}$$

Substituting Eq. (2.2.19) into Eq. (2.2.18) would result in equation below

$$\varepsilon^0(\ddot{x}_0 + x_0) + \varepsilon(\ddot{x}_1 + x_1 + x_0^2) + \varepsilon^2(\ddot{x}_2 + x_2 + 2x_0x_1) + \cdots \tag{2.2.20}$$

Then, the following set of equations are found:

$$\begin{cases} \varepsilon^0: \ddot{x}_0 + x_0 = 0 \\ \varepsilon^1: \ddot{x}_1 + x_1 = -x_0^2 \\ \varepsilon^2: \ddot{x}_2 + x_2 = -2x_0x_1 \end{cases} \tag{2.2.21}$$

The initial guess for the solution could be in the form of

$$x_0(t) = a\cos t \tag{2.2.22}$$

Substituting Eq. (2.2.22) into the second row of Eq. (2.2.21), one obtains the following equation:

$$\ddot{x}_1 + x_1 = -a^2\cos^2 t = \frac{-a^2}{2}(1 + \cos 2t) \tag{2.2.23}$$

Considering the given two initial conditions, one could obtain x_1 as

$$x_1(t) = a^2\left\{\frac{\cos t}{3} - \left(\frac{1}{2} + \frac{1}{6}\cos 2t\right)\right\} \tag{2.2.24}$$

and for the second-order approximation, it will result in the following expression:

$$\varepsilon^2 : \ddot{x}_2 + x_2 = -2a^3\left\{\frac{1}{6} + \frac{1}{6}\cos 2t - \frac{1}{2}\cos t + \frac{1}{12}(\cos 3t + \cos t)\right\} \tag{2.2.25}$$

Finally, the final form of Eq. (2.2.26) is obtained as

$$\varepsilon^2 : \ddot{x}_2 + x_2 = -2a^3\left\{\overbrace{\frac{-5}{12}\cos t}^{\text{SecularTerm}} + \overbrace{\frac{1}{6} + \frac{1}{6}\cos 2t + \frac{1}{12}\cos 3t}^{\text{Non-Secular Term}}\right\} \tag{2.2.26}$$

If the straightforward procedure is implemented for the higher orders, the terms including the two factors $t^m\cos(\omega t + \alpha)$ and $t^m\cos(\omega t + \alpha)$ will then appear. They are called "secular" terms. Accordingly, the expansion of Eq. (2.2.26) is not periodic and the solution grows without bound as t tends to infinity. It must be noted that the uniformly expansion is valid as t increases. In order to increase the accuracy of the method and also overcome its shortcoming, the following modification is applied to SEM by assuming the following nonlinear differential equation [5]:

$$\ddot{x} + x + \varepsilon x^3 = 0 \text{ with initial conditions of } x(0) = A, \text{ and } \dot{x}(0) = 0 \tag{2.2.27}$$

It is assumed that the period of oscillation is $T = \frac{2\pi}{\omega}$, in which ω is a constant that must be identified. Thus, the following linearized equation is constructed:

$$\ddot{x} + \omega^2 x = 0 \tag{2.2.28}$$

Finally, one could rewrite Eq. (2.2.27) in the following form:

$$\ddot{x} + \omega^2 x + \varepsilon(x^3 + \eta x) = 0 \quad x(0) = A, \quad \text{and } \dot{x}(0) = 0 \tag{2.2.29}$$

where

$$\omega^2 + \varepsilon\eta = 1 \tag{2.2.30}$$

Substituting Eq. (2.2.19) into Eq. (2.2.29) and equating the coefficients of similar powers of ε yield

$$\ddot{x}_0 + \omega^2 x_0 = 0 \quad x_0(0) = A, \quad \dot{x}_0(0) = 0 \tag{2.2.31}$$

and

$$\ddot{x}_1 + \omega^2 x_1 + x_0^3 + \eta x_0 = 0 \quad x_1(0) = 0, \quad \dot{x}_1(0) = 0 \tag{2.2.32}$$

Solving Eq. (2.2.31) would result in

$$x_0 = A\cos\omega t \tag{2.2.33}$$

Therefore, Eq. (2.2.32) can be rewritten in the following form:

$$\ddot{x}_1 + \omega^2 x_1 + \left(\frac{3}{4}A^2 + \eta\right)A\cos\omega t + \frac{1}{4}A^3\cos 3\omega t = 0 \tag{2.2.34}$$

By neglecting the secular term, one obtains

$$\eta = -\frac{3}{4}A^2 \tag{2.2.35}$$

and accordingly,

$$x_1 = -\frac{A^3}{32\omega^2}(\cos\omega t - \cos 3\omega t) \tag{2.2.36}$$

The first-order approximation is

$$x = A\cos\omega t - \frac{\varepsilon A^3}{32\omega^2}(\cos\omega t - \cos^3\omega t) \tag{2.2.37}$$

Finally, its natural frequency can be obtained in the following form:

$$\omega = \sqrt{1 + \frac{3}{4}\varepsilon A^2} \tag{2.2.38}$$

2.2.2 *Lindstedt–Poincaré Perturbation Method (LPPM)*

In order to identify the weakness of SEM, one could apply the Lindstedt–Poincaré perturbation method to the following nonlinear differential equation:

$$\ddot{x} + f(x) = 0 \tag{2.2.39}$$

Considering $x = x_0$ as the singular point and defining the new variable as $u = x - x_0$, then one could rewrite Eq. (2.2.39) in the following form:

$$\ddot{u} + f(u + x_0) = 0 \tag{2.2.40}$$

Using the Taylor's expansion around x_0, the following new form of function f is obtained:

$$\begin{aligned} f(u + x_0) &= f(x_0) + f'(x_0)u + \frac{f''(x_0)}{2!}u^2 + \cdots = f'(x_0)u + \frac{f''(x_0)}{2!}u^2 + \cdots \\ &= \sum_{n=1}^{\infty} \alpha_n u^n \end{aligned} \tag{2.2.41}$$

Accordingly, Eq. (2.2.41) will have the following general form:

$$\ddot{u} + \sum_{n=1}^{\infty} \alpha_n u^n = 0 \tag{2.2.42}$$

By assuming $\omega_0^2 = \alpha_1$, one could obtain the successive form of Eq. (2.2.42):

$$\ddot{u} + \omega_0^2 u + \sum_{n=2}^{\infty} \alpha_n u^n = 0 \tag{2.2.43}$$

Based on the method of Lindstedt–Poincaré perturbation, the following expansion is defined:

$$u(\tau) = \varepsilon u_1(\tau) + \varepsilon^2 u_2(\tau) + \cdots \tag{2.2.44}$$

and

$$\tau = (\omega_0 + \omega_1 \varepsilon + \omega_2 \varepsilon^2 + \cdots)t \tag{2.2.45}$$

It is well-known that

$$\frac{\mathrm{d}}{\mathrm{d}t} = \frac{\mathrm{d}}{\mathrm{d}\tau} \cdot \frac{\mathrm{d}\tau}{\mathrm{d}t} = (\omega_0 + \omega_1 \varepsilon + \omega_2 \varepsilon^2 + \cdots) \frac{\mathrm{d}}{\mathrm{d}\tau} \tag{2.2.46}$$

and

$$\frac{\mathrm{d}^2}{\mathrm{d}t^2} = \frac{\mathrm{d}}{\mathrm{d}\tau} \cdot \frac{\mathrm{d}\tau}{\mathrm{d}t} = (\omega_0 + \omega_1 \varepsilon + \omega_2 \varepsilon^2 + \cdots)^2 \frac{\mathrm{d}^2}{\mathrm{d}\tau^2} \tag{2.2.47}$$

By substituting Eqs. (2.2.44) and (2.2.45) into Eq. (2.2.43), the following equation is obtained.

$$\begin{aligned} &(\omega_0 + \omega_1 \varepsilon + \omega_2 \varepsilon^2 + \cdots)^2 (\varepsilon \ddot{u}_1 + \varepsilon \ddot{u}_2 + \cdots) + \omega_0^2 (\varepsilon u_1 + \varepsilon^2 u_2 + \cdots) \\ &+ \sum_{n=2}^{\infty} \alpha_n (\varepsilon u_1 + \varepsilon^2 u_2 + \cdots)^n = 0 \end{aligned} \tag{2.2.48}$$

and the following set of equations is obtained by equating the coefficients of ε^i to zero:

$$\begin{cases} \varepsilon^1 \colon \ddot{u}_1 + u_1 = 0 \\ \varepsilon^2 \colon \omega_0^2 \ddot{u}_2 + \omega_0^2 u_2 = -2\omega_0 \omega_1 \ddot{u}_1 - \alpha_2 u_1^2 \end{cases} \tag{2.2.49}$$

The initial guess for the first part of Eq. (2.2.49) is

$$x_1(t) = a\cos(\tau + \beta) = a\cos\phi \tag{2.2.50}$$

Substituting Eq. (2.2.50) into the second part of Eq. (2.2.49) yields

$$\ddot{x}_2 + x_2 = \underbrace{2\frac{\omega_1}{\omega_0} a\cos\phi}_{\text{Secular Term}} - \frac{\alpha^2 a^2}{2\omega_0^2}(1 + \cos 2\phi) \tag{2.2.51}$$

By equating the secular term to zero, the following equation is obtained:

$$\ddot{x}_2 + x_2 = -\frac{\alpha^2 a^2}{2\omega_0^2}(1 + \cos 2\phi) \tag{2.2.52}$$

The solution of the above equation is

$$x_2 = -\frac{\alpha^2 a^2}{2\omega_0^2}\left(1 - \frac{1}{3}\cos 2\phi\right) \tag{2.2.53}$$

Substituting Eq. (2.2.53) in the third-order expansion as

$$\begin{aligned} \ddot{x}_3 + x_3 &= \frac{2\omega_2 a}{\omega_0}\cos\phi + 2\alpha_2^2\frac{a^3}{2\omega_0^4}\cos\phi\left(1 - \frac{1}{3}\cos 2\phi\right) - \alpha_3\frac{a^3}{\omega_0^2}\cos^3\phi \\ &= \underbrace{\cos\phi\left\{\frac{2\omega_2 a}{\omega_0} + \frac{\alpha_2^2 a^3}{\omega_0^4}\left(1 - \frac{1}{6}\right) - \frac{3}{4}\frac{\alpha_3 a^3}{\omega_0^2}\right\}}_{\text{Secular Term}} - \frac{\alpha_2^2 a^3}{6\omega_0^4}\cos 3\phi - \frac{1}{4}\frac{\alpha_3 a^3}{\omega_0^2}\cos 3\phi \end{aligned} \tag{2.2.54}$$

By omitting the secular term, one obtains ω_2 as

$$\omega_2 = \omega_0 a^2\left(\frac{9\alpha_3\alpha_1 - 10\alpha_2^2}{24\omega_0^4}\right) \tag{2.2.55}$$

Therefore, the third-order approximation is

$$x_3 = \frac{-a^3}{96\omega_0^2}\left(\frac{2\alpha_2^2}{\omega_0^2} + 3\alpha_3\right)\cos 3\phi$$

Finally, the approximate solution of Eq. (2.2.43) is found as

$$\begin{aligned} x(\tau) &= \varepsilon x_1 + \varepsilon^2 x_2 + \varepsilon^3 x_3 + \cdots = \varepsilon a\cos\phi - \varepsilon^2\frac{\alpha^2 a^2}{2\omega_0^2}\left(1 - \frac{1}{3}\cos 2\phi\right) \\ &\quad - \frac{\varepsilon^3 a^3}{96\omega_0^2}\left(\frac{2\alpha_2^2}{\omega_0^2} + 3\alpha_3\right)\cos 3\phi + \cdots \end{aligned} \tag{2.2.56}$$

and it is known that

$$\phi = \tau + \beta = (\omega_0 + \varepsilon^2 a^2\frac{9\alpha_3\alpha_1 - 10\alpha_2^2}{24\omega_0^3} + \cdots)t + \beta_0 \tag{2.2.57}$$

Substituting Eq. (2.2.57) into Eq. (2.2.56), the following equation is obtained:

$$x(\tau) = \varepsilon a \cos\left(\left(\omega_0 + \varepsilon^2 a^2 \frac{9\alpha_3\alpha_1 - 10\alpha_2^2}{24\omega_0^3} + \cdots\right)t + \beta_0\right) - \varepsilon^2 \frac{\alpha^2 a^2}{2\omega_0^2}\left(1 - \frac{1}{3}\cos 2\left(\left(\omega_0 + \varepsilon^2 a^2 \frac{9\alpha_3\alpha_1 - 10\alpha_2^2}{24\omega_0^3} + \cdots\right)t + \beta_0\right)\right) - \frac{\varepsilon^3 a^3}{96\omega_0^2}\left(\frac{2\alpha_2^2}{\omega_0^2} + 3\alpha_3\right)\cos 3\left(\left(\omega_0 + \varepsilon^2 a^2 \frac{9\alpha_3\alpha_1 - 10\alpha_2^2}{24\omega_0^3} + \cdots\right)t + \beta_0\right) + \cdots \tag{2.2.58}$$

The two unknown parameters a and β_0 are obtained by applying the two initial condition.

In 2002, a modified version of the Lindstedt–Poincaré perturbation method (LPPM) was proposed by He [6]. In order to illustrate LPPM, the following non-linear differential equation is considered:

$$\ddot{x} + \varepsilon x^3 = 0 \quad x(0) = a, \quad \text{and } \dot{x}(0) = 0 \tag{2.2.59}$$

Rewrite the above equation in the following form:

$$\ddot{x} + 0x + \varepsilon x^3 = 0 \quad x(0) = a, \quad \text{and } \dot{x}(0) = 0 \tag{2.2.60}$$

Then one could expand the constant coefficient "zero" as

$$0 = \omega^2 + \varepsilon c_1 + \varepsilon^2 c_2 + \cdots \tag{2.2.61}$$

Substituting Eqs. (2.2.5) and (2.2.61) into Eq. (2.2.60) and applying the perturbation procedure, the following equation is obtained:

$$\ddot{x}_0 + \omega^2 x_0 = 0 \quad x_0(0) = a, \quad \text{and } \dot{x}_0(0) = 0 \tag{2.2.62}$$

and

$$\ddot{x}_1 + \omega^2 x_1 + c_1 x_0 + x_0^3 = 0 \quad x_1(0) = 0, \quad \text{and } \dot{x}_1(0) = 0 \tag{2.2.63}$$

By solving Eq. (2.2.62), one could have the following equation as the initial solution:

$$x_0 = a\cos\omega t \tag{2.2.64}$$

Substituting Eq. (2.2.63) into Eq. (2.2.63), the following equation is obtained:

$$\ddot{x}_1 + \omega^2 x_1 + a(c_1 + \frac{3}{4}a^2)\cos\omega t + \frac{1}{4}a^3\cos 3\omega t = 0 \quad (2.2.65)$$

Eliminating the secular term requires that

$$c_1 = -\frac{3}{4}a^2 \quad (2.2.66)$$

and the following frequency is obtained as the first approximate solution:

$$\omega = \frac{\sqrt{3}}{2}\varepsilon^{1/2}a^2 \quad (2.2.67)$$

2.2.3 *Multiple Time-Scales Method (MTSM)*

Multiple time-scales method (MTSM) is one of the most well-known and highly applicable perturbation-based approaches [1, 2, 7, 8]. In order to fully examine this method, let us first consider the following nonlinear differential equation:

$$\ddot{x} + f(x) = 0 \quad (2.2.68)$$

where $f(x) = \sum_{n=1}^{\infty} \alpha_n x^n$. Based on the MTSM, the timescaling parameter is defined as follows:

$$T_n = \varepsilon^n t \quad n = 0, 1, 2, \ldots \quad (2.2.69)$$

Accordingly,

$$T_0 = t,\ T_1 = \varepsilon t,\ T_2 = \varepsilon^2 t, \ldots \quad (2.2.70)$$

The solution of Eq. (2.2.68) is perturbed around ε as

$$x(t) = \varepsilon x_1(T_0, T_1, T_2, \ldots) + \varepsilon^2 x_2(T_0, T_1, T_2, \ldots) + \cdots \quad (2.2.71)$$

where ε shows the order of solution.

Table 2.3 illustrates few basic notations, which have been frequently used in MSTM.

In order to solve Eq. (2.2.68), one should substitute Eq. (2.2.71) in Eq. (2.2.68) to get

Table 2.3 Few basic notations used in MSTM

Mathematical symbols and operators
$\frac{\partial}{\partial T_0} = D_0$, $\frac{\partial}{\partial T_1} = D_1$ and $\frac{\partial}{\partial T_2} = D_2, \ldots$
$\frac{\partial}{\partial t} = \frac{\partial}{\partial T_0} \cdot \frac{\partial T_0}{\partial t} + \frac{\partial}{\partial T_1} \cdot \frac{\partial T_1}{\partial t} + \frac{\partial}{\partial T_2} \cdot \frac{\partial T_2}{\partial t} + \cdots = D_0 + \varepsilon D_1 + \varepsilon^2 D_2 + \cdots$
$\frac{\partial^2}{\partial t^2} = (D_0 + \varepsilon D_1 + \varepsilon^2 D_2 + \cdots)(D_0 + \varepsilon D_1 + \varepsilon^2 D_2) = D_0^2 + 2D_0 D_1 \varepsilon + (D_1^2 + 2D_0 D_2)\varepsilon^2 + \cdots$
$(A + \bar{A})^2 = A^2 + 2A\bar{A} + \bar{A}^2 = A^2 + A\bar{A} + C.C.$
$(A + \bar{A})^3 = A^3 + 3A^2\bar{A} + 3A\bar{A}^2 + \bar{A}^3 = A^3 + 3A^2\bar{A} + C.C.$
$(A + \bar{A})(B + \bar{B}) = AB + A\bar{B} + \bar{A}B + \bar{A}\bar{B} = AB + A\bar{B} + C.C.$

$$
\begin{aligned}
&(D_0^2 + 2D_0 D_1 \varepsilon + (D_1^2 + 2D_0 D_2)\varepsilon^2 + \cdots)(\varepsilon x_1 + \varepsilon^2 x_2 + \cdots) \\
&+ \sum_{n=1}^{\infty} \alpha_n (\varepsilon x_1 + \varepsilon^2 x_2 + \cdots)^n = 0
\end{aligned} \tag{2.2.72}
$$

Equating the coefficients of similar powers of ε yields

$$\varepsilon^1 : D_0^2 x_1 + \omega_0^2 x_1 = 0 \tag{2.2.73}$$

and

$$\varepsilon^2 : D_0^2 x_2 + \omega_0^2 x_2 = -2D_0 D_1 x_1 - \alpha_2 x_1^2 \tag{2.2.74}$$

One assumes $\alpha_1 = \omega_0^2$ which is equivalent to the linearized natural frequency. Thus, the initial solution of the problem, considering Eq. (2.2.73), is defined as

$$x_1(T_0, T_1, T_2, \ldots) = a(T_1, T_2, \ldots)\cos(\omega_0 T_0) + b(T_1, T_2, \ldots)\sin(\omega_0 T_0) \tag{2.2.75}$$

In order to simplify the solution procedure, the following form of the solution is considered:

$$x_1 = \underbrace{A(T_1, T_2, \ldots)}_{\text{A complex number}} e^{i\omega_0 T_0} + \bar{A}(T_1, T_2, \ldots)e^{-i\omega_0 T_0} \tag{2.2.76}$$

Also, one may convert it in the following form:

$$x_1 = \underbrace{A(T_1, T_2, \ldots)}_{\text{A complex number}} e^{i\omega_0 T_0} + C.C. \tag{2.2.77}$$

Substituting Eq. (2.2.77) into Eq. (2.2.74) leads to

$$\begin{aligned} D_0^2x_2+\omega_0^2x_2 &= -2D_0D_1(Ae^{i\omega_0T_0})+C.C.-\alpha_2(A^2e^{2i\omega_0T_0}+A\bar{A})+C.C. \\ &= \underbrace{-2i\omega_0(D_1A)e^{i\omega_0T_0}}_{\text{Secular Term}}-\alpha_2(A^2e^{2i\omega_0T_0}+A\bar{A})+C.C. \end{aligned} \tag{2.2.78}$$

Eliminating the secular terms results in

$$D_1A=0\Rightarrow A=A(T_2,T_3) \tag{2.2.79}$$

The above equation indicates that the parameter A is not a function of T_1. Thus, one could obtain

$$x_2=\alpha_2\frac{A^2}{3\omega_0^2}e^{2i\omega_0T_0}-\alpha_2\frac{A\bar{A}}{\omega_0^2}+C.C. \tag{2.2.80}$$

The third-order expansion is

$$\varepsilon^3: D_0^2x_3+\omega_0^2x_3=-2D_0D_1x_2-(D_1^2+2D_0D_2)x_1-2\alpha_2x_1x_2-\alpha_3x_1^3 \tag{2.2.81}$$

Substituting the approximate solution into Eq. (2.2.81) results in

$$\begin{aligned} D_0^2x_3+\omega_0^2x_3 = &-\overbrace{2i\omega_0(D_2A)e^{i\omega_0T_0}}^{\text{Secular Term}}-2\alpha_2\left\{\begin{array}{l} \dfrac{\alpha_2A^3e^{3i\omega_0T_0}}{3\omega_0^2}-\overbrace{\dfrac{\alpha_2A^2\bar{A}}{\omega_0^2}e^{i\omega_0T_0}}^{\text{Secular Term}} \\ +\overbrace{\dfrac{\alpha_2A^2\bar{A}}{3\omega_0^2}e^{i\omega_0T_0}}^{\text{Secular Term}}-\overbrace{\dfrac{\alpha_2A\bar{A}^2}{\omega_0^2}e^{-i\omega_0T_0}}^{\text{Secular Term}} \end{array}\right\} \\ &-\alpha_3(A^3e^{3i\omega_0T_0}+\overbrace{3A^2\bar{A}e^{i\omega_0T_0}}^{\text{Secular Term}})+C.C. \end{aligned} \tag{2.2.82}$$

It must be noted that $-\alpha_2\frac{\bar{A}^2A}{\omega_0^2}e^{-i\omega_0T_0}$ is also a secular term since its complex conjugate exists in the $C.C.$ term, and it is equivalent to $-\alpha_2\frac{A^2\bar{A}}{\omega_0^2}e^{-i\omega_0T_0}$.

Eliminating the secular term would lead to

$$-2i\omega_0D_2A+\frac{10\alpha_2}{3\omega_0^2}A^2\bar{A}-3\alpha_3A^2\bar{A}=0 \tag{2.2.83}$$

It is known that

$$A = \frac{1}{2} a e^{i\beta} \tag{2.2.84}$$

Substituting Eq. (2.2.84) into Eq. (2.2.83) results in

$$-2i\omega_0 \left(\frac{1}{2} a' + \frac{1}{2} i\beta' a \right) e^{i\beta} + \left(\frac{10\alpha_2^2 - 9\alpha_3\omega_0^2}{3\omega_0^2} \right) \frac{1}{8} a^3 e^{i\beta} = 0 \tag{2.2.85}$$

The following set of equation is obtained by equating to zero the imaginary parts and the real parts of the above equation:

$$(\textbf{Imaginary Part}) \quad \omega_0 a' = 0 \Rightarrow a' = 0 \rightarrow a \neq f(T_1, T_2) a = \text{cte} = a_0 \tag{2.2.86}$$

and

$$(\textbf{Real Part}) \quad \beta' a_0 \omega_0 = \left(\frac{9\alpha_3\omega_0^2 - 10\alpha_2^2}{24\omega_0^2} \right) a_0^3 \tag{2.2.87}$$

From Eq. 2.2.87 one has

$$\beta = \left(\frac{9\alpha_3\omega_0^2 - 10\alpha_2^2}{24\omega_0^2} \right) a_0^2 T_2 + \beta_0 \tag{2.2.88}$$

where $T_2 = \varepsilon^2 T$. Therefore, the following form of solution is

$$x_1(t) = a_0 \cos(\omega_0 T_0 + B) = a_0 \cos\left\{ \omega_0 (1 + \frac{9\alpha_3\alpha_1 - 10\alpha_2^2}{24\alpha_1^2} \varepsilon^2 a_0^2 + \cdots) t + \beta_0 \right\} \tag{2.2.89}$$

It is evident from the above equation that the MSTM would have a similar solution to that of the LPPM.

The next few examples will focus on more complicated nonlinear problems and clearly illustrate the capability of MTSM for analyzing different types of nonlinear dynamical systems.

2.2.3.1 Vibration of Cantilever Beam Carrying an Intermediate Lumped Mass [9]

The governing differential equation for the first mode of vibration of a cantilever beam carries an intermediate lumped mass that is developed by Sadri, Younesian, and Esmailzadeh [9]

$$\ddot{x}+x=\varepsilon(-\alpha_1x^3-\alpha_2x^5-\beta_1x^2\ddot{x}-\beta_1x\dot{x}^2-\beta_2x^4\ddot{x}-\beta_2x^3\dot{x}^2-\mu x+F_0\cos\Omega\tau) \tag{2.2.90}$$

Considering the following expansion term for x:

$$x(\tau,\varepsilon)=x_0(T_0,T_1)+\varepsilon x_1(T_0,T_1)+\cdots \tag{2.2.91}$$

Substituting the above equation into Eq. (2.2.90) and equating the same power of ε lead to

$$\varepsilon^0: D_0^2x_0+x_0=0 \tag{2.2.92}$$

$$\begin{aligned}\varepsilon^1: D_0^2x_1+x_1=&-2D_0D_1x-\mu D_0x-\alpha_1x_0^3-\alpha_2x_0^5-\beta_1x_0^2D_0^2x_0\\&-\beta_1x_0^2[D_0x_0]^2-\beta_2x_0^4[D_0^2x_0]-\beta_2x_0^3[D_0x_0]^2+1/2F_0(e^{i\Omega\tau}+C.C.)\end{aligned} \tag{2.2.93}$$

The solution of Eq. (2.2.92) is $x_0=A(T_1)e^{iT_0}+C.C.$ and by substituting it in Eq. (2.2.93) leads to

$$\begin{aligned}D_0^2x_1+x_1=&\frac{1}{2}e^{i\Omega T_0}F_0-\alpha_1A^3e^{3iT_0}-3\alpha_1A^2\bar{A}e^{iT_0}-\alpha_2A^5e^{5iT_0}-5\alpha_2A^4\bar{A}e^{3iT_0}\\&-10\alpha_2A^3\bar{A}^2e^{iT_0}-i\mu Ae^{iT_0}+2\beta_1A^3e^{3iT_0}+2\beta_1A^2\bar{A}e^{iT_0}\\&+6\beta_2A^4\bar{A}e^{3iT_0}+2\beta_2A^5\bar{A}e^{5iT_0}+6\beta_2A^4\bar{A}e^{3iT_0}\\&+8\beta_2A^3\bar{A}^2e^{iT_0}-2iA'e^{iT_0}+C.C.\end{aligned} \tag{2.2.94}$$

Considering the primary resonance case, one considers the following excitation frequency Ω:

$$\Omega=1+\varepsilon\sigma \tag{2.2.95}$$

where σ is called the “detuning parameter”. Eliminating the secular terms yields

$$-3\alpha_1A^2\bar{A}-10\alpha_2A^3\bar{A}^2-i\,\mu A+2\beta_1A^2\bar{A}+8\beta_2A^3\bar{A}^2-2iA'+\frac{1}{2}F_0e^{i\sigma T_1}=0 \tag{2.2.96}$$

As was mentioned before, the secular terms must be eliminated since they will cause an unbounded growth of the response of the model. Assuming $\gamma=\sigma T_1-\beta$ and by separating the real and the imaginary parts would result in the following expression:

$$\frac{1}{2}F_0\cos\gamma - \frac{3}{8}\alpha_1 a^3 - \frac{5}{16}\alpha_2 a^5 + \frac{1}{4}\beta_1 a^3 + \frac{1}{4}\beta_2 a^5 + a(\sigma - \gamma') = 0 \quad (2.2.97)$$

$$\frac{1}{2}F_0\sin\gamma - \frac{1}{2}a\mu - a' = 0 \quad (2.2.98)$$

Finally, it is possible to obtain the steady-state response of the system by equating the two parameters a' and γ' to zero.

Interested readers are encouraged to find detailed information on this example in Ref. [9].

2.2.3.2 Nonlinear Viscoelastic Plates Subjected to Subsonic Flow and External Loads

The nonlinear vibration of a viscoelastic plate subjected to both a subsonic flow and external loads was investigated by Younesian and Norouzi [10]. They developed the governing differential equation by using the Galerkin method to investigate the nonlinear vibration of viscoelastic plates:

$$\ddot{x} + \varepsilon C\dot{x} + \omega_0^2 x + \varepsilon N x^3 + \varepsilon S x^2 \dot{x} = \varepsilon f_0 \cos\Omega t \quad (2.2.99)$$

Substituting Eq. (2.2.91) into Eq. (2.2.99) and separating the terms with the same power of ε, the following equation is obtained:

$$\varepsilon^0 : D_0^2 x_0 + x_0 = 0 \quad (2.2.100)$$

The solution of the above equation is

$$x_0(T_0, T_1) = A(T_1)e^{i\omega_0 T_0} + C.C. \quad (2.2.101)$$

Furthermore, one could have

$$\varepsilon^1 : D_0^2 x_1 + x_1 = -2D_0 D_1 x_0 - C D_0 x_0 - N x_0^3 - S x_0^2 D_0 x_0 + f_0 \cos\Omega t + C.C. \quad (2.2.102)$$

Initially, the nonresonant case is investigated, in which the frequency of the external force is assumed to be chosen far from the natural frequency of the linear system. Accordingly, by eliminating the secular terms for this case would result in

$$-2i\omega_0 D_1 A - CAi\omega_0 - N(3A^2\bar{A}) - S(A^2\bar{A}i\omega_0) = 0 \quad (2.2.103)$$

Substituting Eq. (2.2.84) into Eq. (2.2.103) and, by separating the real and the imaginary parts of the equation, the following system of differential equations will be obtained:

$$\begin{cases} \text{Re}: a\omega_0 \dfrac{d\beta}{dT_1} - \dfrac{3}{8}Na^3 = 0 \Rightarrow \begin{cases} a & = & 0 \\ \frac{d\beta}{dT_1} & = & \frac{3}{8}\frac{Na^2}{\omega_0} \end{cases} \\ \text{Im}: -\dfrac{da}{dT_1}\omega_0 - \dfrac{1}{2}aC\omega_0 - \dfrac{1}{8}a^3 S\omega_0 = 0 \end{cases} \tag{2.2.104}$$

Solving the above differential equations, the following solutions are determined for a and β:

$$a(T_1) = \left(\frac{4(\frac{C}{S})a_0^2 e^{-CT_1}}{a_0^2(1 - e^{-CT_1}) + \frac{4C}{S}} \right)^{1/2} \tag{2.2.105}$$

and

$$\beta(T_1) = \frac{3N}{2\omega_0 S} \ln \left(\frac{(1 - e^{-CT_1}) + 4\frac{C}{Sa_0^2}}{4\frac{C}{Sa_0^2}} \right)^{1/2} + \beta_0 \tag{2.2.106}$$

The following equation as the solution of Eq. (2.2.99) is obtained after eliminating the secular terms:

$$x_1(T_0, T_1) = \frac{a^3}{32\omega_0^2}(N\cos 3\phi - S\omega_0 \sin 3\phi) + \Lambda \cos \omega_0 T_0 \tag{2.2.107}$$

where

$$\phi = \omega_0 T_0 + \beta, \Lambda = \frac{f_0}{2(\omega_0^2 - \Omega^2)} \tag{2.2.108}$$

and the approximate solution of the problem is determined as [10]

$$x(T_0, T_1) = x_0(T_0, T_1) + \varepsilon x_1(T_0, T_1) + O(\varepsilon^2) \tag{2.2.109}$$

$$\begin{aligned} x(t) = a\cos(\omega_0 t + \beta) + \frac{\varepsilon a^3}{32\omega_0^2}(N\cos(3\omega_0 t + 3\beta) \\ - S\omega_0 \sin(3\omega_0 t + 3\beta)) + 2\varepsilon\Lambda \cos \Omega t \end{aligned} \tag{2.2.110}$$

In order to find the primary resonance of the equation, the following equation is assumed:

$$\Omega = \omega_0 + \varepsilon\sigma \tag{2.2.111}$$

Rewriting Eq. (2.2.102) in the following form:

$$\varepsilon^1 : D_0^2 x_1 + \omega_0^2 x_1 = -2i\omega_0 D_1 A e^{i\omega_0 T_0} - CAi\omega_0 e^{i\omega_0 T_0} - N(A^3 e^{3i\omega_0 T_0} + 3A^2\bar{A} e^{i\omega_0 T_0}) - S(i\omega_0 A^3 e^{3i\omega_0 T_0} + A^2\bar{A} i\omega_0 e^{i\omega_0 T_0}) + \frac{f_0}{2} e^{i\omega_0 T_0} e^{i\sigma T_1} + C.C. \tag{2.2.112}$$

Then the secular terms were eliminated to obtain the equation below

$$-2i\omega_0 \frac{dA}{dT_1} - CAi\omega_0 - 3NA^2\bar{A} - Si\omega_0 A^2\bar{A} + \frac{f_0}{2} e^{i\sigma T_1} = 0 \tag{2.2.113}$$

Substituting Eq. (2.2.84) into Eq. (2.2.113) and by separating the real and the imaginary parts, the following set of differential equations is obtained:

$$\begin{cases} \mathrm{Re} : \dfrac{da}{dT_1} = -\dfrac{1}{2} aC - S\dfrac{a^3}{3} + \dfrac{f_0}{2\omega_0} \sin\gamma \\ \mathrm{Im} : \dfrac{d\gamma}{dT_1} - \sigma a = -\dfrac{3Na^2}{8} + \dfrac{f_0}{2a\omega_0} \cos\gamma \end{cases} \tag{2.2.114}$$

where $\gamma = \sigma T_1 - \beta$. After setting the two parameters to zero ($a' = \gamma' = 0$), one obtains

$$\left(\frac{1}{2} aC + S\frac{a^3}{3}\right)^2 + \left(\frac{3Na^3}{8\omega_0} - \sigma a\right)^2 = \left(\frac{f_0}{2\omega_0}\right)^2 \tag{2.2.115}$$

After few mathematical simplifications, the following relationship between the detuning parameter and the steady-state amplitude is obtained:

$$\sigma = \frac{3Na^2}{8\omega_0^2} \pm \left[\left(\frac{f_0}{2a\omega_0}\right)^2 - \left(\frac{C}{2} + \frac{Sa^2}{3}\right)^2\right]^{1/2} \tag{2.2.116}$$

2.2.3.3 Nonlinear Vibration of Variable Speed Rotating Viscoelastic Beams

Younesian and Esmailzadeh studied the nonlinear vibration of variable speed rotating viscoelastic beam [11, 12]. In this part of the chapter, their model is considered as an example for the utilization of MTSM for solving a nonlinear ordinary differential equation. The following nonlinear differential equation describes the vibration of a variable speed rotating viscoelastic beam [11]:

$$
\begin{aligned}
&\ddot{x}+\left[12.36+\Omega^2(1.193+1.987\beta)\right]x+12.362\eta\dot{x}\\
&+(22.319-5.088\Omega^2)x^3+4.265x\dot{x}^2-17.963\eta x\dot{x}\\
&-320.058x^5-52.724\eta x^2\dot{x}-18.19x^3\dot{x}^2+76.612\eta x^3\dot{x}=0
\end{aligned}
\tag{2.2.117}
$$

The above differential equation is rewritten in the following form:

$$
\ddot{x}+\left[12.36+\ \Omega^2(1.193+1.987\beta)\right]x=\varepsilon f(x,\dot{x}) \tag{2.2.118}
$$

where

$$
f(x,\dot{x})=-\begin{bmatrix} 12.362\eta\dot{x} \\ +(22.319-5.088\Omega^2)x^3+4.265x\dot{x}^2-17.963\eta x\dot{x} \\ -320.058x^5-52.724\eta x^2\dot{x}-18.19x^3\dot{x}^2+76.612\eta x^3\dot{x} \end{bmatrix} \tag{2.2.119}
$$

Substituting Eq. (2.2.91) into Eq. (2.2.118) would lead to

$$
D_0^2x_0+\left[12.362+\ \Omega^2(1.193+1.987\beta)\right]x_0=0 \tag{2.2.120}
$$

$$
D_0^2x_1+\left[12.362+\ \Omega^2(1.193+1.987\beta)\right]x_1=-2D_0D_1x_0+f(x_0,D_0x_0) \tag{2.2.121}
$$

Considering Eq. (2.2.101) as the solution of Eq. (2.2.120) and by substituting Eq. (2.2.101) into Eq. (2.2.121), and neglecting the secular terms, the following mathematical relation is obtained:

$$
\begin{aligned}
2iDA_1=\frac{1}{2\pi}\int_0^{2\pi/\hat{\Omega}} &f\left[A\exp(i\hat{\Omega}T_0)+\bar{A}\exp(-i\hat{\Omega}T_0),i\hat{\Omega}A\exp(i\hat{\Omega}T_0)-i\hat{\Omega}\bar{A}\exp(-i\hat{\Omega}T_0)\right]\\
&\times\exp(-i\hat{\Omega}T_0)\mathrm{d}T_0
\end{aligned}
\tag{2.2.122}
$$

In order to solve the above differential equation, the following function A is considered:

$$
A(T_1)=\frac{1}{2}a(T_1)\exp(i\gamma(T_1)) \tag{2.2.123}
$$

Substituting Eq. (2.2.123) into Eq. (2.2.122) results in

$$
i(a'+ia\gamma')=\tfrac{1}{2\pi\hat{\Omega}}\int_0^{2\pi}f\left[a\cos(\phi),-a\hat{\Omega}\sin(\phi)]\times\exp(-i\phi)\mathrm{d}\phi\right]
$$

After few mathematical works, the following vibration amplitude equation is obtained:

$$a(T_1) = 0.095a(0)\sqrt{\frac{\exp(-12.36\eta T_1)}{0.009 + 0.119\eta\ \exp(-12.36\eta T_1)}} \tag{2.2.124}$$

Detailed solution procedure for the case of time varying rotating beams is given in Ref. [1].

2.2.3.4 Resonance Analyses of Clamped–Clamped Microbeams

As another example, part of a recently published article by Younesian et al. [13], in which they studied the nonlinear vibration of a clamped–clamped microbeam, is discussed here. Considering the clamped–clamped boundary conditions and by using the Galerkin method, the following nonlinear ordinary differential equation is obtained:

$$\ddot{x} + \omega_0^2 x = \varepsilon\left[-\mu\dot{x} - \frac{a_1}{a_3}x^4\ddot{x} - \frac{\alpha_2}{\alpha_3}x^2\ddot{x} - \frac{\alpha_5}{\alpha_3}x^3 - \frac{\alpha_6}{\alpha_3}x^5 - \frac{\alpha_7}{\alpha_3}x^7 + F_0\cos\Omega t\right] \tag{2.2.125}$$

where ε is a small dimensionless parameter. Substituting Eq. (2.2.91) into Eq. (2.2.125) and equating the terms with the same power of ε would give

$$\varepsilon^0 : D_0^2 x_0 + \omega_0^2 x_0 = 0 \tag{2.2.126}$$

$$\begin{aligned}\varepsilon^1 : D_0^2 x_1 + \omega_0^2 x_1 = &- 2D_0 D_1 x_0 - \mu D_0 x_0 - \frac{\alpha_5}{\alpha_3}x_0^3 - \frac{\alpha_6}{\alpha_3}x_0^5 \\ &- \frac{\alpha_7}{\alpha_3}x_0^7 - \frac{\alpha_2}{\alpha_3}x_0^2[D_0^2 x_0] - \frac{\alpha_1}{\alpha_3}x_0^4[D_0^2 x_0] + \frac{1}{2}F_0(e^{i\Omega t} + C.C.)\end{aligned} \tag{2.2.127}$$

From the above equation, the parameter x_0 is determined as [13]

$$x_0 = A(T_1)e^{i\omega_0 T_0} + C.C. \tag{2.2.128}$$

The following equation is obtained after substituting Eq. (2.2.128) into Eq. (2.2.127):

$$\varepsilon^1 : D_0^2 x_1 + \omega_0^2 x_1 = \frac{1}{\alpha_3}\left[\begin{array}{r} -2ie^{i\omega_0 T_0}\omega_0\alpha_3 A' - A^3\alpha_5 e^{3i\omega_0 T_0} - A^5\alpha_6 e^{5i\omega_0 T_0} - \alpha_7 A^7 e^{7i\omega_0 T_0} \\ +\frac{1}{2}\alpha_3 F_0 e^{i\Omega T_0} - 3\alpha_5\bar{A}A^2 e^{i\omega_0 T_0} - 5\alpha_6\bar{A}A^4 e^{3i\omega_0 T_0} \\ -7\bar{A}A^6\alpha_7 e^{5i\omega_0 T_0} - 10\bar{A}^2A^3\alpha_6 e^{i\omega_0 T_0} - 21\bar{A}^2A^5\alpha_7 e^{3i\omega_0 T_0} \\ -35\bar{A}^3A^4\alpha_7 e^{i\omega_0 T_0} - i\mu\omega_0 A\alpha_3 e^{i\omega_0 T_0} + \alpha_2\omega_0^2 A^3 e^{3i\omega_0 T_0} \\ +\alpha_1 A^5\omega_0^2 e^{5i\omega_0 T_0} + 3\alpha_2\bar{A}A^2 e^{i\omega_0 T_0}\omega_0^2 + 5\alpha_1\bar{A}A^4 e^{3i\omega_0 T_0}\omega_0^2 \\ +10\alpha_1\bar{A}^2A^3 e^{i\omega_0 T_0}\omega_0^2 + C.C. \end{array}\right] \quad (2.2.129)$$

Considering the primary resonance as $\Omega = \omega_0 + \varepsilon\sigma$, and by eliminating the secular terms, the following equation is obtained:

$$-2i\omega_0\alpha_3 A' - 3\alpha_5\bar{A}A^2 - 10\bar{A}^2A^3\alpha_6 - 35\bar{A}^3A^4\alpha_7 - i\mu\omega_0 A\alpha_3 + 3\alpha_2\bar{A}A^2\omega_0^2 + 10\alpha_1\bar{A}^2A^3\omega_0^2 + \frac{1}{2}\alpha_3 F_0 e^{i\sigma T_1} = 0 \quad (2.2.130)$$

Substituting Eq. (2.2.84) into Eq. (2.2.130), and by separating the real and the imaginary parts, the following set of differential equations is obtained [13]:

$$-\frac{3\alpha_5 a^3}{8} - 5\frac{a_6 a^5}{16} - 35\frac{a_7 a^7}{128} + \frac{1}{2}\alpha_3 F_0\cos\gamma + \sigma a\omega_0\alpha_3 + \frac{3\alpha_2\omega_0^2 a^3}{8} + \frac{5\alpha_1 a^5\omega_0^2}{16} - a\alpha_3\omega_0\gamma' = 0 \quad (2.2.131)$$

and

$$\frac{1}{2}F_0\sin\gamma - \frac{1}{2}\mu a\omega_0 - \omega_0 a' = 0 \quad (2.2.132)$$

The steady-state response of the system can be obtained by setting the two parameters to zero $a' = \gamma' = 0$. The above set of equations can then be solved to determine σ in terms of the parameter a.

Interested readers can find further analyses of several nonlinear mechanical systems using the multiple time-scales method (MTSM) that were investigated by the authors and are compiled in the “Practice Problems” of this chapter.

2.2.4 Bogoliubov–Krylov Averaging Method (BKAM)

The Bogoliubov–Krylov averaging method (BKAM) is a powerful and useful method of solving nonlinear differential equations presented in the following form [2, 14–16]:

$$\ddot{x} + \omega_0^2 x = f(x, \dot{x}) \tag{2.2.133}$$

The main concept of BKAM is to expand the solution of the nonlinear differential equation around its linear solution, so that the linear solution of the above equation would be

$$x(t) = a\cos(\omega_0 t + \beta) \tag{2.2.134}$$

where a and β are the two constant parameters. In the nonlinear case, the solution of the equation has the following form:

$$x(t) = a(t)\cos(\omega_0 t + \beta(t)) \tag{2.2.135}$$

Taking derivative with respect to time would lead to

$$\dot{x}(t) = \dot{a}\cos\phi - a\omega_0 \sin\phi - a\dot{\beta}\sin\phi \tag{2.2.136}$$

It is first assumed that $a\omega_0 \sin\phi - a\dot{\beta}\sin\phi \approx 0$.
Accordingly, the second derivative of $x(t)$ with respect to time is

$$\ddot{x}(t) = -\dot{a}\omega_0 \sin\phi - a\omega_0^2 \cos\phi - a\omega_0\dot{\beta}\cos\phi \tag{2.2.137}$$

Substituting the above equation into Eq. (2.2.133) results in

$$\begin{aligned} &-\dot{a}\omega_0 \sin\phi - a\omega_0^2 \cos\phi - a\omega_0\dot{\beta}\cos\phi \\ &+ a\omega_0^2 \cos\phi = f(a\cos\phi, -a\omega_0 \sin\phi) \end{aligned} \tag{2.2.138}$$

Therefore, the two sets of differential equations are obtained as

$$\begin{cases} -\dot{a}\omega_0 \sin\phi - a\omega_0\dot{\beta}\cos\phi = f(a\cos\phi, -a\omega_0 \sin\phi) \\ \dot{a}\cos\phi - a\dot{\beta}\sin\phi = 0 \end{cases} \tag{2.2.139}$$

Solving the above two sets of differential equations will lead to

$$-\dot{a} = -\frac{1}{\omega_0} f(a\cos\phi, -a\omega_0 \sin\phi)\sin\phi \tag{2.2.140}$$

and

$$\dot{\beta} = -\frac{1}{a\omega_0} f(a\cos\phi, -a\omega_0 \sin\phi)\cos\phi \tag{2.2.141}$$

It is assumed that there is a slowly time variation in the time gradients. Accordingly, one can write the following relationships:

$$\dot{a} \approx \dot{a}_{\text{ave}} \text{ and } \dot{\beta} \approx \dot{\beta}_{\text{ave}} \tag{2.2.142}$$

Thus, it is possible to have the following expressions for one period of oscillation:

$$\underset{0-2\pi}{\dot{a}_{\text{ave}}} = \frac{1}{2\pi}\int_0^{2\pi} \dot{a}\mathrm{d}\phi \text{ and } \underset{0-2\pi}{\dot{\beta}_{\text{ave}}} = \frac{1}{2\pi}\int_0^{2\pi} \dot{\beta}\mathrm{d}\phi \tag{2.2.143}$$

Accordingly, the following set of equations could be written for one period of oscillation:

$$\begin{cases} \dot{a}_{\text{ave}} = \dfrac{-1}{2\pi\omega_0}\displaystyle\int_0^{2\pi} f(a\cos\phi, -a\omega_0\sin\phi)\sin\phi\mathrm{d}\phi \\ \dot{\beta}_{\text{ave}} = \dfrac{-1}{2\pi\omega_0 a}\displaystyle\int_0^{2\pi} f(a\cos\phi, -a\omega_0\sin\phi)\cos\phi\mathrm{d}\phi \end{cases} \tag{2.2.144}$$

The above two equations are the principal formulas of the Bogoliubov–Krylov averaging method (BKAM).

2.2.4.1 Linear Differential Equation

For a dynamic system, its governing differential equation could have the following form:

$$\ddot{x} + \omega_0^2 x = -2\xi\omega_0\dot{x} \tag{2.2.145}$$

Based on the Bogoliubov–Krylov averaging method, and by substituting Eq. (2.2.144) into Eq. (2.2.145), the following formulation can be obtained:

$$\dot{a} = \frac{-1}{2\pi\omega_0}\int_0^{2\pi}(-2\xi\omega_0)(-a\omega_0 \sin\phi\cos\phi)\mathrm{d}\phi \tag{2.2.146}$$

and

$$\dot{a} = \frac{-\xi\omega_0}{\pi}\frac{2\pi}{2}a = -\xi\omega_0 a \tag{2.2.147}$$

Therefore, one could deduce the following equation:

$$a(t) = a_0 \mathrm{e}^{-\xi\omega_0 t};\ a_0:\ \text{Initial condition} \tag{2.2.148}$$

and

$$\dot{\beta} = \frac{-1}{2\pi\omega_0 a}\int_0^{2\pi}(-2\xi\omega_0)(-a\omega_0 \sin\phi\cos\phi)\mathrm{d}\phi = 0 \rightarrow \beta = \beta_0 \tag{2.2.149}$$

Thus, the following equation is presented as the solution of Eq. (2.2.145):

$$x = a_0 \mathrm{e}^{-\xi\omega_n t}\cos(\omega_0 t + \beta_0) \tag{2.2.150}$$

2.2.4.2 Coulomb Friction

The oscillation of a single-degree-of-freedom system in the presence of Coulomb damping (dry friction) can be described by the following differential equation:

$$\ddot{x} + \omega_0^2 x = \begin{cases} -f_f & \dot{x} > 0 \\ f_f & \dot{x} < 0 \end{cases} \tag{2.2.151}$$

Based on the Bogoliubov–Krylov averaging method (BKAM), one could have the following integration for Eq. (2.2.151):

$$\dot{a} = \frac{-1}{2\pi\omega_0}\left\{\int_0^{\pi} f_f \sin\phi\mathrm{d}\phi + \int_{\pi}^{2\pi} -f_f \sin\phi\mathrm{d}\phi\right\} = \frac{-2f_f}{\pi\omega_0} \tag{2.2.152}$$

Performing the integration of the above equation with respect to time leads to

$$a(t) = \frac{-2f_f}{\pi\omega_0}t + a_0 \tag{2.2.153}$$

and

$$\dot{\beta} = \frac{-1}{2\pi\omega_0 a}\left\{\int_0^{\pi} f_f \cos\phi \mathrm{d}\phi + \int_{\pi}^{2\pi} f_f \cos\phi \mathrm{d}\phi = 0\right\} \to \beta = \beta_0 \tag{2.2.154}$$

Accordingly, the solution of Eq. (2.2.151) is given as

$$x = \left(\frac{-2f_f t}{\pi\omega_0} + a\right)\cos(\omega_0 t + \beta_0) \tag{2.2.155}$$

2.2.4.3 Van der Pol Equation

The well-known van der Pol equation is mathematically described by the following equation [17]:

$$\ddot{x} + \omega_0^2 x = \dot{x}(1 - x^2) \tag{2.2.156}$$

A comprehensive review of the van der Pol equation will be carried out in Chap. 6. Based on the Bogoliubov–Krylov averaging method, one could develop the following expression:

$$\dot{a} = \frac{-1}{2\pi\omega_0}\int_0^{2\pi} (-a\omega_0 \sin^2\phi)(1 - a^2\cos^2\phi)\mathrm{d}\phi = -\frac{1}{8}a(a^2 - 4) \tag{2.2.157}$$

Equation (2.2.157) can be rewritten in the following form:

$$\frac{\mathrm{d}a}{a(a^2 - 4)} = -\frac{1}{8}\mathrm{d}t \to \frac{1}{8}\mathrm{Ln}\left(1 - \frac{4}{a^2}\right) = -\frac{1}{8}t + t_0 \tag{2.2.158}$$

and therefore

$$a^2 = \frac{4}{1 - \left(1 - \frac{4}{a_0^2}\right)\mathrm{e}^{-t}} \tag{2.2.159}$$

Moreover, based on Eq. (2.2.144), the following formulation can be developed:

$$\dot{\beta} = \frac{-1}{2\pi a\omega_0}\int_0^{2\pi} (-a\omega_0 \sin\phi)(1 - a^2\cos^2\phi)\cos\phi \mathrm{d}\phi = 0 \to \beta = \beta_0 = \text{Constant} \tag{2.2.160}$$

2.2.4.4 Rayleigh Equation

The well-known Rayleigh equation is described by the following differential equation [2]:

$$\ddot{x} + \omega_0^2 x = \dot{x} - \dot{x}^3 \tag{2.2.161}$$

Based on the Bogoliubov–Krylov averaging method, the following equation can be developed:

$$\dot{a} = \frac{-1}{2\pi\omega_0}\int_0^{2\pi} (-a\omega_0 \sin^2\phi + (a\omega_0)^3 \sin^4\phi)\mathrm{d}\phi = \frac{a}{2}(1 - \frac{3}{4}a^2\omega_0^2) \tag{2.2.162}$$

The differential equation of Eq. (2.2.162) can be rewritten in the following form:

$$\mathrm{d}t = \frac{2\,\mathrm{d}a}{a\left(1 - \frac{3}{4}a^2\omega_0^2\right)} \tag{2.2.163}$$

By carrying out the integration and also considering the initial condition, it would lead to

$$a^2 = \frac{\frac{3\omega_0^2}{4}}{1 + (1 - \frac{4}{3\omega_0^2 a_0^2})\mathrm{e}^{-t}} \tag{2.2.164}$$

Moreover, based on Eq. (2.2.144), the following formulation could be developed:

$$\dot{\beta} = \frac{-1}{2\pi a\omega_0}\int_0^{2\pi} -a\omega_0 \sin\phi\cos\phi + (a\omega_0)^3 \sin^3\phi\cos\phi \mathrm{d}\phi = 0 \to \beta = \beta_0 = \text{Constant} \tag{2.2.165}$$

2.2.4.5 Generalized van der Pol Type Oscillator

Kovacic and Mickens analyzed a generalized form of the van der Pol type oscillator with the following differential equation form [18]:

$$\ddot{x} + \mathrm{sgn}(x)|x|^{\alpha} = \varepsilon f(x, \dot{x}) \tag{2.2.166}$$

in which $f(x, \dot{x}) = (1 - |x|^{\beta})|\dot{x}|^{\gamma}\mathrm{sgn}(\dot{x})$.

Based on the Bogoliubov–Krylov averaging method, the following equation is derived.

$$\dot{a} = -\frac{2\varepsilon}{\pi c(\alpha+3)|a|^{\alpha-1/2}}\int_0^{2\pi} (1 - |a\cos\phi|^{\beta})|-a\omega\sin\phi|^{\gamma}\mathrm{sgn}(-a\omega\sin\phi)\sin\phi\, \mathrm{d}\phi \tag{2.2.167}$$

and

$$a\dot{\beta} = -\frac{\varepsilon}{\pi c|a|^{\alpha-1/2}}\int_0^{2\pi} (1 - |a\cos\phi|^{\beta})|-a\omega\sin\phi|^{\gamma}\mathrm{sgn}(-a\omega\sin\phi)\cos\phi\, \mathrm{d}\phi \tag{2.2.168}$$

Integrating the above two equations would result in getting interesting results [18]

$$\dot{a} = \frac{4\varepsilon c^{\gamma-1}}{\pi(\alpha+3)} \frac{|a|^{\gamma\frac{\alpha+1}{2}-\frac{\alpha-1}{2}}\Gamma(1+\frac{\gamma}{2})\left[\sqrt{\pi}\,\Gamma(\frac{3+\beta+\gamma}{2}) - |a|^{\beta}\,\Gamma(\frac{1+\beta}{2})\Gamma(\frac{3+\gamma}{2})\right]}{\Gamma(\frac{3+\gamma}{2})\Gamma(\frac{3+\beta+\gamma}{2})} \tag{2.2.169}$$

$$\dot{\beta} = 0 \tag{2.2.170}$$

2.3 Parametric Excitation and Hill's Equation

The focus of this section will be on the differential equations with the time-dependent coefficients as indicated in the following equation:

$$\ddot{x} + P_1(t)\dot{x} + P_2(t)x = 0 \tag{2.3.1}$$

where P_1 and P_2 are periodic functions with the period of T. The complete second-order differential equation of Eq. (2.3.1) is the well-known Hill's equation. In general case, there is no analytical solution for this equation. The above equation can be altered to the form of Eq. (2.3.3) by using the change of variable shown below

$$u = xe^{\frac{1}{2}\int_0^t -P_1(\tau)\mathrm{d}\tau} \tag{2.3.2}$$

and the differential equation is now converted into a second-order incomplete differential equation

$$\ddot{u} + P(t)u = 0 \tag{2.3.3}$$

There are many physical systems that can be mathematically modeled in the form of Eq. (2.3.3).

2.3.1 Floquet Theorem

Initially, one could assume that $x_1(t)$ and $x_2(t)$ are two fundamental solutions of Eq. (2.3.1). It can be said that any other solution of Eq. (2.3.1) in the form a linear combination of $x_1(t)$ and $x_2(t)$ can exist as indicated below

$$x(t) = C_1x_1(t) + C_2x_2(t) \tag{2.3.4}$$

For the prescribed time $(t + T)$, the differential equation would become

$$\ddot{x}(t+T) + P_1(t+T)\dot{x}(t+T) + P_2(t+T)x(t+T) = 0 \tag{2.3.5}$$

Since both coefficients P_1 and P_2 are periodic functions, then Eq. (2.3.5) can be written as

$$\ddot{x}(t+T) + P_1(t)\dot{x}(t+T) + P_2(t)x(t+T) = 0 \tag{2.3.6}$$

It can be seen from Eq. (2.3.6) that one could arrive at another Hill's equation by substituting the term $(t + T)$ in Eq. (2.3.1). Accordingly, it can be deduced that $x_1(t + T)$ and $x_2(t + T)$ are the solutions of Hill's equation. Thus, it can be written as

$$x_1(t+T) = a_{11}x_1(t) + a_{12}x_2(t) \quad x_2(t+T) = a_{21}x_1(t) + a_{22}x_2(t) \tag{2.3.7}$$

or in the state equation form

$$\{x\}_{t+T} = [A]\{x\}_t \tag{2.3.8}$$

Using the change of variables as $x = PV$ would lead to

$$x_{t+T} = Ax_t \rightarrow V_{t+T} = P^{-1}x_{t+T} = \underbrace{P^{-1}AP}_{\begin{bmatrix} \lambda_1 & 0 \\ 0 & \lambda_2 \end{bmatrix}} V \tag{2.3.9}$$

where P is the corresponding modal matrix of A. Accordingly, one could write

$$\begin{aligned} V_1(t+T) &= \lambda_1 V_1(t) \\ V_2(t+T) &= \lambda_2 V_2(t) \end{aligned} \tag{2.3.10}$$

Based on the results obtained for λ_1 and λ_2, different types of the solution of Eq. (2.3.1) can be determined. The detailed information about the relationship of λ_i and the type of solution of Eq. (2.3.1) is presented in Table 2.3.

For the case of $P^{-1}AP = \begin{bmatrix} \lambda & 1 \\ 0 & \lambda \end{bmatrix}$, one could write

$$\begin{aligned} V_1(t+T) &= \lambda_1 V_1(t) + V_2(t) \\ V_2(t+T) &= \lambda_2 V_2(t) \end{aligned} \tag{2.3.11}$$

In order to reach a periodic solution for the above case, one should adjust the value of $\lambda = \pm 1$. The abovementioned theorem provides a useful tool for analyzing the solution type for Eq. (2.3.1). However, if one is interested to solve these types of equations, then it is appropriate to use either the method of strained parameters or the multiple time-scales method (MTSM).

2.3.2 *Mathieu Equation*

As fully discussed in Chap. 1, there are many physical systems, which can be mathematically described by the well-known Mathieu equation. In this section, the method of strained parameters will be utilized to solve the Mathieu equation.

The Mathieu equation has the following general form:

$$\ddot{x} + (\delta + 2\varepsilon \cos 2t)x = 0 \tag{2.3.12}$$

Table 2.4 Solution type for Hill's equation

λ_i	Solution type
$\lambda_{1,2} = 1$	Periodic solution for x or V
$\lambda_{1,2} = -1$	Periodic solution, $T' = 2T$
$\lvert\lambda_{1,2}\rvert > 1$	Instability
$\lvert\lambda_{1,2}\rvert < 1$	Damped solution

The solution of the above equation is expanded as

$$x(t;\varepsilon) = x_0(t) + \varepsilon x_1(t) + \varepsilon^2 x_1(t) + \cdots \tag{2.3.13}$$

where

$$\delta = \delta_0 + \varepsilon\delta_1 + \varepsilon^2\delta_2 + \cdots \tag{2.3.14}$$

Substituting Eqs. (2.3.13) and (2.3.14) into Eq. (2.3.12) and equating the same power of ε^i will then lead to

$$\begin{aligned} \ddot{u}_0 + \delta_0 u_0 &= 0 \\ \ddot{u}_1 + \delta_0 u_1 &= -\delta_1 u_0 - 2u_0 \cos 2t \\ \ddot{u}_2 + \delta_0 u_2 &= -\delta_2 u_0 - -\delta_1 u_1 - 2u_1 \cos 2t \end{aligned} \tag{2.3.15}$$

The solution of the first part of Eq. (2.3.15) could easily be obtained as

$$u_0 = a\cos\sqrt{\delta_0}t + b\sin\sqrt{\delta_0}t \tag{2.3.16}$$

Assuming the special case of $\delta_0 = 0$ would result in

$$u_0 = a \tag{2.3.17}$$

and therefore,

$$\ddot{u}_1 = -a\delta_1 - 2a\cos 2t \tag{2.3.18}$$

For the case of $\delta_1 = 0$, the equations would become

$$u_1 = -2a\cos 2t \tag{2.3.19}$$

In the same way as the first two cases, it can be assumed that $\delta_2 = 0$, and thus leads to

$$\ddot{u}_2 = -\frac{1}{2}a(1 + \cos 4t) \tag{2.3.20}$$

Interested readers could find many useful references about Mathieu equation given in Chap. 1.

2.4 Practice Problems

P.2.1. Nonlinear vibration of a carbon nanotube with the consideration of an added particle is mathematically described by the following nonlinear partial differential equation [19]:

$$EI\frac{\partial^4 w}{\partial x^4}+(\rho A+m_p\delta(x-x_0))\frac{\partial^2 w(x,t)}{\partial t^2}+\mu\frac{\partial w(x,t)}{\partial t}+k_1 w(x,t)+k_3 w^3(x,t)$$
$$-k_p\frac{\partial^2 w}{\partial x^2}-F(x,t)+F_{ff}-\left(\frac{EA}{2L}\int_0^L\left(\frac{\partial w}{\partial x}\right)^2\mathrm{d}x\right)\frac{\partial^2 w(x,t)}{\partial x^2}-(ea)^2$$
$$\frac{\partial^2}{\partial x^2}\left(\begin{array}{c}(\rho A+m_p\delta(x-x_0))\frac{\partial^2 w(x,t)}{\partial t^2}+\mu\frac{\partial w(x,t)}{\partial t}+k_1 w(x,t)\\ +k_3 w^3(x,t)-k_p\frac{\partial^2 w}{\partial x^2}-F(x,t)+F_{ff}-\left(\frac{EA}{2L}\int_0^L\left(\frac{\partial w}{\partial x}\right)^2\mathrm{d}x\right)\frac{\partial^2 w(x,t)}{\partial x^2}\end{array}\right)$$

where $F(x,t)=F_0\cos\omega t\delta(x-x_0)$ and $F_{ff}=m_f\left(2v\frac{\partial^2 w(x,t)}{\partial x\partial t}+v^2\frac{\partial^2 w(x,t)}{\partial x^2}+\frac{\partial^2 w(x,t)}{\partial t^2}\right)$.

P.2.1.a. Implement the Galerkin procedure to obtain its nonlinear ordinary differential equation considering the simply supported boundary conditions.
P.2.1.b. Determine the singular point(s) of the nonlinear differential equation obtained.
P.2.1.c. Analyze the primary resonance of the considered system with and without considering of the added particle using MTSM.
P.2.1.d. Discuss the effect of the added particle on the jump phenomenon in accordance with the results obtained based on MTSM.

P.2.2. Forced vibration of a fluid conveying carbon nanotube considering thermal effect is mathematically described by the following partial differential equation: [20].

$$(EI^*)\frac{\partial^4 w}{\partial x^4}+(1-(e_0a)^2\frac{\partial^2}{\partial x^2})\Gamma=0$$

where

$$\Gamma=\rho A\frac{\partial^2 w(x,t)}{\partial t^2}+\mu\frac{\partial w(x,t)}{\partial t}+k_1 w(x,t)+k_3 w^3(x,t)+H_\tau\frac{\partial^2 w}{\partial x^2}-k_p\frac{\partial^2 w(x,t)}{\partial x^2}$$
$$-F(x,t)+F_{ff}-\left(\frac{EA^*}{2L}\int_0^L\left(\left(\frac{\partial w}{\partial x}\right)^2+N_{tt}\right)\mathrm{d}x\right)\frac{\partial^2 w(x,t)}{\partial x^2}$$

and $F_{ff}=m_f\left(2V\frac{\partial^2 w(x,t)}{\partial x\partial t}+V^2\frac{\partial^2 w(x,t)}{\partial x^2}+\frac{\partial^2 w(x,t)}{\partial t^2}\right)$

Definition of the parameters in the above equation can be found in Ref. [20]. Assume that the velocity of fluid changes by the following time-dependent equation:

$$V = V_0(1 + \cos\omega t)$$

P.2.2.a. Determine the nonlinear ordinary differential equation of the system using the Galerkin procedure (Consider the clamped–clamped boundary conditions).
P.2.2.b. By neglecting the nonlinear part of the nonlinear differential equation obtained, determine the stable zone of the solution considering different initial fluid velocities.

P.2.3. Nonlinear vibration of a nanoplate considering an added mass is mathematically described by the following equation [21]:

$$D\nabla^4 w(x,y,t) = \left(1 - (e_0 a)^2\nabla^2\right)\left(\left(-\rho h\frac{\partial^2 w(x,y,t)}{\partial t^2} - m_p\delta(x - x_0)\delta(y - y_0)\frac{\partial^2 w(x,y,t)}{\partial t^2}\right)\right.$$
$$\left. + f(x,y,t) - \mu\dot{w}(x,y,t) - k_1 w(x,y,t) - k_3 w^3(x,y,t)\right)$$

P.2.3.a. Determine the nonlinear ordinary differential equation of the system using the Galerkin procedure (Consider the simply supported boundary conditions).
P.2.3.b. Obtain the solution of the nonlinear ordinary differential equation using the BKAM and MTSM.
P.2.3.c. Compare the solution obtained from part ***b*** with those obtained from the LPPM and SEM.

P.2.4. The size-dependent vibration of an actuated electrical micro-cantilever-based MEMS is mathematically described by the following partial differential equation [22]:

$$\rho A\frac{\partial^2\omega}{\partial t^2} + c_d\frac{\partial w}{\partial t} + (EI + \mu A l^2)\frac{\partial^4\omega}{\partial t^4} = \frac{3}{2}EA\left(\frac{\partial^2\omega}{\partial x^2}\right)^2 + \frac{\varepsilon b[V_{DC} + V_{AC}\cos\omega t]^2}{2(d - w)^2}$$

P.2.4.a. Develop the nonlinear ordinary differential equation of the system using the Galerkin procedure.
P.2.4.b. Using MTSM, obtain the solution of the nonlinear ordinary differential equation and investigate the case of primary resonance for the vibration of the system.

P.2.5. The size-dependent parametric dynamics of imperfect microbeams are described by the following partial differential equation [23]:

$$\rho A\frac{\partial^2 w}{\partial t^2}+(EI+\mu A l^2)\frac{\partial^4 w}{\partial x^4}-\frac{EA}{L}\left(\frac{\partial^2 w}{\partial x^2}+\frac{\partial^2 w_0}{\partial t^2}\right)$$

$$=\int_0^L\left[\frac{\partial w}{\partial x}\frac{\mathrm{d}w_0}{\mathrm{d}x}+\frac{1}{2}\left(\frac{\partial w}{\partial x}\right)^2\right]\mathrm{d}x$$

$$+(P_0+P_1\cos(\omega t))\left(\frac{\partial^2 w}{\partial x^2}+\frac{\mathrm{d}^2 w_0}{\mathrm{d}x^2}\right)+c\frac{\partial w}{\partial t}$$

P.2.5.a. Determine the nonlinear ordinary differential equation of the system using the Galerkin procedure (Consider the simply supported boundary conditions).
P.2.5.b. Using the method of strained parameters, investigate the stability of the system.

P.2.6. The vibration of a Timoshenko beam subjected to an axial load is described by the following differential equation [24]:

$$\rho A\frac{\partial^2 w}{\partial t^2}+\left(\frac{\mu A l^2}{4}\right)\left(\frac{\partial^4 w}{\partial x^4}-\frac{\partial^3 \phi}{\partial x^3}\right)-\frac{3}{2}EA\frac{\partial^2 w}{\partial x^2}\left(\frac{\partial^2 w}{\partial t^2}\right)^2-\mu A\left(\frac{\partial^2 w}{\partial x^2}-\frac{\partial \phi}{\partial x}\right)=+P\frac{\partial^2 w}{\partial x^2}+c_d\frac{\partial w}{\partial t}$$
$$-F(x)\cos(\omega t)$$

$$\rho I\frac{\partial^2 \phi}{\partial t^2}-EI\frac{\partial^2 \phi}{\partial x^2}+\mu A\left(\frac{\partial w}{\partial x}+\phi\right)+\frac{\mu A l^2}{4}\left(\frac{\partial^3 w}{\partial x^3}-\frac{\partial^2 \phi}{\partial x^2}\right)+c_r\frac{\partial \phi}{\partial t}=0$$

P.2.6.a. Develop the nonlinear ordinary differential equation of the system using the Galerkin procedure (Consider the general boundary conditions).
P.2.6.b. Using MTSM, determine the solution of the nonlinear ordinary differential equation and investigate the case of primary resonance for the vibration of the system.

P.2.7. Use the method of multiple time-scales to find the transition curve(s) of the differential equation described below [25]:

$$\ddot{x}+\left[\delta+\sum_{m=1}^{\infty}\varepsilon^m f_m(t)\right]x=0$$

where $f_m(t)=f_m(t+\pi)$ and express f_m as a Fourier series of the form:

$$f_m=\sum_{n=1}^{\infty}(\alpha_{mn}\cos 2nt+\beta_{mn}\sin 2nt)$$

Hint: $x=\sum_{n=0}^{2}\varepsilon^n u_n(T_0,T_1,T_2)+O(\varepsilon^2)$ and $\delta=\delta_0+\varepsilon\delta_1+\varepsilon^2\delta_2+O(\varepsilon^2)$

P.2.8. The nonlinear acoustic propagation in the two-dimensional duct is described by the following partial differential equation [26]:

$$\phi_{xx} + \phi_{yy} - \frac{1}{c^2}\left[\phi_{tt} + \frac{\partial}{\partial t}(\phi_x^2 + \phi_y^2) + \phi_{xx}\phi_x^2 + \phi_{yy}\phi_y^2 + 2\phi_x\phi_y\phi_{xy}\right] = 0$$

where $c^2 = (1-\gamma)\left[\phi_t + \frac{1}{2}(\phi_x^2 + \phi_y^2)\right] + 1$ and with the boundary conditions of $\phi_y(x, 0, t) = 0$ will lead to

$$\begin{aligned} &-\phi_t - \frac{1}{2}(\phi_x^2 + \phi_y^2) + \frac{1}{2}\phi_t^2 + \frac{1}{2}\phi_t(\phi_x^2 + \phi_y^2) - \frac{1}{6}(2-y)\phi_t^3 \\ &= \left[R_0 + R_2\phi_y^2 + (\chi_0 + \chi_2\phi_y^2)\frac{\partial}{\partial t}\right]\phi_y + 0(\phi^4); \quad \text{at } y = 1 \end{aligned}$$

Use the method of multiple time-scales to obtain the approximate solution for the above equation with the corresponding boundary conditions.

P.2.9. Investigate the response of a self-excited oscillator, as described by the following equation, under several harmonic excitations using the multiple time-scales method (MTSM) [27]:

$$\ddot{x} + x = \varepsilon\left(\dot{x} - \frac{\dot{x}^3}{3}\right) + \sum_{n=1}^{N} P_n \cos(\Omega_n t + \theta_n)$$

P.2.10. The nonlinear vibration of a parametrically excited buckled beam is mathematically presented by the following partial differential equation:

$$m\frac{\partial^2 w}{\partial t^2} + c\frac{\partial w}{\partial t} + EI\frac{\partial^4 w}{\partial x^4} + \left[P - F\cos\Omega t - \frac{EA}{2l}\int_0^l \left(\frac{\partial w}{\partial x}\right)^2\right]\frac{\partial^2 w}{\partial x^2} = 0$$

P.2.10.a. Using the Galerkin method, determine its ordinary differential equation by considering a simply supported boundary condition.
P.2.10.b. Develop the transition curve(s) using the method of strained parameters.
P.2.10.c. Determine the transition curve(s) by employing MTSM and compare the solution obtained with those found in P.2.10.a.

P.2.11. Nonlinear oscillation of a viscoelastic rectangular plate is described mathematically by the following governing equation [28]:

$$D\left(1 + \eta\frac{\partial}{\partial t}\right)\nabla^2\nabla^2 w + \rho h w_{tt} - \phi_{yy}w_{xx} - \phi_{xx}w_{yy} + 2\phi_{xy}w_{xy} = \rho h A\omega^2 \cos\omega t$$

and the following compatibility equation:

$$\nabla^2\nabla^2\phi = Eh(w_{xy}^2 - w_{xx}w_{yy})$$

P.2.11.a. Determine the nonlinear ordinary differential equation of the above equation considering the first mode of the vibration.
P.2.12.a. Obtain the primary resonance of the viscoelastic rectangular plate.

P.2.12. Nonlinear vibration of a micro-machined asymmetric resonator is presented by the following nonlinear differential equation [29, 30]:

$$w_{tt} + \mu w_{tt}(\xi, t)\delta(x-\xi) - \mu\eta^2 w_{xtt}(\xi, t)\delta(x-\xi) - 2Pw_{xx} + w_{xxxx}$$
$$= \frac{1}{2r^2}(1 - 2r^2P)\left[w_{xx}\int_0^1 w_x^2 \mathrm{d}x\right] + f\delta(x-\xi)$$

where $f = f_0 \cos \Omega t$.
P.2.12.a. Determine the nonlinear ordinary differential equation of the above system considering the first mode of vibration.
P.2.12.b. Obtain the primary resonance of the micro-machined asymmetric resonator.

P.2.13. Find an approximate solution for the van der Pol oscillator coupled by the fractional derivatives using MTSM [31]:

$$\ddot{x} - \varepsilon(1-x^2)\dot{x} + x = \varepsilon\gamma D^{\alpha}(y-x)$$
$$\ddot{y} - \varepsilon(1-y^2)\dot{y} + y = \varepsilon\gamma D^{\alpha}(x-y)$$

P.2.14. Vibration of a slender beam subjected to both horizontal and vertical excitations is defined by the following nonlinear differential equation [32]:

$$\ddot{x} + 2\mu\dot{x} + \Omega_0^2(1 - \varepsilon\cos\omega_v t)x + \beta x^3$$
$$+ kx(\dot{x}^2 + x\ddot{x}) = \bar{\delta}\cos(\omega_H t + \psi)$$

P.2.14.a. Obtain the primary resonance of the above equation.
P.2.14.b. Determine the transition curve(s) of the above equation.

References

1. Nayfeh AH (2011) Introduction to perturbation techniques. Wiley
2. Nayfeh AH, Mook DT (1995) Nonlinear oscillations. Wiley-VCH Verlag GmbH
3. Nayfeh AH, Pai PF (2008) Linear and nonlinear structural mechanics. Wiley
4. Mickens RE (2010) Truly nonlinear oscillations: harmonic balance Iteration, and averaging methods. World Scientific, Singapore
5. He J-H (1999) Modified straightforward expansion. Meccanica 34(4):287–289

6. He J-H (2002) Modified lindstedt-poincare methods for some strongly non-linear oscillations —part I: expansion of a constant. Int J Nonlinear Mech 37(2):2002
7. Nayfeh AH (2008) Perturbation methods. Wiley
8. Holmes MH (2013) Introduction to perturbation methods. Springer, New York
9. Sadri M, Younesian D, Esmailzadeh E (2016) Nonlinear harmonic vibration and stability analysis of a cantilever beam carrying an intermediate lumped mass. Nonlinear Dyn 84 (3):1667–1682
10. Younesian D, Norouzi H (2015) Frequency analysis of the nonlinear viscoelastic plates subjected to subsonic flow and external loads. Thin-Walled Struct 92:65–75
11. Younesian D, Esmailzadeh E (2010) Non-linear vibration of variable speed rotating viscoelastic beams. Nonlinear Dyn 60(1–2):193–205
12. Younesian D, Esmailzadeh E (2011) Vibration suppression of rotating beams using time-varying internal tensile force. J Sound Vib 330(2):308–320
13. Younesian D, Sadri M, Esmailzadeh E (2014) Primary and secondary resonance analyses of clamped-clamped micro-beams. Nonlinear Dyn 76(4):1867–1884
14. Kovacic I, Brennan MJ (2001) The duffing equation: nonlinear oscillators and their behaviour. Wiley
15. Cveticanin L (2015) Generalization of the Krylov-Bogoliubov method for nonlinear oscillators. Appl Mech Mater 801:3–11
16. Mitropolsky IA (1967) Averaging method in non-linear mechanics. Int J Nonlinear Mech 2 (1):69–96
17. Shohat J (1944) On van der Pol's and related non-linear differential equations. J Appl Phys 15 (7):568
18. Kovacic I, Mickens RE (2012) A generalized van der Pol type oscillator: investigation of the properties of its limit cycle. Math Comput Model 55(3–4):645–653
19. Jamshidifar H, Askari H, Fidan B (2016) Parameter identification and adaptive control of carbon nanotube resonators. Asian J Control 19(2):1–10
20. Askari H, Esmailzadeh E (2017) Forced vibration of fluid conveying carbon nanotubes considering thermal effect and nonlinear foundations. Compos Part B Eng 113(1):31–43
21. Askari H, Jamshidifar H, Fidan B (2017) High resolution mass identification using nonlinear vibrations of nanoplates. Measurement 101:166–174
22. Farokhi H, Ghayesh MH (2016) Size-dependent behaviour of electrically actuated microcantilever-based MEMS. Int J Mech Mater Des 12(3):301–315
23. Farokhi H, Ghayesh MH (2016) Size-dependent parametric dynamics of imperfect microbeams. Int J Eng Sci 99:39–55
24. Ghayesh MH, Farokhi H (2015) Post-buckling dynamics of Timoshenko microbeams under axial loads. Int J Dyn Control 3(4):403–415
25. Nayfeh AH (1972) Characteristic exponents and stability of Hill's equation. J Appl Mech Trans ASME 39 Ser E(4):1156–1158
26. Nayfeh AH, Tsai M-S (1974) Nonlinear acoustic propagation in two-dimensional ducts. J Acoust Soc Am 55(6):1166–1172
27. Asfar KR, Nayfeh AH, Mook DT (1981) Response of self-excited oscillators to multifrequency excitations. J Sound Vib 79(4):589–604
28. Esmailzadeh E, Jalali MA (1999) Nonlinear oscillations of viscoelastic rectangular plates. Nonlinear Dyn 18(4):311–319
29. Hassanpour PA, Esmailzadeh E, Cleghorn WL, Mills JK (2010) Nonlinear vibration of micromachined asymmetric resonators. J Sound Vib 329(13):2547–2564
30. Hassanpour PA, Cleghorn WL, Esmailzadeh E, Mills JK (2007) Vibration analysis of micro-machined beam-type resonators. J Sound Vib 308(1):287–301
31. Suchorsky MK, Rand RH (2012) A pair of van der Pol oscillators coupled by fractional derivatives. Nonlinear Dyn 69(1):313–324
32. Chiba M, Shimazaki N, Ichinohe K (2014) Dynamic stability of a slender beam under horizontal–vertical excitations. J Sound Vib 333(5):1442–1472

Chapter 3
Energy Balance Methods

Abstract Most of semi-analytical techniques, based on Energy Balance method (EBM), are presented in this chapter and the fundamental of Energy Balance method and its modified version, based on the Galerkin and Petrov–Galerkin approaches, are discussed. The first-, second- and the third-order Hamiltonian approaches are comprehensively illustrated and the Rational Energy Balance method, as a potent technique for solving nonlinear oscillators, is presented. The applications of the above-mentioned techniques for solving a large number of nonlinear differential equations with applications in structural vibrations, nano systems and fluid sloshing have been fully discussed. More than twenty examples, based on the recently published works in the area of nonlinear oscillations, are selected to reiterate the important potential of Energy Balance method in analyzing the nonlinear systems.

Keywords Energy balance method · Petrov–Galerkin approach
Hamiltonian approach · Rational energy balance method · Nano systems
Fluid sloshing

Energy Balance Method (EBM) is a powerful and widely used analytical technique to obtain the approximate periodic solutions of varieties of the nonlinear type oscillators. Almost all the solution methods that are based on the energy balance are discussed in this chapter. He [1] in 2002 proposed the energy balance method and his technique was initially developed using the collocation method. Other researchers including Ganji et al. [2–4], Momeni et al. [5], Younesian et al. [6–9] and Askari et al. [10–12], Yazdi et al. [13] applied the energy balance solution procedure accordingly to a variety of problems. Younesian et al. considered an additional flexibility in the solution procedure and modified the energy balance using the least square method (LSM) in conjunction with the Galerkin approach. The fundamental work on the energy balance method was subsequently modified by He [14] in 2010 which he referred to as the Hamiltonian approach. Further improvement on EBM was carried out by Cveticanin et al. [15] who modified the fundamental approach by means of the Ritz method.

E. Esmailzadeh et al., *Analytical Methods in Nonlinear Oscillations*, Solid Mechanics and Its Applications 252, https://doi.org/10.1007/978-94-024-1542-1_3

Initially, the fundamentals and methodology of the energy balance method are presented in this chapter. The sequence of the modifications based on the weighted residual method is discussed. Detailed analyses of the rational EBM using a rational approximate solution are also presented. Different applications of the energy balance method in nonlinear oscillation and multi-body dynamics are finally discussed.

Part I: Approach

3.1 Fundamentals of the Energy Balance Method

The fundamental approach, detailed analyses and the features of the EBM are presented in this section and an associated variational principle is established for a nonlinear system. Subsequently, an associating Hamiltonian function is developed and the angular frequency is obtained using the collocation method. The validity of the results is confirmed even for the strongly nonlinear systems. The detailed analysis of the collocation method is presented in the following section.

3.1.1 Collocation Method

In the collocation method, one would essentially require that the residual R be vanished at a selected number of points x^i in the solution domain

$$R_N(x^i, \{c\}, \{\phi\}, f) = 0, \quad (i = 1, 2, \ldots, N) \tag{3.1.1}$$

The fundamental equation given above can be reconstructed in the following form

$$\int_{\Omega} \delta(x - x^i) R_N(x, \{c\}, \{\phi\}, f) \mathrm{d}x = 0, \quad (i = 1, 2, \ldots, N) \tag{3.1.2}$$

in which δ stands for the Dirac Delta function. In the collocation method, one may choose as many collocation points as the quantity of the trial functions. In general, these points can either be distributed uniformly across the solution domain or condensed in a pre-defined critical region.

3.1.2 Solution Procedure

Let us consider a general nonlinear oscillator governed in the most general form [1]:

$$\ddot{u} + f(u(t)) = 0 \tag{3.1.3}$$

where u and t represent the respective dimensionless generalized coordinate and time. The associating variational principle can be developed as:

$$V(u) = \int_0^t \left\{ -\frac{1}{2}\dot{u}^2 + F(u) \right\} \mathrm{d}t, \tag{3.1.4}$$

in which, $F(u) = \int f(u)\mathrm{d}u$.

The corresponding Hamiltonian expression is therefore given by

$$H = \frac{1}{2}\dot{u}^2 + F(u) = F(A) \tag{3.1.5}$$

followed by the consequent residual of

$$R(t) = \frac{1}{2}\dot{u}^2 + F(u) - F(A) \tag{3.1.6}$$

The two initial conditions assumed in this case are

$$u(0) = A \text{ and } \dot{u}(0) = 0 \tag{3.1.7}$$

The associate trial function can then be expressed as a simple harmonic motion

$$u(t) = A\cos\omega t \tag{3.1.8}$$

The corresponding residual from Eq. (3.1.6) can be constructed as [1]:

$$R(t) = \frac{1}{2}\omega^2 A^2 \sin^2 \omega t + F(A\cos\omega t) - F(A) \tag{3.1.9}$$

It must be noted if the exact solution had been selected as the initial guess, accordingly, it would be mathematically possible to set the value of R be zero. Since the trial function is only an approximate one, the residual R cannot be independent of t and set to zero [1].

Based on Ref. [1], the residual vanishes exactly at the specific collocation points over the entire interval period. Therefore, the collocation at $\omega t = \frac{\pi}{4}$ gives [1]:

$$\omega = \mathrm{Limit}_{\omega t \to \pi/4} \sqrt{\frac{2(F(A) - F(A\cos\omega t))}{A^2 \sin^2 \omega t}} \tag{3.1.10}$$

3.2 Modified Energy Balance Method: Galerkin Approach

The conventional energy balance method has been modified by Younesian et al. using the Galerkin Approach. The average measure of the error vanishes in this combined method as described briefly below [16].

3.2.1 *Galerkin Approach*

The Galerkin method is recognized as a special case of the Petrov–Galerkin method for which the approximate and weighting functions are assumed to be identical. Hence, the expression for the Galerkin integral is given by

$$\int_{\Omega} \phi_i R_N(x, \{c\}, \{\phi\}, f)\mathrm{d}x = 0, \quad (i = 1, 2, \ldots, N) \tag{3.2.1}$$

The above integral equations will coincide with a set of nonlinear algebraic equations with the unknown vector $\{c\}$.

3.2.2 *Modified Petrov–Galerkin Approach*

A further modification is applied by Younesian and Askari [6–9] through the Galerkin approach so that three types of the periodic test functions are chosen as the weighting functions [6–9].

$$1.\ \phi = A\cos\omega t \tag{3.2.2}$$

$$2.\ \phi = A\cos^2\omega t \tag{3.2.3}$$

$$3.\ \phi = A\cos^{\alpha}\omega t \tag{3.2.4}$$

They have considered further flexibility in the solution procedure and presented the possibility of applying any optimization method to minimize the error measure. According to the Galerkin approach, the proposed test functions are employed in order to eliminate the following residual:

$$\tilde{R}(t) = \int_0^{\frac{T}{4}} \left(\frac{1}{2}\omega^2 A^2 \sin^2\omega t + F(A\cos\omega t) - F(A)\right)\phi\,\mathrm{d}t \tag{3.2.5}$$

3.3 Modified Energy Balance Method: Least Square Method

In this section, we would like to introduce a new kind of EBM being constructed on the basis of the least square method. The least square method is based on the idea of minimizing the integral of the square of residual, i.e.

$$I(c_1, c_2, \ldots, c_N) = \int_{\Omega} R_N^2(x, \{c\}, \{\phi\}, f)\mathrm{d}x \tag{3.3.1}$$

By implementing $\delta I = 0$, one could arrive at

$$\int_{\Omega} \frac{\partial R_N}{\partial c_i} R_N(x, \{c\}, \{\phi\}, f)\mathrm{d}x = 0. \tag{3.3.2}$$

According to Eq. (3.2.1), a new weighting function can be defined as $\frac{\partial R(A,\omega,t)}{\partial A}$ so that

$$\begin{aligned}\tilde{R}(t) &= \int_0^{\frac{T}{4}} \left(\frac{1}{2}\omega^2 A^2 \sin^2 \omega t + F(A\cos \omega t) - F(A)\right) \frac{\partial R(A, \omega, t)}{\partial A}\, \mathrm{d}t \\ &= 0\end{aligned} \tag{3.3.3}$$

3.4 Hamiltonian Approach

The terms $\frac{1}{2}\dot{u}^2$ and $F(u)$ are represented by the respective kinetic and potential energies in the expression given by Eq. (3.1.5) and therefore, the functional Eq. (3.1.5) is known as the Lagrangian function in the dynamical system. The corresponding Hamiltonian functions can then be expressed as [14]:

$$H = \frac{1}{2}\dot{u}^2 + F(u) = \text{constant} = H_0 \tag{3.4.1}$$

or equivalently one can write

$$\frac{1}{2}\dot{u}^2 + F(u) - H_0 = 0 \tag{3.4.2}$$

Equation (3.4.1) implies that the total energy of the system remains unchanged during the oscillation and it is known as the energy conservation principle. In the energy balance method, $u = A\cos\omega t$ is employed as the trial function in Eq. (3.4.2). The consequent residual is then derived as:

$$R(t) = \frac{1}{2}\omega^2 A^2 \sin^2 \omega t + F(A\cos\omega t) - H_0 \tag{3.4.3}$$

One can arrive at the frequency-amplitude relationship for a nonlinear oscillator by collocating at a specific point, say $\omega t = \frac{\pi}{4}$, and then eliminating the consequent residual as: $R\left(t = \frac{\pi}{4\omega}\right) = 0$.

It must be noted the collocation point plays an important role in the accuracy of collocation method. In order to overcome the shortcomings of the conventional energy balance method, He [14] suggested an innovative approach, which is based on the Hamiltonian function. This approach has been proposed based on the properties of the invariant H_0:

$$\frac{\partial H_0}{\partial A} = 0 \tag{3.4.4}$$

By introducing a new function, $\tilde{H}(u)$ defined in the integral form of

$$\tilde{H}(u) = \int_0^{T/4} \left\{\frac{1}{2}\dot{u}^2 + F(u)\right\} dt = \frac{1}{4} T H_0 \tag{3.4.5}$$

one could arrive at the following expression

$$\frac{\partial \tilde{H}}{\partial T} = \frac{1}{4} H_0 \tag{3.4.6}$$

Substituting Eq. (3.4.6) into Eq. (3.4.4) yields

$$\frac{\partial}{\partial A}\left(\frac{\partial \tilde{H}}{\partial T}\right) = 0 \tag{3.4.7}$$

Based on He's paper, this corresponds to

$$\frac{\partial}{\partial A}\left(\frac{\partial \tilde{H}}{\partial (1/\omega)}\right) = 0 \tag{3.4.8}$$

From the above generic equation, one can readily obtain the approximate frequency-amplitude relationship for a nonlinear oscillator [14].

3.5 Modified Hamiltonian Approach

Another alternative method for calculating higher-order Hamiltonian is to assume a higher order cosine approximation term. In some cases, a higher order expansion can lead to an improved approximation in the nonlinear systems [17–19].

The trial function is assumed in the form of [19]:

$$u(t) = A_1 \cos \omega t + A_3 \cos 3\omega t + \cdots + A_{2n-1} \cos(2n-1)\omega t \tag{3.5.1}$$

or in the case of

$$u(t) = A_1 \cos \omega t - A_2 \cos 2\omega t + A_3 \cos 3\omega t + \cdots + A_n \cos(n\omega t) \tag{3.5.2}$$

Attempt is made here to introduce the second and third orders of the Hamiltonian approach in the following two sections [19].

3.5.1 Second-Order Hamiltonian Approach

In order to improve the accuracy of the conventional Hamiltonian approach, one can consider the following periodic solution as [17–19]:

$$u(t) = A_1 \cos \omega t + A_3 \cos 3\omega t \tag{3.5.3}$$

such that the initial condition is expressed as:

$$A = A_1 + A_3 \tag{3.5.4}$$

Substituting Eq. (3.5.3) into Eq. (3.4.5), one can obtain [17–19]:

$$\begin{aligned} \tilde{H}(u) &= \int_0^{T/4} \left\{ \frac{1}{2} (A_1 \omega \sin \omega t + 3A_3 \omega \sin 3\omega t)^2 + F(A_1 \cos \omega t + A_3 \cos 3\omega t) \right\} \mathrm{d}t \\ &= \frac{1}{4} T H_0 \end{aligned} \tag{3.5.5}$$

According to the Hamiltonian procedure explained earlier one requires that

$$\frac{\partial}{\partial A_1} \left(\frac{\partial \tilde{H}}{\partial \left(\frac{1}{\omega}\right)} \right) = 0 \tag{3.5.6}$$

and

$$\frac{\partial}{\partial A_3}\left(\frac{\partial \tilde{H}}{\partial\left(\frac{1}{\omega}\right)}\right)=0 \tag{3.5.7}$$

The expression for the natural frequency can then be obtained through the solvability condition of the above set of algebraic equations [17–19].

3.5.2 Third-Order Hamiltonian Approach

Using the same pattern presented for the second-order Hamiltonian approach, one could assume the following expression as the trial function in this case [17–19]:

$$u = A_1\cos(\omega t)+A_3\cos(3\omega t)+A_5\cos(5\omega t) \tag{3.5.8}$$

The initial condition must be satisfied by:

$$A = A_1 + A_3 + A_5 \tag{3.5.9}$$

The corresponding Hamiltonian integration is obtained as:

$$\tilde{H}(u)=\int_0^{\frac{T}{4}}\left[\frac{1}{2}(A_1\omega\sin\omega t+3A_3\omega\sin 3\omega t+5A_5\omega\sin 5\omega t)^2 + F(A_1\cos\omega t+A_3\cos 3\omega t+A_5\cos 5\omega t)\right]dt \tag{3.5.10}$$

The consequent set of generic equations are derived in the third order approximation by the following set of equations [17–19]:

$$\frac{\partial}{\partial A_1}\left(\frac{\partial \tilde{H}}{\partial\left(\frac{1}{\omega}\right)}\right)=0 \tag{3.5.11}$$

$$\frac{\partial}{\partial A_3}\left(\frac{\partial \tilde{H}}{\partial\left(\frac{1}{\omega}\right)}\right)=0 \tag{3.5.12}$$

and

$$\frac{\partial}{\partial A_5}\left(\frac{\partial \tilde{H}}{\partial\left(\frac{1}{\omega}\right)}\right)=0 \tag{3.5.13}$$

3.6 Rational Energy Balance Method

A useful alternative procedure to calculate a second-order energy balance method, i.e., the rational approximation is studied in this section. This technique was first introduced by Sfahani et al. [20, 21] in 2009. The main advantage of the rational approximation is that it provides an implicit inclusion of all the harmonics contributing to the periodic solutions. This rational trial function is given by the expression

$$u(t) = \frac{A_1 \cos \omega t}{1 + B_1 \cos 2\omega t} \tag{3.6.1}$$

where A_1 and B_1 are the unknown constants, which must be determined accordingly. For any particular application, they are determined as functions of the initial conditions and mathematical structure of the oscillatory system. Based on the Fourier expansion, a general rational approximation is developed in the following section.

3.6.1 Fourier Expansion

The rational trial function, given by Eq. (3.6.1), is periodic in terms of the variable θ with a period of 2π. It can therefore, be expanded by the Fourier series

$$u(t) = \sum_{k=1}^{\infty} a_k \cos(2k+1)\theta \tag{3.6.2}$$

The corresponding coefficients are then expressed as

$$a_k = \frac{1}{\pi} \int_{-\pi}^{\pi} \left[\frac{A_1 \cos \theta}{1 + B_1 \cos 2\theta} \right] \cos(2k+1)\theta \, d\theta \tag{3.6.3}$$

The integral can be readily be evaluated by the following generic integration equation

$$\int_{0}^{\pi} \frac{\cos(n\phi) d\phi}{1 + b \cos \phi} = \frac{\pi(-1)^n}{\sqrt{1-b^2}} \left[\frac{1 - \sqrt{1-b^2}}{b} \right] \tag{3.6.4}$$

Accordingly, the unknown coefficients a_k are obtained as

$$a_k = \frac{(-1)^K A}{\sqrt{1-B_1^2}}\left[\frac{1-\sqrt{1-B_1^2}}{B_1}\right]^K \left[\frac{B_1-1+\sqrt{1-B_1^2}}{B_1}\right] \tag{3.6.5}$$

The rational trial function can now be utilized for each of the pre-defined approximation methods, i.e., the conventional Energy Balance Method (EBM), modified EBM using Galerkin approach, Hamiltonian approach and the least square method.

The detailed analysis of the rational EBM application in the form of a numerical example has been presented in the following section.

Part II: Applications

In order to clarify the details of the pre-discussed solution procedures further, several numerical examples are presented in this section which are based on recently published articles using the illustrated methods. The case-study problems are selected to cover a wide variety of well-known applications in nonlinear dynamics and oscillations.

3.7 Generalized Duffing Equation

Duffing equation is one of the most referred differential equation in nonlinear dynamics and vibrations. It can be generally applied to varieties of dynamical systems consist of nonlinear elastic restoring force with either hardening or softening characteristics. The generalized form of the Duffing oscillator is characterized by [6, 22]:

$$\ddot{u} + \alpha_1 u + \alpha_3 u^3 + \alpha_5 u^5 + \cdots + \alpha_n u^n = 0 \tag{3.7.1}$$

with the initial conditions of $u(0) = A$ and $\dot{u}(0) = 0$.

Younesian et al. [6], used the energy balance method (EBM) to solve the above equation. According to Sect. 3.1.2 and Ref. [6], the corresponding variational principle can be obtained as

$$V(u) = \int_0^t \left(-\frac{1}{2}\dot{u}^2 + \frac{u^2}{2} + \frac{\alpha_3 u^4}{4} + \frac{\alpha_5 u^6}{6} + \cdots + \frac{\alpha_n u^{n+1}}{n+1}\right) \mathrm{d}t \tag{3.7.2}$$

and the Hamiltonian function given by Eq. (3.7.1.) is consequently presented as [6]:

$$\begin{aligned} H &= \frac{\dot{u}^2}{2} + \frac{u^2}{2} + \frac{\alpha_3 u^4}{4} + \frac{\alpha_5 u^6}{6} + \cdots + \frac{\alpha_n u^{n+1}}{n+1} \\ &= \frac{A^2}{2} + \frac{\alpha_3 A^4}{4} + \frac{\alpha_5 A^4}{4} + \cdots + \frac{\alpha_n A^{n+1}}{n+1} \end{aligned} \tag{3.7.3}$$

in which, A stands for the initial amplitude.

The following residual can be derived using the trial function of $u(t) = A\cos\omega t$.

$$
\begin{aligned}
R(t) &= \frac{1}{2}A^2\omega^2\sin^2\omega t + \frac{1}{2}A^2\cos^2\omega t + \frac{1}{4}\alpha_3 A^4\cos^4\omega t + \frac{1}{6}\alpha_5 A^6\cos^6\omega t \\
&\quad + \cdots + \frac{1}{n+1}\alpha_n A^{n+1}\cos^{n+1}\omega t - \left(\frac{A^2}{2} + \frac{\alpha_3 A^4}{4} + \frac{\alpha_5 A^6}{6} + \frac{\alpha_7 A^8}{8}\right. \\
&\quad \left. + \cdots + \frac{\alpha_n A^{n+1}}{n+1}\right) = 0
\end{aligned}
\tag{3.7.4}
$$

According to Sect. 3.1.2, employing the conventional EBM, one could arrive at:

$$
\omega = \frac{1}{\sin\omega t}\sqrt{\frac{1}{2} + 2\sum_{\substack{n=2k+1 \\ k=1,2,3,\ldots,m}}^{n} \frac{\alpha_n A^{n-1}}{n+1}\left(1 - \cos^{n+1}\omega t\right)} \quad \text{when } \omega t \to \frac{\pi}{4} \tag{3.7.5}
$$

Based on the discussion given in Sect. 3.2.2, the nonlinear natural frequency can eventually be obtained as

$$
\omega = \sqrt{1 + \sum_{\substack{n=2k+1 \\ k=1,2,3,\ldots,m}}^{n} \frac{3\alpha A^{n-1}}{n+1}\left(1 - \frac{(2n)!}{(2n+1)!}\right)} \tag{3.7.6}
$$

or equivalently in the form of

$$
\omega = \sqrt{1 + \sum_{\substack{n=2k+1 \\ k=1,2,3,\ldots,m}}^{n} \alpha_n A^{n-1}\left(\frac{3\times 5\times\cdots\times(n-1)}{2\times 4\times\cdots\times(n+1)}\right)} \tag{3.7.7}
$$

It is possible to consider the conventional Duffing equation being characterized as

$$
\ddot{u} + u + \alpha_3 u^3 = 0 \tag{3.7.8}
$$

The closed-form solution can be obtained using the Eqs. (3.7.5) and (3.7.7) [6].

$$
\omega = \sqrt{1 + 0.75\times A^2} \tag{3.7.9}
$$

Alternatively, if one uses the Younesian modification, presented in Sect. 3.2.2, and follows the same procedure one could obtain

$$\omega = \sqrt{1 + 0.7 \times A^2} \tag{3.7.10}$$

The ratio between the approximate and exact values of the frequency in the conventional EBM is given by [6]:

$$\frac{\omega_{\text{app}}}{\omega_{\text{ex}}} = \frac{\sqrt{1 + 0.75\alpha_3 A^2}}{2\pi\left(4\int_0^{\pi/2} \frac{d\theta}{\sqrt{1 + 0.5(1 + \sin^2\theta)\alpha_3 A^2}}\right)^{-1}} \tag{3.7.11}$$

where ω_{app} and ω_{ex} represent the approximate and exact frequencies, respectively.

Consequently, the frequency ratio is evaluated for the extreme cases of having large and infinitesimal values of the oscillation amplitude as [6]:

$$\lim_{A \to \infty} \frac{\omega_{\text{app}}}{\omega_{\text{ex}}} = 1.022 \tag{3.7.12}$$

and

$$\lim_{A \to 0} \frac{\omega_{\text{app}}}{\omega_{\text{ex}}} = 1 \tag{3.7.13}$$

Similarly, the ratio between the approximate and exact values of the frequency in the Modified EBM is obtained as [6]:

$$\frac{\omega_{\text{app}}}{\omega_{\text{ex}}} = \frac{\sqrt{1 + 0.7\alpha_3 A^2}}{2\pi\left(4\int_0^{\pi/2} \frac{d\theta}{\sqrt{1 + 0.5(1 + \sin^2\theta)\alpha_3 A^2}}\right)^{-1}} \tag{3.7.14}$$

Consequently, the frequency ratio is evaluated for the extreme cases of large and infinitesimal values of the oscillation amplitude as:

$$\lim_{A \to \infty} \frac{\omega_{\text{app}}}{\omega_{\text{ex}}} = 1.014 \tag{3.7.15}$$

and

$$\lim_{A \to 0} \frac{\omega_{\text{app}}}{\omega_{\text{ex}}} = 1 \tag{3.7.16}$$

The applied modification seems to be effective and can reduce the asymptotic final error from 2.2 to 1.4%. Figure 3.1 shows the relative error based on the general EBM and Modified EBM for Eq. (3.7.8).

Fig. 3.1 Relative errors versus initial amplitudes

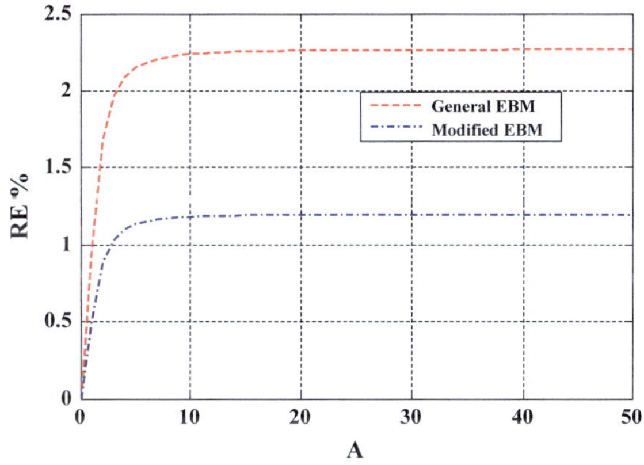

Furthermore, a higher-order Duffing oscillator, as a generalized case, has been analyzed where the nonlinear differential equation is shown as

$$\ddot{u} + u + \alpha_3 u^3 + \alpha_5 u^5 = 0 \tag{3.7.17}$$

The same solution procedure with the same trial function is employed and the general solution is obtained using the conventional EBM (Eq. 7.1.7) as [6]:

$$\omega = \sqrt{1 + \frac{3}{4} \times \alpha_3 \times A^2 + \frac{7}{12} \times \alpha_5 \times A^4} \tag{3.7.18}$$

The general solution based on the Hamiltonian approach (Sect. 3.4) is obtained as

$$\omega = \sqrt{1 + \frac{3}{4} \times \alpha_3 \times A^2 + \frac{5}{8} \times \alpha_3 \times A^4} \tag{3.7.19}$$

Employing the Younesian modification, as explained in Sect. 3.2.2, one arrives at

$$\omega = \sqrt{1 + \frac{7}{10} \times \alpha_3 \times A^2 + 0.5502 \times \alpha_3 \times A^4} \tag{3.7.20}$$

Using the conventional EBM, the frequency ratio and the corresponding asymptotic final values for the large and infinitesimal oscillations are obtained as [6]:

$$\frac{\omega_{\text{app}}}{\omega_{\text{ex}}} = \frac{\sqrt{1 + \frac{3}{4}\alpha_3 A^2 + \frac{7}{12}\alpha_5 A^4}}{2\pi\left(4\int_0^{\pi/2} \frac{\mathrm{d}\theta}{\sqrt{1+\frac{1}{2}\left(1+\sin^2\theta\right)\alpha_3 A^2 + \frac{1}{3}\left(1+\sin^2\theta+\sin^4\theta\right)\alpha_5 A^4}}\right)^{-1}} \tag{3.7.21}$$

$$\lim_{A\to\infty} \frac{\omega_{\text{app}}}{\omega_{\text{ex}}} = 1.022 \tag{3.7.22}$$

and

$$\lim_{A\to 0} \frac{\omega_{\text{app}}}{\omega_{\text{ex}}} = 1 \tag{3.7.23}$$

The comparable frequency ratios are obtained using the Hamiltonian approach [6]:

$$\frac{\omega_{\text{app}}}{\omega_{\text{ex}}} = \frac{\sqrt{1 + \frac{3}{4}\alpha_3 A^2 + \frac{5}{8}\alpha_5 A^4}}{2\pi\left(4\int_0^{\pi/2} \frac{\mathrm{d}\theta}{\sqrt{1+\frac{1}{2}\left(1+\sin^2\theta\right)\alpha_3 A^2 + \frac{1}{3}\left(1+\sin^2\theta+\sin^4\theta\right)\alpha_5 A^4}}\right)^{-1}} \tag{3.7.24}$$

$$\lim_{A\to\infty} \frac{\omega_{\text{app}}}{\omega_{\text{ex}}} = 1.058 \tag{3.7.25}$$

and

$$\lim_{A\to 0} \frac{\omega_{\text{app}}}{\omega_{\text{ex}}} = 1 \tag{3.7.26}$$

Eventually, one can conclude from the results obtained that the Younesian modification provides far more precise results as [6]:

$$\frac{\omega_{\text{app}}}{\omega_{\text{ex}}} = \frac{\sqrt{1 + \frac{7}{10}\alpha_3 A^2 + 0.5502\alpha_5 A^4}}{2\pi\left(4\int_0^{\pi/2} \frac{\mathrm{d}\theta}{\sqrt{1+\frac{1}{2}\left(1+\sin^2\theta\right)\alpha_3 A^2 + \frac{1}{3}\left(1+\sin^2\theta+\sin^4\theta\right)\alpha_5 A^4}}\right)^{-1}} \tag{3.7.27}$$

$$\lim_{A\to\infty} \frac{\omega_{\text{app}}}{\omega_{\text{ex}}} = 0.0074 \tag{3.7.28}$$

and

$$\lim_{A\to 0} \frac{\omega_{\text{app}}}{\omega_{\text{ex}}} = 1 \tag{3.7.29}$$

Fig. 3.2 Relative errors versus initial amplitude

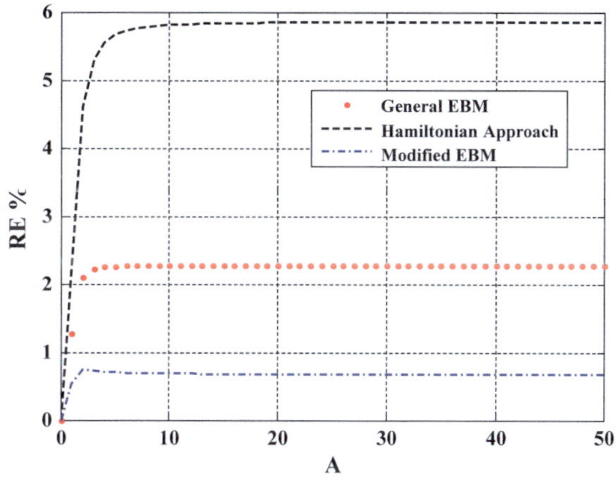

Three additional applications of the energy balance method used in the dynamic analysis of Duffing oscillators are reviewed in the next sections.

Figure 3.2 represents the variations of the relative error with changes in the initial condition for Eq. (3.7.17).

3.7.1 Nonlinear Oscillations of Single-Walled Carbon Nanotubes

Free vibration of an embedded single-walled carbon nanotube, shown schematically in Fig. 3.3, is characterized by the following nonlinear partial differential equation [23]:

$$\mathrm{EI}\frac{\partial^4 w}{\partial x^4} + \rho A\frac{\partial^2 w}{\partial t^2} + kw = \left[\frac{\mathrm{EA}}{2l}\int\limits_0^l \left(\frac{\partial w}{\partial x}\right)\mathrm{d}x\right]\frac{\partial^2 w}{\partial x^2} \tag{3.7.30}$$

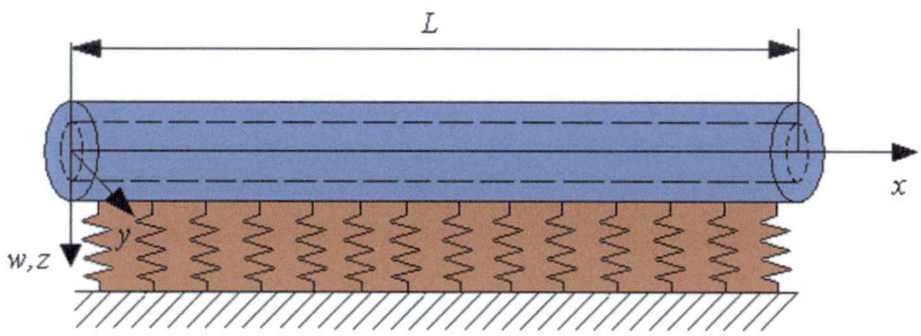

Fig. 3.3 A typical nanotube embedded in an elastic medium [24]

According to the Winkler model, the interaction pressure per unit axial length between the outermost tube and the surrounding medium, denoted by $\{-kw\}$, is described by the material elastic constant k. The elastic rigidity of a single tube is EI and the parameters A and ρ are the respective cross-sectional area and the mass density per unit volume. Assuming that the nanotube is simply-supported at both ends and hence, the boundary conditions can be written as [24]:

$$\text{at}\, x = \big|_0^l : \text{with the deflection}\, w = 0\, \text{and the bending moment}\, M = 0 \qquad (3.7.31)$$

In order to satisfy the boundary conditions, the deflection could be approximated as

$$w(x,t) = W(t)\sin\left(\frac{\pi x}{l}\right) \qquad (3.7.32)$$

Substituting Eq. (3.7.32) into (3.7.30) one could have:

$$\frac{\mathrm{d}^2 W}{\mathrm{d}t^2} + \left(\frac{\pi^4}{l^4}\frac{\mathrm{EI}}{\rho A} + \frac{k}{\rho A}\right)W + \frac{\pi^4}{4l^4}\frac{E}{\rho}W^3 = 0 \qquad (3.7.33)$$

Finally, the governing equation of free oscillation of a single-walled nanotube is summarized as [24]:

$$\frac{\mathrm{d}^2 W}{\mathrm{d}t^2} + \omega^2 W + \alpha W^3 = 0,\ W(0) = W_0,\ \dot{W}(0) = 0 \qquad (3.7.34)$$

in which,

$$\begin{cases} \omega^2 = \dfrac{\pi^4 \mathrm{EI}}{l^4 \rho A} + \dfrac{k}{\rho A} \\ \alpha = \dfrac{\pi^4 E}{4 l^4 \rho} \end{cases} \qquad (3.7.35)$$

It can be seen that Eq. (3.7.34) coincides with the standard Duffing equation and based on Sect. 2.7 the frequency response is found in a straightforward manner.

The physical parameters and material constants are selected to be:

$E = 1.1\,\mathrm{TPa}$; the inside diameter $d_0 = 2.32\,\mathrm{nm}$; and $\rho = 1.3 \times 10^3\,\mathrm{kg/m^3}$ [25].

The variations of the frequency ratio, i.e., the nonlinear natural frequency/linear natural frequency, with respect to the initial amplitude, for different outside diameter d_1 and material constants k, are shown in Figs. 3.4 and 3.5, respectively.

It can be seen that the frequency responses have a hardening characteristic behavior and they are quite correlated with the exact solutions even for the case of large-amplitude oscillations [25]. Furthermore, it can be observed that the

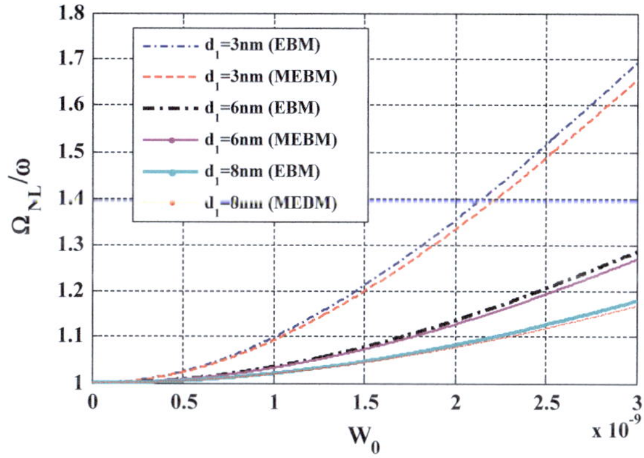

Fig. 3.4 Effect of the outer diameter on the frequency-amplitude response (*EBM* Energy Balance Method, and *MEBM* Modified Energy Balance Method)

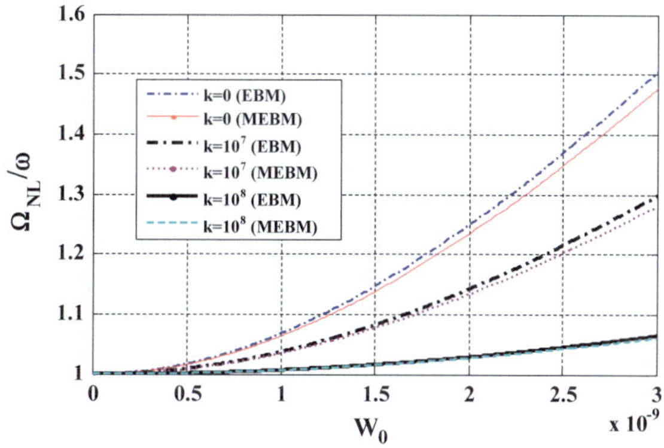

Fig. 3.5 Effect of material constant *k* on the frequency-amplitude response (*EBM* Energy Balance Method, and *MEBM* Modified Energy Balance Method)

sensitivity of the frequency responses with respect to the initial amplitudes are enhanced for the smaller values of parameters d_1 and k.

3.7.2 Nonlinear Oscillations of Rectangular Plates

A two-dimensional elastic medium, i.e., an elastic plate resting on a nonlinear elastic foundation is studied in this part. The governing equation of motion is developed in the form shown below [26]:

$$D\left(\frac{\partial^4 w}{\partial x^4}+\frac{\partial^4 w}{\partial x^2 \partial y^2}+\frac{\partial^4 w}{\partial y^2}\right)+\rho h\frac{\partial^2 w}{\partial t^2}+k_1 w+k_3 w^3=0 \tag{3.7.36}$$

in which, $D=\frac{Eh^3}{12(1-\upsilon^2)}$ represents the flexural rigidity of the plate, and E and υ are the respective elastic modulus and Poisson's ratio. The parameters h, a, and b are denoted as the thickness, length and the width of the rectangular plate, respectively, and k_1 and k_3 are the corresponding constants of the foundation material [26].

The approximate solution, based on the fundamental mode shape, can be obtained as [26]

$$w(x, y, t) = \sin\left(\frac{\pi}{a}x\right)\sin\left(\frac{\pi}{b}y\right)P(t) \tag{3.7.37}$$

Applying the principle of the orthogonality of the mode shapes, one could arrive at

$$\ddot{P}(t) + \omega^2 P(t) + \alpha P^3(t) = 0 \tag{3.7.38}$$

in which,

$$\begin{cases} \omega^2 = \dfrac{D\left[\left(\frac{\pi}{a}\right)^2 + \left(\frac{\pi}{b}\right)^2\right] + k_1}{\rho h} \\ \alpha = \dfrac{9k_3}{16\rho h} \end{cases} \tag{3.7.39}$$

The frequency-amplitude relationship for a rectangular plate is evaluated for three different values of the aspect ratio and their corresponding results have been illustrated in Figs. 3.6 and 3.7.

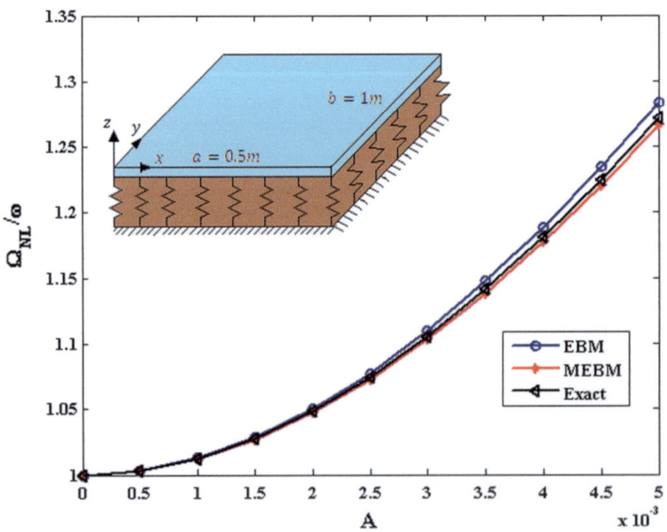

Fig. 3.6 Variations of frequency ratio with initial amplitudes for a rectangular plate $a = 0.5$ and $b = 1$

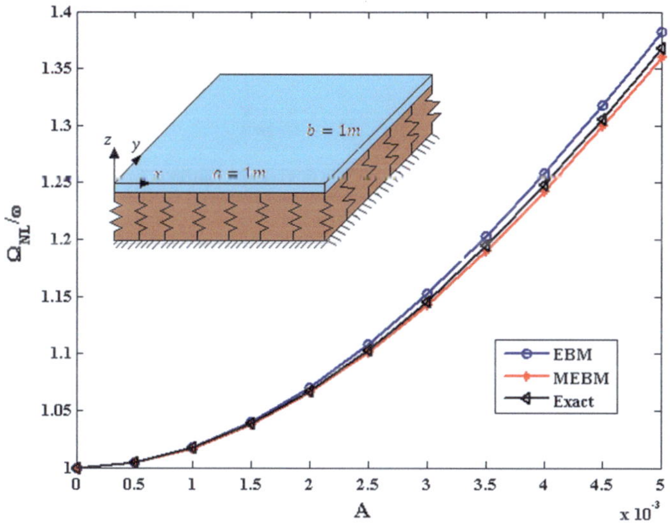

Fig. 3.7 Variations of frequency ratio with initial amplitudes for a rectangular plate $a = 1$ and $b = 1$

The conventional and modified EBM are utilized to solve the nonlinear governing equation of motion, using Sects. 3.1.2 and 3.2.2. The reliability of both solution methods is guaranteed in comparison with those from the exact solution [26].

3.8 Nonlinear Dynamic Buckling of an Elastic Column

According to the Vito modeling [27], a column can be dynamically modeled by two identical light rigid links supported by two nonlinear springs attached at their pivoted joints as shown in Fig. 3.8.

The governing differential equation of motion is then derived as:

$$m\,\ddot{u} + \left(k_1 - \frac{2P}{l}\right)u + \left(k_3 - \frac{2P}{3l^3}\right)u^3 + \cdots = 0 \tag{3.8.1}$$

where the total force of the springs in the general form is given by

$$F_{\text{Spring}} = k_1 u + k_3 u^3 + \cdots \tag{3.8.2}$$

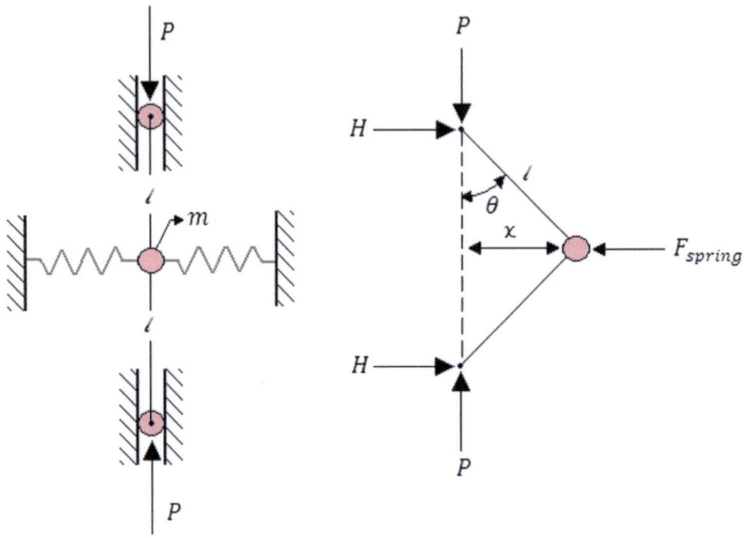

Fig. 3.8 Dynamic buckling analysis of elastic columns: an equivalent analogy [18]

Equation 3.8.1 can be represented by the generalized form of Duffing equation [18].

$$\ddot{u} + \alpha_1 u + \alpha_3 u^3 + \cdots = 0 \tag{3.8.3}$$

Neglecting the terms of the higher-order of 3, the frequency-amplitude relationship can be obtained using both the second-order Hamiltonian approach (SHA) and the third-order Hamiltonian approach (THA). The expressions for the frequency-amplitude responses, are presented as [18]:

$$\omega_{\mathrm{SHA}} = \sqrt{\alpha_1 + 0.72058\alpha_3 A^2} \tag{3.8.4}$$

and

$$\omega_{\mathrm{THA}} = \sqrt{\alpha_1 + 0.717896\alpha_3 A^2} \tag{3.8.5}$$

For the limiting case of the nonlinear buckling forces, one could set to zero the left-hand side of Eqs. 3.8.4 and 3.8.5 [18].

$$P_{\mathrm{cr-SHA}} = \frac{(k_1 + 0.72058k_3 A^2)}{\frac{2}{l} + \frac{1.44116A^2}{3l^3}} \tag{3.8.6}$$

and

$$P_{\mathrm{cr-THA}} = \frac{(k_1 + 0.71789k_3A^2)}{\frac{2}{l} + \frac{1.43578A^2}{3l^3}} \tag{3.8.7}$$

The frequency-amplitude responses for two different elastic columns have been derived and their numerical results are illustrated in Figs. 3.9 and 3.10. It can be observed that the accuracy of the HA is consistently enhanced in the higher-order expansions [18].

3.9 Vibrations of Cracked Rectangular Plate

The equation of vibration of a cracked rectangular plate is developed by Diba et al. [28]. The authors studied the free vibration of the cracked plate modeled by using the following equation:

$$D\left(\frac{\partial^4 w}{\partial x^4} + 2\frac{\partial^4 w}{\partial x^2 \partial y^2} + \frac{\partial^4 w}{\partial y^4}\right) = -\rho h\frac{\partial^2 w}{\partial t^2} + f_x\frac{\partial^2 w}{\partial x^2} + \frac{\partial^2 \bar{M}_y}{\partial y^2} + \bar{f}_y\frac{\partial^2 w}{\partial y^2} \tag{3.9.1}$$

where D is the flexural rigidity, w is the displacement in z-direction, f_x is the in-plane force per unit length in the x-direction, $\bar{M}_y$, is the in-plane moment per unit length, and $\bar{f}_y$ is the in-plane force per unit length in the y-direction. The moment $\bar{M}_y$ and the in-plane force $\bar{f}_y$ are caused by the crack in the plate.

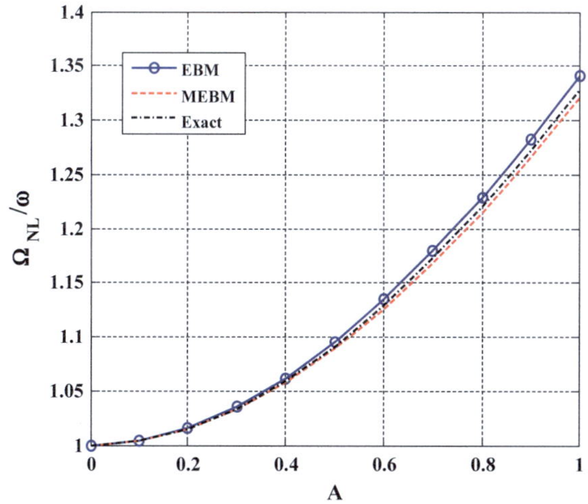

Fig. 3.9 Nonlinear/linear frequency ratio variations with initial amplitude (case 1) $m = 50$; $L = 10$; $K_1 = 500$; $K_3 = 500$; and $P = 150$

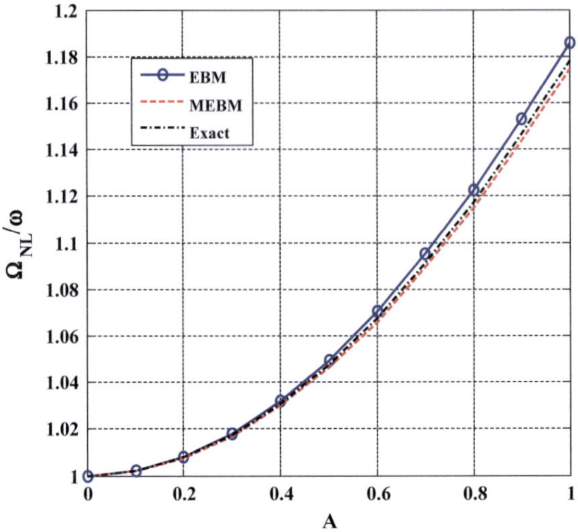

Fig. 3.10 Nonlinear/linear frequency ratio variations with initial amplitude (case 2) $m = 1$; $L = 1$; $K_1 = 10$; $K_3 = 5$; and $P = 1$

The pertinent formulations for $\bar{M}_y$ and $\bar{f}_y$ are defined by Diba and Esmailzadeh [28]:

$$\bar{f}_y = -\frac{2a}{(6\alpha_{tb}^o + \alpha_{tt}^o)(1 - \upsilon^2)h + 2a} f_y \tag{3.9.2}$$

$$\bar{M}_y = -\frac{2a}{3\left(\frac{\alpha_{bt}^o}{6} + \alpha_{bb}^o\right)(3 + \upsilon)(1 - \upsilon)h + 2a} M_y \tag{3.9.3}$$

where a is the half-crack length, f_y represents the in-plane force per unit length at the other-side of the plate, M_y refers to the bending moment per unit length at the other-side of the plate.

The coefficients α_{tt}^o, α_{bb}^o and $\alpha_{bt}^o = \alpha_{tb}^o$ are the non-dimensional bending compliance, the stretching compliance, and the bending-stretching compliance coefficients at the crack center, respectively [1, 3]. Applying the Galerkin procedure, the nonlinear ordinary differential equation is obtained in the form of a Duffing equation:

$$m\ddot{\psi}(t) + k\psi(t) + g\psi^3(t) = 0 \tag{3.9.4}$$

One could easily obtain the nonlinear natural frequency of Eq. (3.9.4). Figures 3.11 and 3.12 show the variations of the frequency ratio for different side lengths [28].

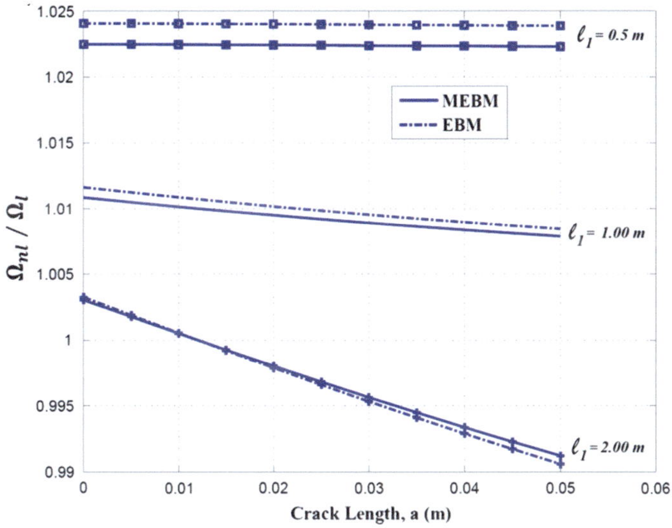

Fig. 3.11 Nonlinear/linear frequency ratio variations with respect to the crack length for the SSSS boundary conditions [28]

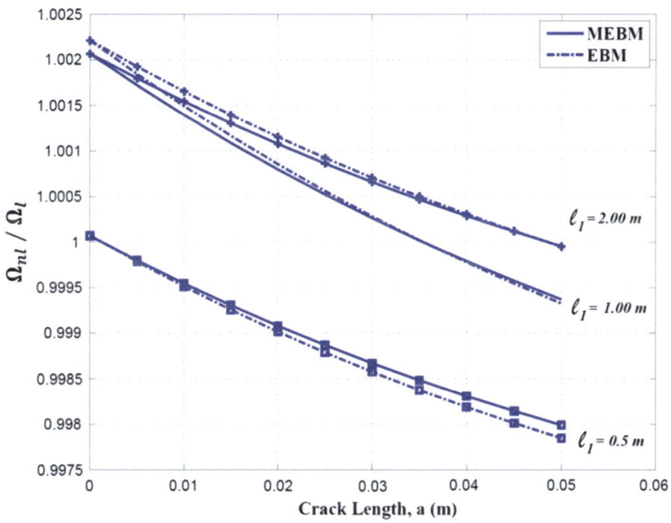

Fig. 3.12 Nonlinear/linear frequency ratio variations with respect to the crack length for the CCSS boundary conditions [28]

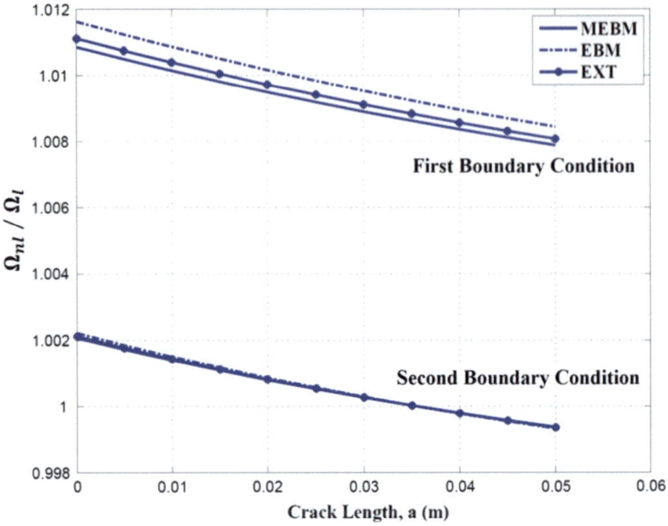

Fig. 3.13 Comparison of the solution results with the half-crack length [28]

A comparison between the results obtained from the modified form of the EBM and those from the general form of the EBM is illustrated in Fig. 3.13.

3.10 Relativistic Oscillator

The relativistic oscillatory motion of a particle is basically recognized when the applied central force is inversely proportional to the distance from its center.

There are many real-world examples, which are physically governed by the relativistic differential equations. This type of motion can also be observed in quantum mechanics.

Classical mechanics examples include the orbital dynamical systems and the oscillation of a light cable with an attached lumped mass [29–33].

The governing equation of a standard relativistic oscillator is represented by

$$\ddot{u} + \frac{u}{\sqrt{u^2+1}} = 0 \text{ with two initial conditions of } u(0) = A \text{ and } \dot{u}(0) = 0 \quad (3.10.1)$$

The corresponding Hamiltonian can then be constructed as [33]

$$H = \frac{1}{2}\dot{u}^2 + \sqrt{1+u^2} \quad (3.10.2)$$

According to the discussions presented in Sect. 3.4, one could have

$$\tilde{H}(u) = \int_0^{\frac{T}{4}} \left\{ \frac{1}{2}\dot{u}^2 + \sqrt{1+u^2} \right\} dt = \int_0^{\frac{T}{4}} \left\{ \frac{1}{2}A^2\omega^2 \sin^2 \omega t + \sqrt{(1+A^2\cos^2 \omega t} \right\} dt$$
$$= \int_0^{\frac{T}{4}} \left\{ \frac{1}{2}\dot{u}^2 + \sqrt{1+u^2} \right\} dt = \int_0^{\frac{T}{4}} \left\{ \frac{1}{2}A^2\omega \sin^2 \omega t + \frac{1}{\omega}\sqrt{(1+A^2\cos^2 t} \right\} dt \tag{3.10.3}$$

and by setting the variations equal to zero yields

$$\frac{\partial}{\partial A}\left(\frac{\partial \tilde{H}}{\partial(\frac{1}{\omega})}\right) = 0 \tag{3.10.4}$$

One could arrive at the frequency-amplitude relationship as given below [33]

$$\omega = \sqrt{\frac{\int_0^{\frac{\pi}{2}} \frac{\cos^2 t}{\sqrt{1+A^2\cos^2 t}}}{\int_0^{\frac{\pi}{2}} \sin^2 t}} \tag{3.10.5}$$

Considering the discussions given in Sect. 3.4, the nonlinear natural frequency is obtained as

$$\omega = \frac{2}{A}\sqrt{\sqrt{A^2+1} - \sqrt{\frac{A^2}{2}+1}} \tag{3.10.6}$$

The exact natural frequency can be determined from the following integral

$$\omega_{ex} = \frac{2\pi}{4\int_0^A \frac{du}{\sqrt{2(\sqrt{1+u^2}-\sqrt{1+A^2})}}} \tag{3.10.7}$$

Although one cannot find an analytical solution for the above integral but its numerical value is a checkpoint reference for any potential comparison [33].

3.11 Plasma Physics Equation

A well-known nonlinear differential equation developed in plasma physics, which describes the motion trajectory of an electron beam injected into a plasma tube, is studied in this section. The beam is applied parallel to the walls of the tube, but the magnetic field bends the path towards the axis. The corresponding differential equation to describe such an oscillatory motion is represented by [34]

$$u'' + \frac{d}{u} = 0 \tag{3.11.1}$$

with the initial conditions given as:

$$u(0) = A \text{ and } \dot{u}(0) = 0 \tag{3.11.2}$$

in which, d stands for the magnetic field constant.

According to Sect. 3.4, the corresponding Hamiltonian can be obtained in a straight forward manner as

$$H = \frac{1}{2}\dot{u}^2 + d \ln u \tag{3.11.3}$$

The resulting Hamiltonian integral is then represented by

$$\tilde{H}(u) = \int_0^{\frac{T}{4}} \left(\frac{1}{2}\dot{u}^2 + d \ln u\right) \mathrm{d}t \tag{3.11.4}$$

Employing the standard cosine trial function, one can arrive at

$$\begin{aligned} \tilde{H}(u) &= \int_0^{\frac{T}{4}} \left(\frac{1}{2}A^2\omega^2 \sin^2 \omega t + d \ln(A \cos \omega t)\right) \mathrm{d}t \\ &= \int_0^{\frac{\pi}{2}} \left(\frac{1}{2}A^2\omega \sin^2 \theta + \frac{d}{\omega} \ln(A \cos \theta)\right) \mathrm{d}\theta \\ &= \frac{\pi}{8}A^2\omega + \frac{d}{\omega}\int_0^{\frac{\pi}{2}} \ln(A \cos \theta) \mathrm{d}\theta \end{aligned} \tag{3.11.5}$$

It is possible to set the expression equal to zero based on the Hamiltonian approach.

$$\frac{\partial}{\partial A}\left(\frac{\partial \tilde{H}}{\partial\left(\frac{1}{\omega}\right)}\right) = -\frac{\pi}{4}A\omega^2 + \frac{d\pi}{2A} = 0 \tag{3.11.6}$$

and eventually, the frequency-amplitude relationship is obtained as [34]

$$\omega = \frac{\sqrt{2d}}{A} \tag{3.11.7}$$

According to Sect. 3.1, the corresponding residual function is represented by

$$R(t) = \frac{1}{2}A^2\omega^2 \sin^2 \omega t + d\ \ln(A\cos\omega t) - d\ \ln(A) \tag{3.11.8}$$

Using the collocation method in accordance with the conventional Energy Balance Method (EBM), one could arrive at [34]:

$$\omega = \frac{2}{A}\sqrt{d\left(\ln(A) - \ln\left(\frac{\sqrt{2}}{2}A\right)\right)} \tag{3.11.9}$$

3.12 Nonlinear Oscillator with Discontinuity

A nonlinear oscillator, which contains a signum function discontinuity, is considered for the analysis in this section. The characteristic equation of the oscillation is given by

$$\ddot{u} + \operatorname{sgn}(u) = 0 \tag{3.12.1}$$

with the two initial conditions of $u(0) = A$, and $\dot{u}(0) = 0$.

The describing variational formulation is obtained as

$$V(u) = \int_0^{T/4} \left(-\frac{1}{2}\dot{u}^2 + u\right) \mathrm{d}t + \int_{T/4}^{T/2} \left(-\frac{1}{2}\dot{u}^2 - u\right) \mathrm{d}t \tag{3.12.2}$$

According to Sect. 3.4 one has

$$\tilde{H}(u) = \int_0^{T/4} \left(\frac{1}{2}\dot{u}^2 + u\right) \mathrm{d}t + \int_{T/4}^{T/2} \left(\frac{1}{2}\dot{u}^2 - u\right) \mathrm{d}t \tag{3.12.3}$$

Using the standard cosine trial function, one could write

$$\begin{aligned}\tilde{H}(u) &= \int_0^{T/4} \left(\frac{1}{2}A^2\omega^2 \sin^2 \omega t + A\cos\omega t\right) \mathrm{d}t + \int_{T/4}^{T/2} \left(\frac{1}{2}A^2\omega^2 \sin^2 \omega t - A\cos\omega t\right) \mathrm{d}t \\ &= A^2\omega\frac{\pi}{4} + \frac{2}{\omega}A\end{aligned} \tag{3.12.4}$$

Accordingly by setting it equal to zero yields

$$\frac{\partial}{\partial A}\left(\frac{\partial \tilde{H}}{\partial\left(\frac{1}{\omega}\right)}\right) = 0 \tag{3.12.5}$$

one could then find the frequency as

$$\omega = \frac{2}{\sqrt{\pi A}} \tag{3.12.6}$$

Results obtained for the changes of the frequency-amplitude expression are quite close to those found from the exact expression given by:

$$\omega_{\text{exact}} = \frac{\pi}{2\sqrt{2A}} \tag{3.12.7}$$

It has been found that a reliable answer (with 98.41% accuracy) is achieved with a simple and straightforward solution procedure carried out by HA.

3.13 Nonlinear Oscillator with Fractional-Power Restoring Force

The governing equation of a nonlinear oscillator with a fractional-power elastic force is considered as another case study. The associating generalized nonlinear differential equation is represented by [35]:

$$\ddot{u} + \sum_i d_i^2 u|u|^{\alpha_i - 1} = 0 \tag{3.13.1}$$

together with the following two initial conditions

$$u(0) = A \text{ and } \dot{u}(0) = 0 \tag{3.13.2}$$

in which, d_i represents a kind of material constant, which is called the rigidity coefficient and α_i denotes a positive fraction or integer. The prescribed system is a conservative type oscillator, which has the Hamiltonian expression given as

$$H = \frac{1}{2}\dot{u}^2 + \sum_i \frac{d_i^2 u^{\alpha_i + 1}}{\alpha_i + 1} = H_0 \quad \text{(being a constant)} \tag{3.13.3}$$

It can be seen from Eq. (3.13.3) that the terms relating to the kinetic and potential energies of the dynamic system have been included.

According to Hamiltonian approach, the corresponding integral is obtained as

$$\bar{H} = \int_0^{T/4} \left(\frac{1}{2}\dot{u}^2 + \sum_i \frac{d_i^2 u^{\alpha_i + 1}}{\alpha_i + 1} \right) \mathrm{d}t \tag{3.13.4}$$

which represents the Hamiltonian integral evaluated between the interval of zero-time to a quarter of the period T, so that

$$H = \frac{\partial \bar{H}}{\partial T} \tag{3.13.5}$$

Employing the standard cosine trial function yields to

$$\bar{H} = \frac{A^2 \omega a_0}{2} + \frac{1}{\omega} \sum_i \frac{d_i^2 A^{\alpha_i + 1} a_i}{\alpha_i + 1} \tag{3.13.6}$$

where

$$a_0 = \int_0^{\pi/2} \sin^2 \theta \mathrm{d}\theta = \frac{\pi}{4} \tag{3.13.7}$$

and

$$a_i = \int_0^{\pi/2} \cos^{\alpha_i + 1} \theta \mathrm{d}\theta = \frac{\pi}{2^{\alpha_i + 1}(\alpha_i + 1) B\left(\frac{\alpha_i + 3}{2}, \frac{\alpha_i + 1}{2}\right)} \tag{3.13.8}$$

The beta function, B included in Eq. (3.13.8) can equally be replaced by the gamma function as

$$B\left(\frac{\alpha_i + 3}{2}, \frac{\alpha_i + 1}{2}\right) \equiv \frac{\Gamma\left(\frac{\alpha_i + 3}{2}\right)\Gamma\left(\frac{\alpha_i + 1}{2}\right)}{\Gamma(\alpha_i + 2)} \tag{3.13.9}$$

The partial derivative of Eq. (3.13.6) with respect to $(1/\omega)$ is

$$\frac{\partial \bar{H}}{\partial (1/\omega)} = -\frac{A^2 \omega^2 a_0}{2} + \sum_i \frac{d_i^2 A^{\alpha_i + 1} a_i}{\alpha_i + 1} \tag{3.13.10}$$

The difference between H_0 and $\partial\bar{H}/(\partial(1/\omega))$ will give the residual function:

$$R(A) = \left(-\frac{A^2\omega^2 a_0}{2} + \sum_i \frac{d_i^2 A^{\alpha_i+1} a_i}{\alpha_i+1}\right) - H_0 \tag{3.13.11}$$

Using the condition for minimization of the residual function $R(A)$, i.e., $d[R(A)]/dA = 0$, one could arrive at

$$A\omega^2 a_0 - \sum_i d_i^2 A^{\alpha_i} a_i = 0 \tag{3.13.12}$$

The approximate value of the frequency of vibration ω_{HA}, after few transformations, is:

$$\omega_{HA} = \sqrt{\frac{\sum_i d_i^2 a_i A^{\alpha_i - 1}}{a_0}} \tag{3.13.13}$$

For the special case, when the residual force is being a linear one, then the expression given by Eq. (3.13.13) transforms into $\omega_{HA} = d_1$, which is a well-known result. To prove the accuracy of this procedure, two different examples are considered.

Example 3.1 In an special case when $\alpha = 1/3$ and $d_1^2 = 1$ the differential equation given by Eq. (3.13.1) is simplified as [35]:

$$\ddot{u} + u^{\frac{1}{3}} = 0 \tag{3.13.14}$$

The approximate value of the natural frequency based on the previously mentioned Hamiltonian approach and in accordance with Eq. (3.13.13), is derived as

$$\omega_{HA} = \frac{1}{A^{\frac{1}{3}}}\left(\frac{3}{2^{4/3}}\frac{\Gamma\left(\frac{7}{3}\right)}{\Gamma\left(\frac{5}{3}\right)\Gamma\left(\frac{2}{3}\right)}\right)^{\frac{1}{2}} = \frac{1.076845}{A^{\frac{1}{3}}} \tag{3.13.15}$$

The approximate solution obtained from Eq. (3.13.15) needs to be compared with those of the exact values of the natural frequency given in Ref. [35]

$$\omega_{ex} = \sqrt{\frac{2\pi}{3}}\frac{\Gamma\left(\frac{5}{4}\right)}{A^{\frac{1}{3}}\Gamma\left(\frac{3}{4}\right)} = \frac{1.070451}{A^{\frac{1}{3}}} \tag{3.13.16}$$

Hence, the relative error value is found to be 0.5973%.

It is worth mentioning that Ozis and Yildirim developed the variational iteration method for Lindstedt-Poincare solving method [36] and obtained the approximate value of the frequency for Eq. (3.13.14) as [36]

$$\omega_{O-Y}(A) = \frac{1.07685}{A^{\frac{1}{3}}} \tag{3.13.17}$$

with a relative error value of 0.6%.

Mickens [37] employed the Harmonic Balance Method for the same example and arrived at the following result.

$$\omega_M(A) = \frac{1.04912}{A^{\frac{1}{3}}} \tag{3.13.18}$$

This indicates a much higher relative error value of 2%.

Comparing the values of the frequency obtained using different methods, one could conclude that the solution procedure using the Hamiltonian approach is far more accurate than that using the harmonic balance method. Furthermore, the solution obtained from the Hamiltonian approach is very similar to the solutions obtained from both the variation iteration method and the Lindstedt-Poincare method.

The time history diagrams for the solution function, $u = A\cos\omega t$ using Hamiltonian approach and the exact methods with the amplitude $A = 1$ is illustrated in Fig. 3.14.

Furthermore, one could see that there are very close agreements between the solutions found from the approximate method with those from the exact method.

Example 3.2 Other types of nonlinear oscillator with a fractional power can be represented by the following differential equation

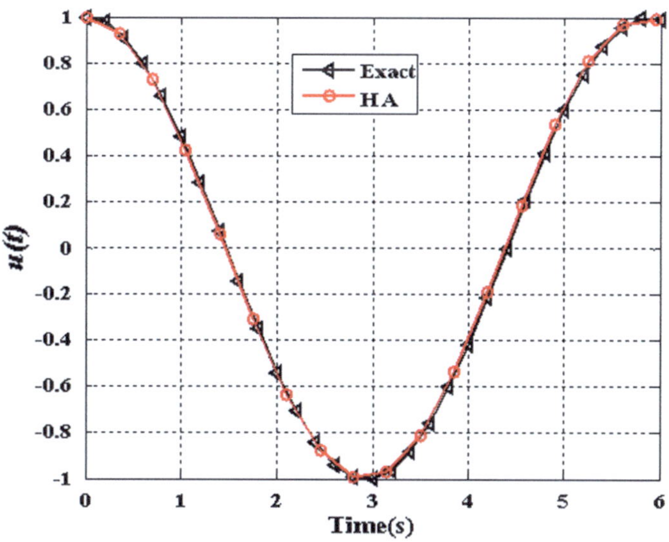

Fig. 3.14 Time history of the solution function for HA and exact methods. $\alpha = 1/3$, $A = 1$

$$\ddot{u} + u^{\frac{2(m-n)+1}{2n+1}} = 0 \quad \text{for } m \geq n \text{ and } m, n = 0, 1, 2, \ldots$$
$$\text{with the initial conditions of} \tag{3.13.19}$$
$$u(0) = A, \ \text{and } \dot{u}(0) = 0$$

The above equation was initially investigated by Hu and Xiong [38]. Younesian et al. [16] studied Eq. (3.13.19) by using the Collocation method and have found:

$$\omega_{\mathrm{EBM}-C} = \frac{2}{A}\left(\frac{2n+1}{2m+2}A^{\frac{2m+2}{2n+1}}\left(1 - \left(\frac{\sqrt{2}}{2}\right)^{\frac{2m+2}{2n+1}}\right)\right)^{\frac{1}{2}} \tag{3.13.20}$$

and by adopting the Petrov–Galerkin approach, they have obtained [16]

$$\omega_{\mathrm{EBM-PG}} = \sqrt{\frac{12n+6}{2m+2}A^{\frac{2m}{2n+1}}\left(1 - \frac{1}{2}\frac{\Gamma\left(\frac{1}{2}\right)\Gamma\left(\frac{2(m+n)+3}{2n+1}+1\right)}{\Gamma\left(\frac{2(m+n)+3}{4n+2}+1\right)}\right)} \tag{3.13.21}$$

For this general case, the approximate solution was initially obtained by Hu and Xiong [38] by applying the harmonic balance method as

$$\omega_{H-X}(A) = \left[\frac{2^{2(2n-m)}\binom{2(m-n)+1}{m-n}}{A^{2(2n-m)}\binom{2n+1}{n}}\right]^{\frac{1}{4n+2}} \tag{3.13.22}$$

The frequency response, according to Sect. 5.1, was found to be [38]

$$\omega_{\mathrm{HVA}} = \frac{1}{A^{\frac{(2n-m)}{2n+1}}}\left(\frac{2}{\pi}\frac{\Gamma\left(\frac{2(m+n)+3}{2(2n+1)}\right)\Gamma\left(\frac{1}{2}\right)}{\Gamma\left(\frac{m+2n+2}{2n+1}\right)}\right)^{\frac{1}{2}} \tag{3.13.23}$$

The solution procedure, which is based on the variational approach, has been explained in detail in Chap. 5. The relative values of the errors derived from different solution methods, as well as, the Hu-Xiong solution are compared and listed in Table 3.1. It can be seen that the relative error is independent of the initial amplitude and the accuracy of different methods for any integer power of resorting force can be sorted as

$$\textbf{Hu-Xiong}\ [38] = \textbf{HA} < \textbf{EBM-C} < \textbf{EBM-PG}$$

Table 3.1 Comparison of relative errors obtained using different methods

m	n	RE_{HA} (%)	$RE_{EBM\text{-}PG}$ (%)	$RE_{EBM\text{-}C}$ (%)	$RE_{H\text{-}X}$ (%)
1	0	3.22	1.24	3.22	3.22
2	0	5.85	1.34	3.26	5.85
3	0	9.56	1.14	1.43	9.56
1	1	0.6	1.21	1.572	1.99
3	2	0.19	0.63	0.88	3.14
4	3	0.41	0.98	1.31	3.47

Accordingly, for any non-integer power one could write

$$\textbf{Hu-Xiong}\ [38] < \textbf{EBM-C} < \textbf{EBM-PG} < \textbf{HA}$$

3.14 Generalized Conservative Oscillatory Systems (Type 1)

A generalized form of the nonlinear differential equation of motion has been considered in this section. This differential equation can cover the Duffing oscillator, the Harmonic Duffing oscillator, and the plasma physics equation as special cases.

The following governing equation of a conservative nonlinear oscillatory system is taken into consideration:

$$\ddot{u} + \gamma u + \Omega u^m + \frac{\alpha u^{n-1}}{u^n + \beta} = 0 \tag{3.14.1}$$

It should be noted that Eq. (3.14.1) can also generate the governing equations of the following oscillatory systems, simultaneously:

For small u :

$$\text{Higher-order Duffing oscillator, i.e., } \ddot{u} + \gamma u + \delta u^{2k-1} = 0 \tag{3.14.2}$$

and

$$\text{For large } u \text{ and } \Omega = \gamma = 0\text{:}$$
$$\text{Plasma Physics equation [26], i.e., } \ddot{u} + \frac{\alpha}{u} = 0 \tag{3.14.3}$$

The variational principle can be applied to Eq. (3.14.1) and the result are obtained as:

$$V(u) = \int_0^t \left(-\frac{1}{2}\dot{u}^2 + \frac{\gamma u^2}{2} + \frac{\Omega u^{m+1}}{m+1} + \frac{\alpha}{n} \ln(u^n + \beta) \right) \mathrm{d}t \tag{3.14.4}$$

The corresponding Hamiltonian can then be presented in the form of

$$H = \frac{\gamma A^2}{2} + \frac{\Omega A^{m+1}}{m+1} + \frac{\alpha}{n} \ln(A^n + \beta) \tag{3.14.5}$$

where A denotes the initial amplitude.

Consequently, the following residual expression can be obtained by using the trial function: $u = A \cos \omega t$.

$$\begin{aligned} R(t) &= \frac{1}{2} A^2 \omega^2 \sin^2 \omega t + \frac{\Omega A^{m+1} \cos^{m+1} \omega t}{m+1} + \frac{\alpha}{n} \ln(A^n \cos^n \omega t + \beta) \\ &+ \frac{\gamma A^2 \cos^2 \omega t}{2} - \left[\frac{\gamma A^2}{2} + \frac{\Omega A^{m+1}}{m+1} + \frac{\alpha}{n} \ln(A^n + \beta) \right] \end{aligned} \tag{3.14.6}$$

Applying few simplifications and employing the collocation method at $\omega t \to \pi/4$, the frequency-amplitude relationship can be obtained as:

$$\begin{aligned} \omega &= \frac{2}{A} \sqrt{\frac{\gamma A^2}{2} (1 - \cos^2 \omega t) + \frac{\Omega A^{m+1}}{m+1} (1 - \cos^{m+1} \omega t)} \\ &+ \frac{\alpha}{n} [\ln(A^n + \beta) - \ln(A^n \cos^n \omega t + \beta)] \end{aligned} \tag{3.14.7}$$

The exact value of the frequency solution of this equation can numerically be obtained from the following natural integral

$$\omega_{\mathrm{ex}} = \frac{2\pi}{\frac{4du}{\sqrt{\gamma(A^2 - u^2) + \frac{2\Omega}{m+1} A^{m+1} + \frac{\alpha}{n} (\ln(A^n + \beta) - \ln(u^n + \beta))}}} \tag{3.14.8}$$

3.15 Generalized Conservative Oscillatory Systems (Type 2)

In studying the generalized conservative oscillatory systems, one has to consider the following nonlinear differential equation of a relativistic harmonic oscillator [8]:

$$\ddot{u} + \delta u + \Omega u^n + \frac{\gamma u^{m-1}}{\sqrt{u^m + \alpha}} = 0; \quad n = 2k - 1,\ m = 2p \tag{3.15.1}$$

with the following two initial conditions

$$u(0) = A \text{ and } \dot{u}(0) = 0 \tag{3.15.2}$$

The variational principle method is applied to Eq. (3.15.1), and eventually that can be obtained as [8]:

$$V(u) = \int_0^t \left(-\frac{1}{2}\dot{u}^2 + \frac{\delta u^2}{2} + \frac{\Omega u^{n+1}}{n+1} + \frac{2\gamma}{m}\sqrt{u^m + \alpha} \right) \mathrm{d}t \tag{3.15.3}$$

The Hamiltonian approach applied to Eq. (3.15.1), is constructed in the form of

$$H = \frac{1}{2}\dot{u}^2 + \frac{\delta u^2}{2} + \frac{\Omega u^{n+1}}{n+1} + \frac{2\gamma}{m}\sqrt{u^m + \alpha} = \frac{\delta A^2}{2} + \frac{\Omega A^{n+1}}{n+1} + \frac{2\gamma}{m}\sqrt{A^m + \alpha} \tag{3.15.4}$$

The standard trial cosine function has been once again utilized and the following residual are finally derived.

$$\begin{aligned} R(t) &= \frac{1}{2}A^2\omega^2 \sin^2 \omega t + \frac{\delta A^2 \cos^2 \omega t}{2} + \frac{\Omega A^{n+1} \cos^{n+1} \omega t}{n+1} \\ &+ \frac{2\gamma}{m}\sqrt{A^m \cos^m \omega t + \alpha} - \left(\frac{\delta A^2}{2} + \frac{\Omega A^{n+1}}{n+1} + \frac{2\gamma}{m}\sqrt{A^m + \alpha} \right) \end{aligned} \tag{3.15.5}$$

The frequency-amplitude relationship could be found in an analytical form by applying the collocation method at $\omega t \to \frac{\pi}{4}$ and after implementing few mathematical simplifications would yield to

$$\omega = \lim_{\omega t \to \frac{\pi}{4}} \left(\frac{2}{A} \sqrt{ \begin{aligned} &\frac{\delta A^2}{2}\left(1 - \cos^2 \omega t\right) + \frac{\Omega A^{n+1}}{n+1}\left(1 - \cos^{n+1} \omega t\right) \\ &+ \frac{2\gamma}{m}\left(\sqrt{A^m + \alpha} - \sqrt{A^m \cos^m \omega t + \alpha}\right) \end{aligned} } \right) \tag{3.15.6}$$

The exact value of the natural frequency can numerically be obtained from the following elliptical integral [8]

$$\omega_{\mathrm{ex}} = \frac{2\pi}{4\int_0^A \frac{\mathrm{d}x}{\sqrt{\delta(A^2 - x^2) + \frac{2\Omega}{(n+1)\left(A^{n+1} - x^{n+1}\right)} + \frac{4\gamma}{m}\left(\sqrt{A^m + \alpha} - \sqrt{x^m + \alpha}\right)}}} \tag{3.15.7}$$

3.16 Duffing Harmonic Oscillator

The governing non-dimensional vibration equation for a Duffing-harmonic oscillator is its most general form given by [33, 39, 40]:

$$\ddot{u}+\frac{u^3}{u^2+1}=0,\quad u(0)=A,\ \dot{u}(0)=0 \tag{3.16.1}$$

For the value of u to be small, the equation will be $\ddot{u}+u^3\approx 0$; and for the large value of u one could show that

$$\ddot{u}+u\approx 0 \tag{3.16.2}$$

According to Sect. 3.4, one may have

$$H=\frac{1}{2}\dot{u}^2+\frac{1}{2}u^2-\frac{1}{2}\log\left(u^2+1\right) \tag{3.16.3}$$

and the Hamiltonian integral will follow as [33]

$$\begin{aligned}\tilde{H}&=\int_0^{\frac{T}{4}}\left\{\frac{1}{2}\dot{u}^2+\frac{1}{2}u^2-\frac{1}{2}\log(u^2+1)\right\}\mathrm{d}t\\&=\int_0^{\frac{T}{4}}\left\{\frac{1}{2}A^2\omega^2\sin^2\omega t+\frac{1}{2}A^2\cos^2\omega t-\frac{1}{2}\log(A^2\cos^2\omega t+1)\right.\\&=\int_0^{\frac{\pi}{2}}\left\{\frac{1}{2}A^2\omega\sin^2 t+\frac{1}{2\omega}A^2\cos^2 t-\frac{1}{2\omega}\log(A^2\cos^2 t+1)\right\}\mathrm{d}t\end{aligned} \tag{3.16.4}$$

Implementing Eq. (3.4.8) would give

$$\omega_{\mathrm{HA}}=\sqrt{\frac{\int_0^{\frac{\pi}{2}}\left\{\cos^2 t-\frac{\cos^2 t}{1+A^2\cos^2 t}\right\}\mathrm{d}t}{\int_0^{\frac{\pi}{2}}\sin^2 t\,\mathrm{d}t}} \tag{3.16.5}$$

The difference between H and $\partial H/(\partial(1/\omega))$ results in the residual equation as:

$$\begin{aligned}R(t)=&\frac{1}{2}A^2\omega^2\sin^2\omega t+\frac{1}{2}A^2\omega^2\cos^2\omega t\\&-\frac{1}{2}\log\left(A^2\cos^2\omega t+1\right)-\left[\frac{1}{2}A^2-\frac{1}{2}\log\left(A^2+1\right)\right]\end{aligned} \tag{3.16.6}$$

Table 3.2 Comparison between the exact and approximate frequencies: Eq. (3.16.1) [33, 39, 40]

A	ω_{ex} [10]	ω_{HEBM}	ω_{HA}	$\frac{\omega_{HEBM}}{\omega_{ex}}$	$\frac{\omega_{HA}}{\omega_{ex}}$
0.01	0.00847	0.00865	0.00865	1.02125	1.02125
0.1	0.08439	0.08626	0.08624	1.02215	1.02192
1	0.63678	0.65164	0.64359	1.02333	1.01069
10	0.99092	0.99314	0.99095	1.00224	1.00003
100	0.99990	0.99993	0.99990	1.00003	1.00000

Solving Eq. (3.16.6) by means of the collocation method at $\omega t \to \frac{\pi}{4}$ will result into

$$\omega = \sqrt{1 + \frac{2}{A^2}\left[\log\left(\frac{A^2}{2} + 1\right) - \log(A^2 + 1)\right]} \tag{3.16.7}$$

A comparison study has been carried out between the exact frequencies and the approximate frequencies, results of which are summarized in Table 3.2.

3.17 Helmholtz Duffing Oscillator

The conservative Helmholtz–Duffing oscillator can be represented by a highly nonlinear second-order differential equation as shown below [10, 41, 42]

$$\begin{aligned}&\ddot{u} + u + (1 - \alpha)u^2 + \alpha u^3 = 0\\ &\text{with the initial conditions of } u(0) = A \text{ and } u(0) = 0\end{aligned} \tag{3.17.1}$$

The nonlinear Helmholtz equation is a well-known nonlinear equation, and has fascinated researchers in the field of applied mathematics and nonlinear vibration, especially those working in the last decade. Several applications of Helmholtz equation are given for the mathematical understanding of many engineering problems. They include the oscillation of the human eardrum, dynamics of ships, particles moving in a cubic potential field, and the oscillation of one-dimensional structural systems with an initial curvature [10, 41, 42].

One can assume that the system oscillates between an asymmetric limit zone [$-b$, a], for positive values of a and b. When either $u = a$ or $u = b$, then $\dot{u} = 0$, since a is given by the initial condition and b is an unknown amplitude needs to be determined.

It has been proved that the behavior of an asymmetric nonlinear oscillator is quite different if in the positive and negative directions. This will lead into studying of the equation in two separate parts, initially for $u = a$ and next part being for $u = b$.

For $u = a$, the variational principle or functional is obtained as:

$$V(u) = \int_0^t \left(-\frac{1}{2}\dot{u}^2 + \frac{u^2}{2} + (1-\alpha)\frac{u^3}{3} + \alpha\frac{u^4}{4} \right) \mathrm{d}t \tag{3.17.2}$$

The associated expression for the Hamiltonian is determined as

$$H = \frac{1}{2}\dot{u}^2 + \frac{u^2}{2} + (1-\alpha)\frac{u^3}{3} + \alpha\frac{u^4}{4} = \frac{a^2}{2} + (1-\alpha)\frac{a^3}{3} + \alpha\frac{a^4}{4} \tag{3.17.3}$$

where a is the initial amplitude in the positive direction. According to the basic idea of energy balance method, explained in Sects. 3.1.1 and 3.3.2, and in a similar procedure explained in Sect. 3.7 and by using $x_a = a\cos(\omega t)$, the following form of the frequency-amplitude relationship is obtained.

$$\omega_{a(\text{Collocation})} = \sqrt{1 + \frac{(4-\sqrt{2})}{3}(1-\alpha)a + \frac{3}{4}\alpha a^2} \tag{3.17.4}$$

and

$$\omega_{a(\text{Petrov-Galerkin})} = \sqrt{1 + \left(2 - \frac{3\pi}{8}\right)(1-\alpha)a + \frac{3}{4}\alpha a^2} \tag{3.17.5}$$

The first-order approximate periodic solution in the negative direction is assumed to be $x_b = b\cos(\omega t)$. The corresponding frequency-amplitude relationship is obtained as [10]

$$\omega_{b(\text{Collocation})} = \sqrt{1 + \frac{(4-\sqrt{2})}{3}(1-\alpha)b + \frac{3}{4}\alpha b^2} \tag{3.17.6}$$

and

$$\omega_{b(\text{Petrov-Galerkin})} = \sqrt{1 + \left(2 - \frac{3\pi}{8}\right)(1-\alpha)b + \frac{3}{4}\alpha b^2} \tag{3.17.7}$$

The first-order approximate solution is then divided into three separate parts, each part is valid for a specific time interval [10, 41, 42]:

$$u(t) = \begin{cases} a\cos(\omega_a t), & \text{for } 0 \le t \le \frac{T_a}{4} \\ b\cos\left(\omega_b\left(t - \frac{T_a}{4} + \frac{T_b}{4}\right)\right), & \text{for } \frac{T_a}{4} \le t \le \frac{T_a}{4} + \frac{T_b}{2} \\ a\cos\left(\omega_a\left(t + \frac{T_a}{2} - \frac{T_b}{2}\right)\right), & \text{for } \frac{T_a}{4} + \frac{T_b}{2} \le t \le T \end{cases} \tag{3.17.8}$$

where the periods are

$$T_a = \frac{2\pi}{\omega_a},\ T_b = \frac{2\pi}{\omega_b} \text{ and } T = \frac{T_a + T_b}{2} \tag{3.17.9}$$

3.18 Autonomous Conservative Oscillatory System

Free vibration of an autonomous conservative nonlinear oscillator, having both the static and inertia cubic nonlinearities, is generally represented by the following equation [17, 43, 44]

$$\ddot{u} + \lambda u + \varepsilon_1 u^2 \ddot{u} + \varepsilon_1 u \dot{u}^2 + \varepsilon_2 u^3 = 0 \tag{3.18.1}$$

with the initial conditions of $u(0) = A$ and $\dot{u}(0) = 0$, in which λ is an integer that can have either values of $\lambda = 1$, or 0, or -1, and also ε_1 and ε_2 are having only positive values.

According to Sect. 3.1.1, the variational principle is considered as:

$$V(u) = \int_0^t \left(-\frac{1}{2}\dot{u}^2\left(1 + \varepsilon_1 u^2\right) + \frac{\lambda u^2}{2} + \frac{\varepsilon_2 u^4}{2} \right) \mathrm{d}t \tag{3.18.2}$$

followed by the Hamiltonian function given as

$$H = \frac{1}{2}\dot{u}^2\left(1 + \varepsilon_1 u^2\right) + \frac{\lambda u^2}{2} + \varepsilon_2 \frac{u^4}{4} = \frac{\lambda A^2}{2} + \varepsilon_2 \frac{A^4}{4} \tag{3.18.3}$$

The residual equation can be obtained using the trial function of $u = A\cos\omega t$.

$$\begin{aligned} R(t) = {} & \frac{1}{2}(A\omega \sin \omega t)^2 \left(1 + \varepsilon_1 (A\cos\omega t)^2\right) \\ & + \frac{\lambda}{2}(A\cos\omega t)^2 + \frac{\varepsilon_1}{4}(A\cos\omega t)^4 - \frac{\lambda A^2}{2} + \varepsilon_2 \frac{A^4}{4} \end{aligned} \tag{3.18.4}$$

By implementing the collocation method at $\omega t \to \frac{\pi}{4}$, one could arrive at

$$\omega = \sqrt{\frac{\lambda + 0.75\varepsilon_2 A^2}{1 + 0.5\varepsilon_1 A^2}} \tag{3.18.5}$$

Using the Petrov–Galerkin modification, implemented by Younesian et al. [6–9], the following condition should be satisfied.

$$\tilde{R}(t) = \int_0^{\frac{T}{4}} R(t) \cos \omega t \, dt = 0 \tag{3.18.6}$$

Therefore, the following frequency-amplitude relationship would then be obtained.

$$\omega = \sqrt{\frac{\lambda + 0.7\varepsilon_2 A^2}{1 + 0.4\varepsilon_1 A^2}} \tag{3.18.7}$$

3.19 Nonlinear Oscillation of Rigid Bar on Semi-circular Surface

The vibration of a uniform rigid bar rocking on a fixed semi-circular surface has been one of the classical examples in the analytical mechanics. The governing equation of the rotational motion is [45]:

$$\ddot{\theta} + \frac{3}{4}\theta^2\ddot{\theta} + \frac{3}{4}\theta\dot{\theta}^2 + \frac{3g\theta}{l}\cos\theta = 0 \tag{3.19.1}$$

with the two initial conditions of $\theta(0) = A$ and $\dot{\theta}(0) = 0$.

Employing the Taylor series expansion, Eq. (3.19.1) is then transformed into:

$$\ddot{\theta} + \frac{3}{4}\theta^2\ddot{\theta} + \frac{3}{4}\theta\dot{\theta}^2 + \frac{3g\theta}{l}\left(1 - \frac{\theta^2}{2} + \frac{\theta^4}{24}\right) = 0 \tag{3.19.2}$$

According to Sect. 3.4, the Hamiltonian form is constructed as

$$H = \frac{1}{2}\dot{\theta}^2\left(1 + \frac{3}{4}\theta^2\right) + \frac{3g}{l}\left(\frac{\theta^2}{2} - \frac{\theta^4}{8} + \frac{\theta^6}{144}\right) \tag{3.19.3}$$

Followed by $\tilde{H}(\theta)$ to be given as

$$\tilde{H}(\theta) = \int_0^{\frac{T}{4}} \left\{\frac{1}{2}\dot{\theta}^2\left(1 + \frac{3}{4}\theta^2\right) + \frac{3g}{l}\left(\frac{\theta^2}{2} - \frac{\theta^4}{8} + \frac{\theta^6}{144}\right)\right\} dt \tag{3.19.4}$$

Substituting $\theta = A\cos\omega t$ into the above residual yields

$$\begin{aligned}
\tilde{H}(\theta) &= \int_0^{\frac{T}{4}} \left\{ \frac{1}{2}A^2\omega^2 \sin^2\omega t \left(1 + \frac{3}{4}A^2\cos^2\omega t\right) \right. \\
&\quad \left. + \frac{3g}{l}\left(\frac{A^2\cos^2\omega t}{2} - \frac{A^4\cos^4\omega t}{8} + \frac{A^6\cos^6\omega t}{144}\right)\right\} \mathrm{d}t \\
&= \int_0^{\frac{\pi}{2\omega}} \left\{ \frac{1}{2}A^2\omega^2 \sin^2\omega t \left(1 + \frac{3}{4}A^2\cos^2\omega t\right) \right. \\
&\quad \left. + \frac{3g}{l}\left(\frac{A^2\cos^2\omega t}{2} - \frac{A^4\cos^4\omega t}{8} + \frac{A^6\cos^6\omega t}{144}\right)\right\} \mathrm{d}t \\
&= \int_0^{\frac{\pi}{2}} \left\{ \frac{1}{2}A^2\omega \sin^2 t \left(1 + \frac{3}{4}A^2\cos^2\omega t\right) \right. \\
&\quad \left. + \frac{3g}{l\omega}\left(\frac{A^2\cos^2\omega t}{2} - \frac{A^4\cos^4\omega t}{8} + \frac{A^6\cos^6\omega t}{144}\right)\right\} \mathrm{d}t \\
&= \frac{1}{2}A^2\omega\left(\frac{\pi}{4} + \frac{3\pi}{64}A^2\right) + \frac{3g}{l\omega}\left(\frac{\pi A^2}{8} - \frac{3\pi}{128}A^4 + \frac{5\pi}{4608}A^6\right)
\end{aligned} \tag{3.19.5}$$

By setting the derivatives with respective to the amplitude A to zero, yields

$$\begin{aligned}
\frac{\partial}{\partial A}\left(\frac{\partial \tilde{H}}{\partial\left(\frac{1}{\omega}\right)}\right) &= -\omega^2\left(\frac{1}{4} + \frac{3}{32}A^2\right) + \frac{3g}{l}\left(\frac{1}{4} - \frac{3}{32}A^2\right) + \frac{3g}{l}\left(\frac{1}{4} - \frac{3}{32}A^2 + \frac{5}{768}A^4\right) \\
&= 0
\end{aligned} \tag{3.19.6}$$

the following frequency-amplitude relationship could then be obtained.

$$\omega = \sqrt{\frac{3g}{l}\frac{(192 - 72A^2 + 5A^4)}{(192 + 72A^2)}} \tag{3.19.7}$$

The exact value of the period of oscillation, T_{Ex} for the rocking motion of the rigid bar can numerically be obtained, using the following integral [45]

$$T_{\mathrm{Ex}} = 4\left(\frac{l}{3g}\right)^{\frac{1}{2}} \int_0^{\frac{\pi}{2}} \left(\frac{(4 + 3A^2\sin^2\varphi)A^2\cos^2\varphi}{8[A\sin A + \cos A - A\sin\varphi\sin(A\sin\varphi) - \cos(A\sin\varphi)]}\right)^{\frac{1}{2}} \mathrm{d}\varphi \tag{3.19.8}$$

3.20 Nonlinear Oscillations of Centrifugal Governor Systems

Centrifugal governors are widely used in automatic control systems, examples of which are in the control of steam and diesel fuel in engines. They are still the major mechanical part of the control systems in a variety of locomotive engines.

Historically, they were invented by James Watt in 1788, known as the starting landmark for the theory of automatic control. The angular motion of the governor, in the absence of friction and inertia of the collar and rods, are defined by the following differential equation [9, 27]:

$$\ddot{\theta} + (1 - \Delta\cos\theta)\sin\theta = 0 \tag{3.20.1}$$

in which

$$\Delta = (\Omega^2 \times r)/g \tag{3.20.2}$$

where, r is the length of the rod and Ω denotes the angular velocity of the rotating support. The Hamiltonian expression given by Eq. (3.20.1) from Sect. 3.1.1 is

$$H = \frac{1}{2}\dot{\theta}^2 - \cos\theta + \frac{1}{4}\Delta\cos 2\theta = -\cos(A) + \frac{1}{4}\Delta\cos(2A) \tag{3.20.3}$$

and the corresponding residual function is derived as

$$R(t) = \frac{1}{2}A^2\omega^2\sin^2\omega t - \cos(A\cos\omega t) + \frac{1}{4}\Delta\cos(2A\cos\omega t) + \cos(A) - \frac{1}{4}\Delta\cos(2A) \tag{3.20.4}$$

Implementing the collocation at $\omega t = \frac{\pi}{4}$ in accordance with the standard procedure in the EBM, one could arrive at

$$\omega = \frac{2}{A}\sqrt{\cos\left(\frac{\sqrt{2}}{2}A\right)\frac{1}{4}\Delta\cos\left(\sqrt{2}A\right) - \cos(A) + \frac{1}{4}\Delta\cos(2A)} \tag{3.20.5}$$

Both the softening and hardening behaviors are observed for different rotating speeds. An interesting bifurcation transition will occur at a specific value of the critical rotating speed Ω^*. The linear behavior is once again preserved for the critical value of the rotating speed. For the rotating speed being exactly matched with Ω^*, the initial condition has no effects on the natural frequency of the system.

3.21 Nonlinear Lateral Sloshing in Partially-Filled Elliptical Tankers

A partially-filled elliptical tanker is schematically illustrated in Fig. 3.15 in which, $2a$ and $2b$ are the respective larger and smaller diameters of the elliptical container.

Accordingly, $2a_{cg}$ and $2b_{cg}$ are the larger and smaller diameter of the center of gravity (CG) contours. The corresponding governing equation is given by [46], [47]:

$$(\bar{a}\cos^2\theta + \bar{b}\sin^2\theta)\ddot{\theta} + 0.5(\bar{b}^2 - \bar{a}^2)\dot{\theta}^2 + g\bar{b}\sin\theta = 0 \tag{3.21.1}$$

where $\alpha = a/b$ and $F = h/2b$ represent the aspect ratio and the fill ratio, respectively, and $\bar{b} = b\gamma$ and $\bar{a} = \bar{b}\gamma$ [6–9], in which,

$$\gamma = 2\left(0.772659 - \frac{1.304727}{\alpha}\right)F^2 + \left(-1.780896 + \frac{1.542048}{\alpha}\right)F + 1 \tag{3.21.2}$$

The corresponding variational principle in accordance with Sect. 3.1.1 is obtained as

$$V = \int_0^t \left[\frac{-1}{2}(\bar{a}\cos^2\theta + \bar{b}\sin^2\theta)\dot{\theta}^2 - g\bar{b}\cos\theta\right] dt \tag{3.21.3}$$

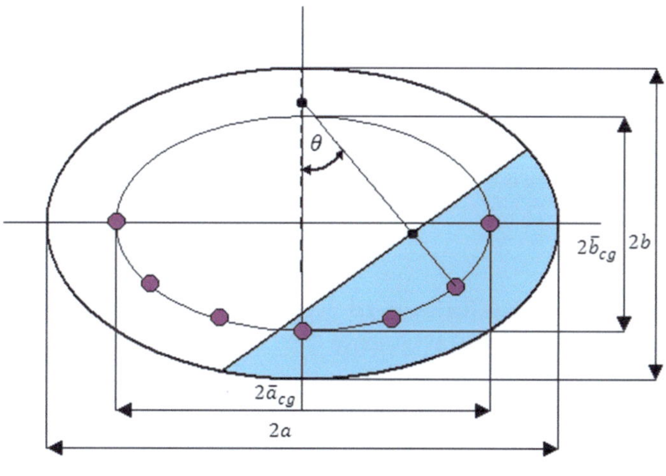

Fig. 3.15 Fluid sloshing in partially-filled elliptical tanker [46, 47]

and consequently, the Hamiltonian expression is determined as

$$H = \frac{-1}{2}\left(\bar{a}\cos^2\theta + \bar{b}\sin^2\theta\right) - g\bar{b}\cos\theta = -g\bar{b}\cos A \tag{3.21.4}$$

Using the initial conditions $\theta(0) = A$ and $\dot{\theta}(0) = 0$, with the trial function $\theta = A\cos\omega t$ being satisfied, the following residual can be reached.

$$\begin{aligned} R(t) = &\frac{1}{2}A^2\omega^2\sin^2\omega t\left(\bar{a}^2\cos^2(A\cos\omega t)\right.\\ &\left.+\bar{b}^2\sin^2(A\cos\omega t)\right) - g\bar{b}\cos(A\cos\omega t) + g\bar{b}\cos A \end{aligned} \tag{3.21.5}$$

Implementing the collocation method at $\omega t = \frac{\pi}{4}$, one can eventually obtain the analytical frequency-amplitude formulation as [46]

$$\omega = \frac{2}{A}\sqrt{\frac{g\bar{b}\left[\cos\left(\sqrt{2}A/2\right) - \cos(A)\right]}{\bar{a}^2\cos^2\left(\sqrt{2}A/2\right) + \bar{b}\sin^2\left(\sqrt{2}A/2\right)}} \tag{3.21.6}$$

In order to improve the correctness of the closed-form solution obtained, one has to integrate the Petrov–Galerkin method with the standard EBM as proposed by Younesian et al. [6–9]. The integral of the residue would vanish with the kernel of the trial solution so that

$$\tilde{R} = \int_0^{\frac{T}{4}} R(t)\cos^2\omega t\,\mathrm{d}t = 0 \tag{3.21.7}$$

The closed-form expression for the frequency-amplitude is hence, given by

$$\omega = \sqrt{\frac{2\int_0^{\frac{\pi}{2}}(g\bar{b}\cos A - g\bar{b}\cos(A\cos\omega t))\cos^2\omega t\,\mathrm{d}t}{\int_0^{\frac{\pi}{2}} A^2\sin^2\omega t\cos^2\omega t(\bar{a}^2\cos^2(A\cos\omega t) + \bar{b}^2\cos^2(A\cos\omega t))\mathrm{d}t}} \tag{3.21.8}$$

3.22 Nonlinear Oscillations of Elevator Cable in a Drum Drive Elevator

The governing equation of the elevator cable in a drum drive elevator is [11]:

$$\ddot{\theta}\left(1 + 2\alpha\sin^2\theta\right) + \alpha\dot{\theta}^2\sin 2\theta + \beta\sin\theta = 0, \quad \text{where:}\alpha = \frac{2m_2}{m_1},\ \beta = \frac{g}{l}(1+\alpha) \tag{3.22.1}$$

The corresponding variational formulation of Eq. (3.22.1) is developed as:

$$V = \int_0^t \left[-\frac{1}{2}\dot{\theta}^2 \left(1 + 2\alpha \sin^2 \theta\right) - \beta \cos \theta \right] \mathrm{d}t \tag{3.22.2}$$

Accordingly, the Hamiltonian function has been constructed as:

$$H = \frac{1}{2}\dot{\theta}^2 \left(1 + 2\alpha \sin^2 \theta\right) - \beta \cos \theta = -\beta \cos A \tag{3.22.3}$$

With the following initial conditions:

$$\theta(0) = A,\ \dot{\theta}(0) = 0 \tag{3.22.4}$$

Based on the initial conditions, the function $\theta = A \cos \omega t$ is considered as the trial function. Therefore, the following residual function is then obtained.

$$R(t) = \frac{1}{2} A^2 \omega^2 \sin^2 \omega t \left(1 + 2\alpha \sin^2 (A \cos \omega t)\right) - \beta \cos(A \cos \omega t) + \beta \cos(A) = 0 \tag{3.22.5}$$

In accordance with the standard EBM procedure and employing the collocation method at $\omega t \rightarrow \frac{\pi}{4}$, the following nonlinear natural frequency is obtained for Eq. (3.22.1):

$$\omega_{\mathrm{EBM}} = \frac{2}{A} \sqrt{\frac{\beta\left(\cos\left(\frac{\sqrt{2}}{2}A\right) - \cos(A)\right)}{1 + 2\alpha \sin^2\left(\frac{\sqrt{2}}{2}A\right)}} \tag{3.22.6}$$

Using the modified form of the energy balance method, we have constructed the following integral using the residual function given in Eq. (3.22.5):

$$\tilde{R} = \int_0^{T/4} R(t) \cos \omega t \,\mathrm{d}t = 0 \tag{3.22.7}$$

Substituting Eq. (3.22.5) into (3.22.7), results into

$$\tilde{R} = \int_0^{\pi/2} \left[\frac{1}{2} A^2 \omega^2 \sin^2 t \left(1 + 2\alpha \sin^2 (A \cos t)\right) - \beta \cos(A \cos(t)) + \beta \cos(A)\right] \cos t \,\mathrm{d}t = 0 \tag{3.22.8}$$

Hence, the following nonlinear frequency for Eq. (3.22.1) is obtained:

$$\omega_{\mathrm{MEBM}} = \frac{1}{A}\sqrt{\frac{2\beta\int_0^{\pi/2}[\cos(A\cos t)-\cos A]\cos t\,\mathrm{d}t}{\int_0^{\pi/2}\left[\sin^2 t\left(1+2\alpha\sin^2(A\cos t)\right)\right]\cos t\,\mathrm{d}t}} \tag{3.22.9}$$

Practice Problems

P.3.1. Use the energy balance method, modified energy balance method, and the higher-order Hamiltonian approach to solve the following nonlinear ordinary differential equations and compare the results obtained with those using the numerical simulation.

P.3.1.a. $\ddot{x}+\alpha x+\beta x^3+\lambda x^5=0$
P.3.1.b. $\ddot{x}+x^{\frac{1}{3}}=0$
P.3.1.c. $\ddot{x}+\frac{x}{x^2+1}=0$
P.3.1.d. $\ddot{x}+\frac{x^3}{x^2+1}=0$
P.3.1.e. $\ddot{x}+\frac{1}{x}=0$
P.3.1.f. $\ddot{x}+\alpha x+\beta x^2+\lambda x^3=0$
P.3.1.h. $\ddot{x}+x|x|=0$

P.3.2. Minimize the relative error of the energy balance method used for Duffing equation using the optimization techniques.
P.3.3. Motion of a particle on a rotating parabola is mathematically formulated by the following nonlinear differential equation [27]:

$$(1+4p^2x^2)\ddot{x}+\Lambda x+4p^2\dot{x}^2x=0$$

P.3.3.a. Use the energy balance method to obtain a closed-form solution for the above equation.
P.3.3.b. Apply the higher-order Hamiltonian approach to obtain the analytical solutions for the above system.
P.3.3.c. Compare the results obtained with those from the numerical simulation.

P.3.4. The governing equation of the vibration of a pendulum attached to a rolling wheel, shown in Fig. 3.16, is given as [27]:

$$(l^2+r^2-2rl\cos\theta)\ddot{\theta}+rl\sin\theta\dot{\theta}^2+gl\sin\theta=0$$

The Fig. 3.16 represents schematically the considered system:

P.3.4.a. Use the energy balance method to obtain a closed-form solution for the above equation.
P.3.4.b. Compare the results obtained with those from the numerical simulation.

Fig. 3.16 Schematic model of a pendulum attached to a rolling wheel [11, 27]

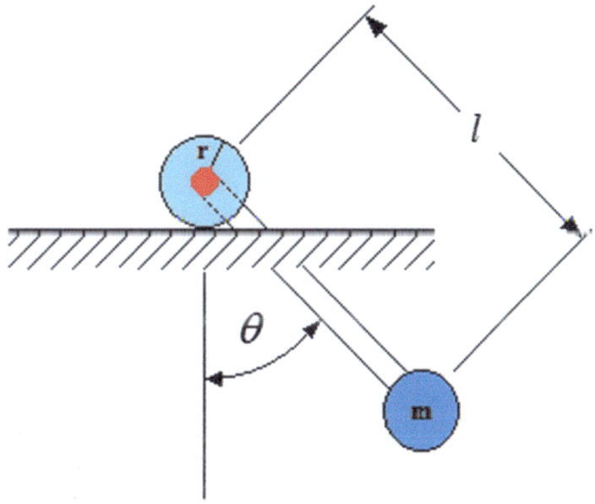

P.3.5. Nonlinear vibration of a nano beam considering the surface effect is mathematically formulated by the following equation:

$$-(\mathrm{EI})^{*}\frac{\partial^4 w}{\partial x^4}-(\rho I)^{*}\frac{\partial^2 w}{\partial t^2}+P\left(\int_0^L\left(\frac{\partial^2 w}{\partial x^2}\right)\right)\left[\frac{\partial^2 w}{\partial x^2}-\mu\frac{\partial^4 w}{\partial x^4}\right]+H\frac{\partial^2 w}{\partial x^2}+\rho\frac{\partial^2 w}{\partial x^2}=0$$

Definition of all parameters in the above equation can be found in Ref. [48].

P.3.5.a. Use the Galerkin approach to obtain the nonlinear differential equation for the principal mode of vibration.
P.3.5.b. Utilize the energy balance method to obtain the natural frequency of the vibration of nanobeam.

P.3.6. Use the modified energy balance method to solve the following nonlinear differential equation:

$$\ddot{u}+u+\alpha u^2\ddot{u}+\alpha u\left(\frac{\mathrm{d}u}{\mathrm{d}t}\right)^2+\beta u^3=0$$

$$u(0)=A,\ \dot{u}(0)=0$$

Compare the results obtained with those from the numerical simulation.
P.3.7. Consider the following nonlinear differential equation:

$$\ddot{u}+u+\alpha u^2\ddot{u}+\alpha u\left(\frac{\mathrm{d}u}{\mathrm{d}t}\right)^2+\alpha u^4\ddot{u}+\alpha u^3\left(\frac{\mathrm{d}u}{\mathrm{d}t}\right)^2+\beta u^3+\alpha u^5=0$$

$$u(0)=A,\ \dot{u}(0)=0$$

P.3.7.a. Use Hamiltonian approach to obtain the closed-form solution for the above equation.
P.3.7.b. Compare the results obtained with those from the exact numerical simulation.

References

1. He J-H (2002) Preliminary report on the energy balance for nonlinear oscillations. Mech Res Commun 29(2–3):107–111
2. Ganji DD, Ranjbar Malidarreh N, Akbarzade M (2009) Comparison of energy balance period with exact period for arising nonlinear oscillator equations : He's energy balance period for nonlinear oscillators with and without discontinuities. Acta Appl Math 108(2), article 353
3. Ganji DD, Esmaeilpour M, Soleimani S (2010) Approximate solutions to van der Pol's damped nonlinear oscillators by means of He's energy balance method. Int J Comput Math 87 (9):2014–2023
4. Ganji SS, Ganji DD, Ganji ZZ, Karimpour S (2009) Periodic solution for strongly nonlinear vibration systems by He's energy balance method. Acta Appl Math 106(1), article 79
5. Momeni M, Jamshidi N, Barari A, Ganji DD (2011) Application of He's energy balance method to Duffing-harmonic oscillators. Int J Comput Math 88(1):135–144
6. Younesian D, Askari H, Saadatnia Z, KalamiYazdi M (2010) Frequency analysis of strongly nonlinear generalized Duffing oscillators using He's frequency–amplitude formulation and He's energy balance method. Comput Math Appl 59(9):3222–3228
7. KalamiYazdi M, Younesian D, Askari H, Saadatnia Z (2011) Free vibration analysis of strongly nonlinear generalized Duffing oscillators using He's variational approach & homotopy perturbation method. Non Sci Lett A 2(1):11–16
8. Younesian D, Askari H, Saadatnia Z, KalamiYazdi M (2012) Analytical approximate solutions for the generalized nonlinear oscillator. Appl Anal 91(5):965–977
9. Yazdi M, Younesian D, Askari H, Saadatnia Z (2011) Periodic solutions for nonlinear oscillation of a centrifugal governor system using the He's frequency-amplitude formulation and He's energy balance method. Nonlinear Sci Lett A 2(3):143–148
10. Askari H, Saadatnia Z, Younesian D, Yildirim A, Kalami-Yazdi M (2011) Approximate periodic solutions for the Helmholtz-Duffing equation. Comput Math Appl 62(10):3894–3901
11. Askari H, Younesian D, Saadatnia Z (2015) Nonlinear oscillations analysis of the elevator cable in a drum drive elevator system. Adv Appl Math Mech 7(1):43–57
12. Askari H, Saadatnia Z, Esmailzadeh E, Younesian D (2014) Multi-frequency excitation of stiffened triangular plates for large amplitude oscillations. J Sound Vib 333(22):5817–5835
13. Yazdi MK, Khan Y, Madani M, Askari H, Saadatnia Z, Yildirim A (2010) Analytical solutions for autonomous conservative nonlinear oscillator. Int J Nonlinear Sci Numer Simul 11(11):979–984
14. He J-H (2010) Hamiltonian approach to nonlinear oscillators. Phys Lett A 374(23):2312–2314
15. Cveticanin L, Kalami-Yazdi M, Saadatnia Z, Askari H (2010) Application of Hamiltonian approach to the generalized nonlinear oscillator with fractional power. Int J Nonlinear Sci Numer Simul 11(12):997–1002
16. Younesian D, Askari H, Saadatnia Z, Yildirim A (2010) Periodic solutions for the generalized nonlinear oscillators containing fraction order elastic force. Int J Nonlinear Sci Numer Simul 11(12):1027–1032

17. Yildirim A, Saadatnia Z, Askari H, Khan Y, KalamiYazdi M (2011) Higher order approximate periodic solutions for nonlinear oscillators with the Hamiltonian approach. Appl Math Lett 24(12):2042–2051
18. Askari H, Saadatnia Z, Yildirim A, Yazdi MK, Khan Y (2013) Application of higher order Hamiltonian approach to nonlinear vibrating systems. J Theor Appl Mech 51(2):287–296
19. Durmaz S, Altay Demirbağ S, Kaya MO (2010) High-order Hamiltonian approach to nonlinear oscillators. Int J Nonlinear Sci Numer Simul 11(8):565–570
20. Sfahani MG, Barari A, Omidvar M, Ganji SS, Domairry G (2011) Dynamic response of inextensible beams by improved energy balance method. Proc Inst Mech Eng Part K J Multi-body Dyn 225(1):66–73
21. Daeichin M, Ahmadpoor MA, Askari H, Yildirim A (2013) Rational energy balance method to nonlinear oscillators with cubic term. Asian-Eur J Math 6(2):1350019
22. Kovacic I, Brennan MJ (2011) Background: on Georg Duffing and the Duffing equation. The Duffing equation. Wiley, USA, pp 1–23
23. Fu YM, Hong JW, Wang XQ (2006) Analysis of nonlinear vibration for embedded carbon nanotubes. J Sound Vib 296(4&5):746–756
24. Askari H, Saadatnia Z, Younesian D, Esmailzadeh E (2013) Large amplitude free vibration analysis of nanotubes using variational and homotopy methods. In: Proceedings, ASME design engineering technical conference, vol 8
25. Pantano A, Boyce MC, Parks DM (2003) Nonlinear structural mechanics based modeling of carbon nanotube deformation. Phys Rev Lett 91(14):145504
26. Younesian D, Askari H, Saadatnia Z, Esmailzadeh E (2011) Analytical solutions for oscillation of rectangular plate on a nonlinear winkler foundation. In: Proceedings, ASME design engineering technical conference, vol 1, parts A & B, pp 755–760
27. Nayfeh AH, Mook DT (1995) Nonlinear oscillations. Wiley, USA
28. Diba F, Esmailzadeh E, Younesian D (2014) Nonlinear vibration analysis of isotropic plate with inclined part-through surface crack. J Nonlinear Dyn 78(4):2377–2397
29. Beléndez A, Pascual C (2007) Harmonic balance approach to the periodic solutions of the (an)harmonic relativistic oscillator. Phys Lett A 371(4):291–299
30. Sun WP, Wu BS, Lim CW (2007) Approximate analytical solutions for oscillation of a mass attached to a stretched elastic wire. J Sound Vib 300(3–5):1042–1047
31. Mickens RE (2010) Truly nonlinear oscillations: harmonic balance, parameter expansions, iteration, and averaging methods. World Scientific, Singapore
32. Mickens RE (1998) Periodic solutions of the relativistic harmonic oscillator. J Sound Vib 212 (5):905–908
33. Yildirim A, Saadatnia Z, Askari H (2011) Application of the Hamiltonian approach to nonlinear oscillators with rational and irrational elastic terms. Math Comput Model 54 (1):697–703
34. Xu L (2011) A Hamiltonian approach for a plasma physics problem. Comput Math Appl 61 (8):1909–1911
35. Cveticanin L, Kalami-Yazdi M, Askari H (2012) Analytical approximations to the solutions for a generalized oscillator with strong nonlinear terms. J Eng Math 77(1):211–223
36. Öziş T, Yıldırım A (2007) Determination of periodic solution for a u1/3 force by He's modified Lindstedt-Poincaré method. J Sound Vib 301(1&2):415–419
37. Mickens RE (2001) Oscillations in an x4/3 potential. J Sound Vib 246(2):375–378
38. Hu H, Xiong Z-G (2003) Oscillations in an x(2m+2)/(2n+1) potential. J Sound Vib 259 (4):977–980
39. Saadatnia Z, Askari H, KalamiYazdi M (2010) Frequency analysis of nonlinear oscillators with rational restoring force via He's energy balance method and He's variational approach. Nonlinear Sci Lett A 1(4):425–430
40. Mickens RE (2001) Mathematical and numerical study of the Duffing-harmonic oscillator. J Sound Vib 244(3):563–567
41. Cao H, Seoane JM, Sanjuán MAF (2007) Symmetry-breaking analysis for the general Helmholtz-Duffing oscillator. Chaos, Solitons Fractals 34(2):197–212

42. Thylwe K-E (1993) Exact quenching phenomenon of undamped driven Helmholtz and Duffing oscillators. J Sound Vib 161(2):203–211
43. Hamdan MN, Shabaneh NH (1997) On the large amplitude free vibrations of a restrained uniform beam carrying an intermediate lumped mass. J Sound Vib 199(5):711–736
44. Abdel-Jaber MS, Al-Qaisia AA, Abdel-Jaber M, Beale RG (2008) Nonlinear natural frequencies of an elastically restrained tapered beam. J Sound Vib 313(3–5):772–783
45. Khan Y, Wu Q, Askari H, Saadatnia Z, Kalami-Yazdi M (2010) Nonlinear vibration analysis of a rigid rod on a circular surface via Hamiltonian approach. Math Comput Appl 15(5):974
46. Younesian D, Askari H, Saadatnia Z, Esmailzadeh E (2011) Analytical solutions for nonlinear lateral sloshing in partially-filled elliptical tankers. In: Proceedings, ASME design engineering technical conference, vol 1, no Parts A and B
47. Salem MI, Mucino VH, Saunders E, Gautam M, Lozano-Guzman A (2009) Lateral sloshing in partially-filled elliptical tanker trucks using a trammel pendulum. Int J Heavy Veh Syst 16 (1–2):207–224
48. Hosseini-Hashemi S, Nazemnezhad R, Rokni H (2015) Nonlocal nonlinear free vibration of nanobeams with surface effects. Eur J Mech A/Solids 52:44–53

Chapter 4
Residual Methods

Abstract The focus of this chapter is on the residual-based methods. Initially, the basic idea behind these methods and full discussions on the main residual methods is presented. The frequency–amplitude formulation (FAF) and its modified versions for solving conservative nonlinear oscillators are discussed in length. The max-min approach, based on the Chengtian's inequality, has also been explained. Eight nonlinear differential equations are solved using the abovementioned techniques to show the capability of the residual methods for analyzing nonlinear oscillations. The considered examples present four types of generalized nonlinear differential equations with their applications in the modeling of structural vibrations and nanoresonators. Eight practice problems are given in the last section to be solved using the residual-based methods.

Keywords Residual method · Frequency–amplitude formulation
Max-min approach · Chengtian's inequality · Structural vibration
Nanoresonators

Part I: Approach

Recently, many mathematicians have modified and proposed several nonlinear analytical approaches based on few mathematical procedures developed from ancient mathematical algorithms. The basic ideas of these novel approaches that have recently been developed are briefly introduced in the first section of this chapter. In accordance with the key reference of these original methods [1], the Ying Buzu Shu method (Literally Method of Surplus and Deficiency) [2], or the method of double false positions, which has been proposed by an Iranian mathematician Khwarizmi (825 AD) is succinctly discussed [3]. The frequency–amplitude formulation (FAF) is fully discussed in the next section and the interconnection of this potent and straightforward technique with the method of surplus and deficiency (S&D) is addressed [4, 5]. Subsequently, different modified approaches based on FAF are presented [6–11].

E. Esmailzadeh et al., *Analytical Methods in Nonlinear Oscillations*, Solid Mechanics and Its Applications 252, https://doi.org/10.1007/978-94-024-1542-1_4

The Geng and Cai alteration of the frequency–amplitude formulation is explicated [5], and then, Ren's and Gui's modification on this method is presented [12]. Further to the discussion and analysis of the frequency–amplitude formulation, one other approach, i.e., max-min technique is elicited from the ancient mathematical algorithms [13, 14]. The concept of the He Chengtian's inequality has been briefly introduced as the basis of the max-min approach [15, 16].

In the last part of this chapter, a number of nonlinear equations are solved using the residual algorithm. The efficiency, accuracy, and performance of the considered methods are evaluated in solving few complicated nonlinear problems [17–20].

4.1 Basic Ideas

The basic ideas, the main references, the fundamentals, and the history of the considered methods are presented in the first part of this chapter.

4.1.1 The Nine Chapters on the Mathematical Art

"Jiu Zhang Suan Shu" or translated as "Nine Chapters on the Mathematical Art" is an old Chinese book on mathematics, which encompasses nine different chapters. The book includes the oldest and the most significant works in the history of Chinese mathematics [2, 21–23]. Indeed, it is one of the great treasures of Chinese mathematics, and the book contains 246 problems in different areas of mathematics. The first chapter is focused on examples related to the land surveying. As one of the examples of the first chapter, an accurate approximation for the value of π has been calculated.

Chapter 2 includes few examples, which are concerned with the exchange of goods. Chapters 3, 4, and 5 provide examples related to the distribution by proportion, short width, and the civil engineering, respectively. Corresponding Chaps. 6–8 of this book relate to the respective subjects of fair distribution of goods, the excess and deficit, and the calculation by square table. The final chapter encompasses 24 problems concerning the subject of right-angled triangles. Interestingly enough, the Nine Chapters on the Mathematical Art is an old practical handbook of mathematics, which provides methods for solving everyday problems concerning different aspects of life [1].

4.1.2 The Ying Buzu Shu, a Method to Approximate Real Roots

Approximation of the real roots of nonlinear equations has always been a fascinating topic among mathematicians. As mentioned before, Chap. 7 of the "Jiu Zhang Suan Shu" introduced the method of Ying Buzu Shu (YBS), which is in fact the method of surplus and deficiency [1]. It is in fact the oldest method for finding the real roots of nonlinear equations.

In order to illustrate the basic idea of YBS, the following example is extracted from the first chapter of the book based on Ref. [1]:

> It is assumed that one can purchase an item with an unknown price. The item is requested by an unknown number of customers. The surplus of 3 dollars is considered if any of them is allocated 8 dollars; and a deficiency of 4 dollars is assumed if each is paid 7 dollars. The problem is to find the exact price and the total number of the costumers.

Figure 4.1 represents the solution procedure of the YBS method. For the purpose of describing the Chinese algorithm in the modern mathematical form, it can be formulated as Eq. (4.1.1) [1]

$$u = \frac{x_2 R_1 - x_1 R_2}{R_1 - R_2} \tag{4.1.1}$$

in which x_1 and x_2 denote the assigned approximate prices, R_1 and R_2 are the deviation with respect to the exact price, and u is the exact price [1].

The algebraic illustration of Eq. (4.1.1) is given as [1]

$$h(u) = 0 \tag{4.1.2}$$

Hence, u_1 and u_2 are assumed to be the approximate solution for Eq. (4.1.2). These two approximate solutions give the residuals of $h(u_1)$ and $h(u_2)$, respectively. In accordance with Eq. (4.1.1), the following form can be defined as the improved approximate solution for Eq. (4.1.2) [1]:

$$u_3 = \frac{u_2 h(u_1) - u_1 h(u_2)}{h(u_1) - h(u_2)} \tag{4.1.3}$$

An alternative procedure to obtain the approximate solution can similarly be given by the following equation based on Ref. [1]:

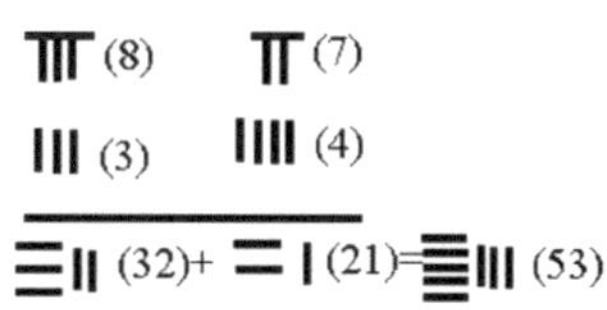

Fig. 4.1 The solution procedure used by ancient Chinese mathematicians. Figure is adopted with permission from Ref. [1]

$$u = \frac{u_1 h_2 h_3}{(h_1 - h_2)(h_1 - h_3)} + \frac{u_2 h_1 h_3}{(h_2 - h_1)(h_2 - h_3)} + \frac{u_3 h_1 h_2}{(h_3 - h_1)(h_3 - h_2)} \tag{4.1.4}$$

where $h_i = h(u_i)$.

4.2 Frequency–Amplitude Formulation

According to the abovementioned mathematical algorithms, a new reliable method has been proposed to solve nonlinear differential equation, which is known as the frequency–amplitude formulation (FAF) [4, 5]. In order to describe the proposed method, the following generalized oscillatory system is considered:

$$\begin{aligned} \ddot{u} + h(u) = 0 \quad \text{with the two initial conditions} \\ u(0) = A \text{ and } \dot{u}(0) = 0 \end{aligned} \tag{4.2.1}$$

One may use the two trial functions in the forms of $u_1 = A\cos t$ and $u_2 = A\cos \omega t$. These two trail functions are the solutions of the following linear oscillator equations [5]:

$$\ddot{u} + \omega_1^2 u = 0, \quad \omega_1^2 = 1 \tag{4.2.2}$$

and

$$\ddot{u} + \omega_2^2 u = 0, \quad \omega_2^2 = \omega^2 \tag{4.2.3}$$

where ω is assumed to be the frequency of the oscillatory system, described by Eq. (4.2.1). The residual functions are then formulated as

$$R_1(t) = -\cos t + h(A\cos t) \tag{4.2.4}$$

and

$$R_2(t) = -\omega^2 \cos \omega t + h(A\cos \omega t) \tag{4.2.5}$$

The original FAF can be obtained from Eq. (4.2.6) [4]:

$$\omega^2 = \frac{\omega_1^2 R_2 - \omega_2^2 R_1}{R_2 - R_1} \tag{4.2.6}$$

or

$$\omega^2 = \frac{\omega_1^2 R_2(\omega t = 0) - \omega_2^2 R_1(t = 0)}{R_2 - R_1} \tag{4.2.7}$$

It is evident from the above formulations that the accuracy of this approach depends significantly on the location choice. In pursuance of overcoming the aforesaid drawback of this method, a weighted residual approach was proposed. The focus of the next chapter will be on the modified version of the frequency–amplitude formulations using the weighted residual methods.

4.2.1 The Method of Weighted Residuals

As mentioned in Sect. 4.2, the method of weighted residuals was used to improve the frequency–amplitude formulation [7]. In accordance with the modified approach, two new residual variables $\tilde{R}_1$ and $\tilde{R}_2$ are defined as

$$\tilde{R}_1 = \frac{4}{T_1}\int_0^{T_1} R_1(t)\cos\left(\frac{2\pi}{T_1}\right)\mathrm{d}t \tag{4.2.8}$$

and

$$\tilde{R}_2 = \frac{4}{T_2}\int_0^{\frac{T_2}{4}} R_2(t)\cos\left(\frac{2\pi}{T_2}\right)\mathrm{d}t \tag{4.2.9}$$

In accordance with Eq. (4.2.6), the following equation is expanded to determine the frequency–amplitude relationship [7]:

$$\omega^2 = \frac{\omega_1^2\tilde{R}_2 - \omega_2^2\tilde{R}_1}{\tilde{R}_2 - \tilde{R}_1} \tag{4.2.10}$$

An alternative combinational method could be proposed in accordance with vanishing of the weighted residual power over a quarter period of oscillation [7].

$$\bar{R} = \int_0^{\frac{T_2}{4}} R_2(t)\cos(\omega_2 t) = 0 \tag{4.2.11}$$

4.2.2 Geng and Cai Modification

A novel analytical method is suggested by Geng and Cai [5], which can not only be used for the weak nonlinear equations but also for the strong ones considering the complete solution domain. This method operates on the basis of an ancient Chinese mathematical method and He's modern interpretation. The following modification for Eq. (4.2.6) is suggested by Gang and Cai [5]:

$$\omega^2 = \frac{\omega_1^2 R_2(t_2) - \omega_2^2 R_1(t_1)}{R_2(t_2) - R_1(t_1)} \tag{4.2.12}$$

in which t_1 and t_2 are the collocation points. Generally speaking they are set as

$$t_1 = \frac{T_1}{12} \quad \text{and} \quad t_2 = \frac{T_2}{12} \tag{4.2.13}$$

where T_1 and T_2 are the two periods of the initial solutions of $u_1(t) = A\cos t$ and $u_2(t) = A\cos \omega t$, respectively.

4.2.3 Ren and Gui Modification

Ren and Gui [12] defined a new phase for the residuals by considering the phase equal to $0.382 \times \frac{\pi}{2} = 0.191\pi$, as the golden fraction of a quarter period. In mathematics, two quantities are in the golden ratio if their ratio is the same as the ratio of their sum to the larger of the two quantities [12]. According to this modification,

$$t_1 = 0.0955T_1, t_2 = 0.0955T_2 \tag{4.2.14}$$

Consequently, the oscillation frequency can be obtained using Eq. (4.2.12) as

$$\omega^2 = \frac{\omega_1^2 R_2(0.0955T_2) - \omega_2^2 R_1(0.0955T_1)}{R_2(0.0955T_2) - R_1(0.0955T_1)} \tag{4.2.15}$$

4.3 Max-Min Approach

The max-min approach is an alternative residual-based method, which generally determines the upper and the lower margins of the exact solution. The maximal and the minimal solution thresholds of a nonlinear problem are determined using the He Chengtian's interpolation. He Chengtian (370–447 AD) [14] proposed a new calendar, which is called the Yuanjia calendar. Chengtian developed an approach to adjust the first day of a calendrical synodic month to the true coincidence of the Moon using the lunar inequality, and accordingly, proposed a new calendar [24].

4.3.1 He Chengtian Inequality

The originality of this inequality comes from a very old mathematical dilemma. The exact interpretation has been brought here without any addition or deletion [14]:

> He Chengtian uses 26/49 as the strong, and 9/17 as the weak ratios. Among the strong and the weak, Chengtian obtains the fractional day (399/752) of the Moon by using the strong factor 15 and the weak factor 1.

The mathematical interpretation of the above statement reads as

$$29\frac{26}{49}\,\text{days} > 1\,\text{Moon} > 29\frac{9}{17}\,\text{days} \tag{4.3.1}$$

Using the weighting factors (15 and 1), He Chengtian (369–447 AD) obtained the fractional day as [14]

$$\text{The fractional day} = \frac{26 \times 15 + 9 \times 1}{49 \times 15 + 17 \times 1} = \frac{399}{752} \tag{4.3.2}$$

and therefore,

$$1\,\text{Moon} = 29\frac{399}{752}\,\text{days}. \tag{4.3.3}$$

Accordingly, Chengtian had employed the following inequality [14]:

$$\frac{a}{b} < u < \frac{d}{c} \tag{4.3.4}$$

where a, b, c, and d are real numbers.

$$\frac{a}{b} < \frac{ma + nd}{mb + nc} < \frac{d}{c} \tag{4.3.5}$$

in which m and n represent the weighting factors.

Therefore, u can approximately be found by

$$u = \frac{ma + nd}{mb + nc} \tag{4.3.6}$$

4.3.2 Application to Nonlinear Oscillators

In order to illustrate the applications of the max-min method in nonlinear systems, one first considers a generalized nonlinear oscillator with its differential equation in the form of [16]

$$\ddot{u} + uh(u) = 0, \quad \text{with the two initial conditions of } u(0) = A, \quad \text{and } \dot{u}(0) = 0 \tag{4.3.7}$$

where $h(u)$ is a nonnegative function of u.

According to the concept of the max-min method, the following trial function is chosen:

$$u(t) = A\cos\omega t \tag{4.3.8}$$

in which ω denotes the unknown frequency that needs to be determined. Also, the following solution, based on the initial conditions and Eq. (4.3.7), is assumed

$$u_1(t) = A\cos(\sqrt{h_{\min}}t) \tag{4.3.9}$$

and accordingly, it will result in the following function:

$$\ddot{u} + uh_{\min} = 0, \quad \text{with the two initial conditions of } u(0) = A, \quad \text{and } \dot{u}(0) = 0 \tag{4.3.10}$$

$h_{\min}$ will represent the minimum value of the function $h(u)$. Furthermore, ω^2 never exceeds the square of the frequency in the solution of Eq. (4.3.10).

Similarly one may assume the following motion:

$$u_1(t) = A\cos\left(\sqrt{h_{\max}}t\right) \tag{4.3.11}$$

For Eq. (4.3.10), the differential equation will be

$$\ddot{u} + uh_{\max} = 0, \quad \text{with the two initial conditions of } u(0) = A, \quad \text{and } \dot{u}(0) = 0 \tag{4.3.12}$$

where $h_{\max}$ is the maximum value of the function $h(u)$. Accordingly, one could also have the following inequality:

$$\frac{h_{\min}}{1} < \omega^2 < \frac{h_{\max}}{1} \tag{4.3.13}$$

In accordance with Chengtain's interpolation [14], one could obtain

$$\omega^2 = \frac{mh_{\min} + nh_{\max}}{m+n} \tag{4.3.14}$$

or

$$\omega^2 = \frac{h_{\min} + kh_{\max}}{1+k} \tag{4.3.15}$$

in which m and n are considered as the weighting factors and $k = \frac{n}{m}$.

Thus, the solution of Eq. (4.3.7) is represented by the following equation [16]:

$$u(t) = A\cos\left(\sqrt{\frac{h_{\min} + kh_{\max}}{1+k}}\right)t \tag{4.3.16}$$

where the value of k can be approximately obtained by various approximate methods [16].

Substituting Eq. (4.3.8) into (4.3.7) leads to the following residual equation:

$$R(t;k) = -\omega^2 A\cos\omega t + (A\cos\omega t)\cdot h(A\cos\omega t) \tag{4.3.17}$$

If Eq. (4.3.16) gives the exact solution, then *R(t; k)* is completely disappeared. In order to minimize the accumulative error, one can set [16]:

$$\int_0^T R(t;k)\cos\left[\sqrt{\frac{h_{\min} + kh_{\max}}{1+k}}\right]t\mathrm{d}t = 0 \tag{4.3.18}$$

Furthermore, the above function can be defined as an optimization problem, and depending on the nonlinear equation, the convex optimization method may be an alternative approach for minimizing the error [25]. In solving the above equation, one could find k in a straightforward way [16]:

$$k = \frac{h_{\max} - h_{\min}}{1 - \sqrt{\frac{A}{\pi}\int_0^{\pi}\cos^2 x\cdot h(A\cos x)\mathrm{d}x}} \tag{4.3.19}$$

The final solution can be obtained by implementing Eqs. (4.3.19) in (4.3.16) [16].

Part II: Applications

In order to discuss about the performance and applications of the residual methods in detail, the solution procedures of few well-known nonlinear differential equations are presented in this section.

4.4 Generalized Duffing Equation

The generalized Duffing equation is solved in this part of the section using the frequency–amplitude formulation.

Accordingly, the following form of the nonlinear differential equations is obtained [20]:

$$\begin{aligned} \ddot{u} + u + \alpha_3 u^3 + \alpha_5 u^5 + \alpha_7 u^7 + \cdots + \alpha_p u^p = 0; \\ u(0) = A, \dot{u}(0) = 0 \end{aligned} \tag{4.4.1}$$

and the trial functions are assumed as $u_1(t) = A\cos t$ and $u_2(t) = A\cos\omega t$. The trial functions would result in the following residuals:

$$R_1(t_1) = \alpha_3 A^3 \cos^3 t + \alpha_5 A^5 \cos^5 t + \cdots + \alpha_p A^p \cos^p t \tag{4.4.2}$$

and

$$\begin{aligned} R_2(t_2) = {} & A\cos\omega t(1-\omega^2) + \alpha_3 A^3 \cos^3 \omega t \\ & + \alpha_5 A^5 \cos^5 \omega t + \cdots + \alpha_p A^p \cos^p \omega t \end{aligned} \tag{4.4.3}$$

By substituting Eqs. (4.4.2) and (4.4.3) into Eq. (4.2.8), one could arrive at

$$\begin{aligned} \tilde{R}_1 = {} & \frac{3}{8}\alpha_3 A^3 + \frac{5}{16}\alpha_5 A^5 + \cdots \\ & + \frac{1\times 3\times 5\times\cdots\times p}{2\times 4\times 6\times\cdots\times p+1}\alpha_p A^p \quad (p = 2k+1, k = 1,2,3,\ldots) \end{aligned} \tag{4.4.4}$$

and

$$\begin{aligned} \tilde{R}_1 = {} & \frac{A}{2}(1-\omega^2) + \frac{3}{8}\alpha_3 A^3 + \frac{5}{16}\alpha_5 A^5 + \cdots \\ & + \frac{1\times 3\times 5\times\cdots\times p}{2\times 4\times 6\times\cdots\times p+1}\alpha_p A^p \end{aligned} \tag{4.4.5}$$

The next step is to substitute Eqs. (4.4.5) and (4.4.6) into Eq. (4.2.10). The following formula as the natural frequency of the generalized Duffing equation is obtained:

$$\omega = \sqrt{1 + 2\sum\nolimits^{p}_{\substack{p=2k+1 \\ k=1,2,3,\ldots,m}} \frac{3\times 5\times 7\times\cdots\times p}{2\times 4\times 6\times\cdots\times p+1}\alpha_p A^{p-1}} \tag{4.4.6}$$

4.5 Generalized Conservative Oscillatory Systems (Type 1)

A generalized form of the nonlinear differential equation of motion is to be investigated. This type of equation can cover the Duffing oscillator, the plasma physics equation, and the harmonic Duffing oscillator as their special cases.

The following governing equation of motion of a conservative nonlinear oscillatory system needs to be fully investigated [26]:

$$\ddot{u} + \Omega u^p + \frac{\lambda u^q}{\mu + \sigma u^r} = 0 \tag{4.5.1}$$

Rewriting the general vibration motion of Eq. (4.5.1) in the following form:

$$(\mu + \sigma u^r)(\ddot{u} + \Omega u^p) + \lambda u^q = 0 \tag{4.5.2}$$

In accordance with the frequency–amplitude formulation, the following first harmonic function as the initial guess is considered:

$$u = A\cos\omega t \tag{4.5.3}$$

Substituting Eq. (4.5.3) into Eq. (4.5.2), the following residual is obtained:

$$\begin{aligned} R(t) = & -\left[\mu A\cos(\omega t) + \sigma A^{r+1}\cos^{r+1}\omega t\right]\omega^2 + \mu\Omega A^p\cos^p\omega t \\ & + \sigma\Omega A^{p+r}\cos^{p+r}(\omega t) + \lambda A^q\cos^q(\omega t) \end{aligned} \tag{4.5.4}$$

According to the modified version of the frequency–amplitude formulation, one could have

$$\int_0^{\frac{T}{4}} R(t)\cos\omega t\mathrm{d}t = 0 \tag{4.5.5}$$

Consequently that will lead to

$$\omega^2 = \frac{\int_0^{\frac{\pi}{2\omega}}\left[\mu\Omega A^p\cos^{p+1}\omega t + \sigma\Omega A^{p+r}\cos^{p+r+1}\omega t + \lambda A^q\cos^{q+1}\omega t\right]}{\int_0^{\frac{\pi}{2\omega}}\left[\mu A\cos^2\omega t + \sigma A^{r+1}\cos^{r+2}\omega t\right]\mathrm{d}t} \tag{4.5.6}$$

Finally, the following equation is obtained to represent the natural frequency of the system [26]:

$$\omega^2 = \frac{\frac{\mu\Omega A^{p-1}\Gamma\left(\frac{p}{2}+1\right)}{\Gamma\left(\frac{p+3}{2}\right)} + \frac{\sigma\Omega A^{p+r-1}\Gamma\left(\frac{p+r}{2}+1\right)}{\Gamma\left(\frac{p+r+3}{2}\right)} + \frac{\lambda A^{q-1}\Gamma\left(\frac{q}{2}+1\right)}{\Gamma\left(\frac{q+3}{2}\right)}}{\frac{\mu\sqrt{\pi}}{2} + \frac{\sigma A^r\Gamma\left(\frac{r+3}{2}\right)}{\Gamma\left(\frac{r}{2}+2\right)}} \tag{4.5.7}$$

4.5.1 Linear Harmonic Oscillator

Choosing the values of the parameters $\Omega = 1, p = 1,$ and $\lambda = 0$ in Eq. (4.5.1), one could obtain the governing equation of a linear harmonic oscillator as

$$\ddot{u} + u = 0 \tag{4.5.8}$$

Equation (4.5.7) can be used to present the exact value, i.e., one to be the natural frequency this equation.

4.5.2 Duffing Harmonic Oscillator

Setting the values of the parameters as $\Omega = 0, \lambda = 1, q = 3, \mu = 1, \sigma = 1$ and $r = 2$ Eq. (4.5.1), the well-known Duffing harmonic equation will then be obtained [26]

$$\ddot{u} + \frac{u^3}{1+u^2} = 0 \tag{4.5.9}$$

The natural frequency of the above equation, using Eq. (4.5.1), can directly be found as

$$\omega^2 = \frac{3A^2}{4+3A^2} \tag{4.5.10}$$

Equation (4.5.10) represents the fact that the frequency–amplitude formulation and the Hamiltonian approach have a similar solution for this type of problem as the first approximation.

4.5.3 Plasma Physics Equation

Equation (4.5.1) has been altered to represent the plasma physics equation by selecting the values of the parameters $\Omega = 0, q = 0, \mu = 1, \sigma = 1$ and $r = 1$.

Subsequently, one could arrive at [27]

$$\ddot{u} + \frac{\lambda}{u} = 0 \tag{4.5.11}$$

According to Eq. (4.5.7), one could obtain Eq. (4.5.12) as a frequency–amplitude relationship of Eq. (4.5.11).

$$\omega = \frac{\sqrt{2\lambda}}{A} \tag{4.5.12}$$

When comparing this with the Hamiltonian approach (HA), it would show that the modified version of the frequency–amplitude formulation (FAF) gives a similar result. To show the accuracy of this method, a comparison with the exact solution has been carried out.

Based on Ref. [27], the exact frequency–amplitude relationship is represented by

$$\omega_{\text{ex}} = \frac{1.2533141}{A} \tag{4.5.13}$$

It has been observed that the relative error is recorded as 12% for this problem when compared with the exact solution.

4.6 Generalized Conservative Oscillator Systems (Type 2)

Based on Ref. [6], the following generalized nonlinear differential equation has been taken into consideration in this section:

$$\begin{aligned} &\ddot{u} + \beta u + u^{2n-1}\sqrt{1+\varepsilon^2 u^{4q}} = 0 \text{ with the two initial conditions} \\ &u(0) = A, \dot{u}(0) = 0 \text{ in which } q = 0, 1, 2, 3, \ldots, k \text{ and } n = 2q \end{aligned} \tag{4.6.1}$$

The parameters q and n represent the orders of the nonlinearity, and β and ε are the arbitrary constant values, which define the linear and nonlinear terms, respectively.

The well-known Ren-Gui and the Geng-Cui approaches have been employed to solve the differential equation Eq. (4.6.2).

Accordingly, $u_1(t) = A\cos\omega_1 t$ and $u_2(t) = A\cos\omega_2 t$ have been assumed as the initial guess. By substituting the initial guess into Eq. (4.6.1), the residual terms are obtained as

$$R_1(t) = A\cos\omega_1 t\left(\left(\beta - \omega_1^2\right) + A^{2n-2}\cos^{2n-2}\omega_1 t\sqrt{1+\varepsilon^2 A^{4q}\cos^{4q}\omega_1 t}\right) \tag{4.6.2}$$

and

$$R_2(t) = A\cos\omega_2 t\left(\left(\beta - \omega_2^2\right) + A^{2n-2}\cos^{2n-2}\omega_2 t\sqrt{1+\varepsilon^2 A^{4q}\cos^{4q}\omega_2 t}\right) \quad (4.6.3)$$

Using the Geng-Cai formulation, the following equation as the required natural frequency of the generalized equation is obtained:

$$\omega_{\text{Geng-Cai}}^2 = \beta + A^{2n-2}\cos^{2n-2}\left(\frac{\pi}{6}\right)\sqrt{1+\varepsilon^2 A^{4q}\cos^{4q}\left(\frac{\pi}{6}\right)} \quad (4.6.4)$$

In accordance with the Ren-Gui's modification for the frequency–amplitude formulation, the following equation is obtained to be the natural frequency of the system [6].

$$\omega_{\text{Ren-Gui}}^2 = \beta + A^{2n-2}\cos^{2n-2}(0.191\pi)\sqrt{1+\varepsilon^2 A^{4q}\cos^{4q}(0.191\pi)} \quad (4.6.5)$$

4.7 Nonlinear Oscillator with Fractional Power

A nonlinear oscillator with the fractional power can be mathematically represented with the following differential equation [28]:

$$\ddot{u} + \sum_i \mathrm{d}_i^2 u|u|^{i-1} = 0 \quad (4.7.1)$$

with the two initial conditions of

$$u(0) = A, \quad \text{and} \quad \dot{u}(0) = 0, \quad (4.7.2)$$

where $i > 0$ will be either a positive integer or a non-integer (fraction), and A is the arbitrary initial displacement. According to their initial conditions, $A\cos\omega t$ is supposed to be the initial guess, and therefore, the following residual term is obtained:

$$R(t) = -A\omega^2\cos\omega t + \sum_i \mathrm{d}_i^2 A\cos\omega t|A\cos\omega t|^{i-1} \quad (4.7.3)$$

In order to find the weighted residual of Eq. (4.7.3), one could develop the following differential equations:

$$\tilde{R} = \int_0^{\frac{T}{4}} \left(-A\omega^2 \cos^2 \omega t + \sum_i \mathrm{d}_i^2 A \cos \omega t |A \cos \omega t|^i \right) \mathrm{d}t \tag{4.7.4}$$

After the integration of Eq. (4.7.5), one could arrive at

$$\tilde{R} = -\frac{\pi}{4} A\omega^2 + \sum_i \mathrm{d}_i^2 A^i \int_0^{\frac{\pi}{2}} \cos^{i+1} \omega t \mathrm{d}t \tag{4.7.5}$$

In order to find the natural frequency, Eq. (4.7.5) has to be vanished, accordingly,

$$\omega = \sqrt{\sum_i \frac{\mathrm{d}_i^2 A^{i-1}}{2^{i-1}(i+1)B(\frac{i+3}{2}, \frac{i+1}{2})}} \tag{4.7.6}$$

in which, the symbol B is the beta function.

4.8 Helmholtz Duffing Oscillator

In this section, the following form of the Helmholtz Duffing oscillator is analyzed. The governing vibration equation of this system is given by [29]

$$\ddot{u} + u + (1 - \alpha)u^2 + \alpha u^3 = 0; \tag{4.8.1}$$
with the two initial conditions of $u(0) = A$, and $\dot{u}(0) = 0$

It is known that an asymmetric nonlinear oscillator has different forms of behavior, in the positive and negative directions. The solution can readily be separated into two different parts:

$$\ddot{u} + u + (1 - \alpha)u^2 \mathrm{sgn}(u) + \alpha u^3 = 0, \quad \text{for } u \geq 0, \tag{4.8.2}$$

and

$$\ddot{u} + u - (1 - \alpha)u^2 \mathrm{sgn}(u) + \alpha u^3 = 0, \quad \text{for } u \leq 0, \tag{4.8.3}$$

The two trial functions $u_1 = a \cos t$ and $u_2 = a \cos \omega t$ are assumed to be in the positive direction. Substituting the trial function into Eq. (4.8.2) would give

$$R_1(t_1) = (1-\alpha)a^2\cos^2 t\,\mathrm{sgn}(a\cos t) + \alpha a^3\cos^3 t \tag{4.8.4}$$

and

$$R_2(t_2) = (1-\omega^2)a\cos\omega t + (1-\alpha)a^2\cos^2\omega t\,\mathrm{sgn}(a\cos\omega t) + \alpha a^3\cos^3\omega t \tag{4.8.5}$$

The weighted residuals can be determined in the following forms of the integrals:

$$\tilde{R}_1 = \frac{4}{T_1}\int_0^{T_1/4} R_1\cos t dt = \frac{4}{3\pi}(1-\alpha)a^2 + \frac{3}{8}\alpha a^3 \tag{4.8.6}$$

and

$$\tilde{R}_2 = \frac{4}{T_1}\int_0^{T_1/4} R_2\cos\omega t dt = \frac{a}{2}(1-\omega^2) + \frac{4}{3\pi}(1-\alpha)a^2 + \frac{3}{8}\alpha a^3 \tag{4.8.7}$$

Substituting the above equations in the frequency equation of $\omega^2 = \frac{\omega_1^2\tilde{R}_2 - \omega_2^2\tilde{R}_1}{\tilde{R}_2 - \tilde{R}_1}$, one would obtain the following equation:

$$\omega_a = \sqrt{1 + \frac{8}{3\pi}(1-\alpha)a + \frac{3}{4}\alpha a^2} \tag{4.8.8}$$

The same procedure for the negative direction results in the following equation for the natural frequency of the targeted equation:

$$\omega_b = \sqrt{1 + \frac{8}{3\pi}(1-\alpha)b + \frac{3}{4}\alpha b^2} \tag{4.8.9}$$

4.9 Relativistic Oscillator

The relativistic oscillator has been previously analyzed in Chap. 3. This equation is solved once again in this section by using the frequency–amplitude formulation (FAF). The governing equation of a standard relativistic oscillator is given by [19, 30]

$$\ddot{u} + \frac{u}{\sqrt{u^2+1}} = 0;$$
$$\text{with initial conditions } u(0) = B, \quad \text{and } \dot{u}(0) = 0 \tag{4.9.1}$$

According to the FAF solution procedure, two trial functions have been selected as

$$u_1(t) = \mathrm{B}\cos t \text{ and } u_2(t) = \mathrm{B}\cos\omega t \tag{4.9.2}$$

The corresponding residual terms are obtained as [19, 30]

$$R_1(t) = -B\cos t + \frac{B\cos t}{\sqrt{1+B^2\cos^2 t}} \tag{4.9.3}$$

and

$$\mathrm{R}_2(t) = -\mathrm{B}\omega^2\cos\omega t + \frac{\mathrm{B}\cos\omega t}{\sqrt{1+\mathrm{B}^2\cos^2\omega t}} \tag{4.9.4}$$

By substituting Eqs. (4.9.3) and (4.9.4) into Eq. (4.2.12), one may get [19, 30]

$$\omega^2 = \frac{-B\omega^2\cos\frac{\omega T_2}{12} + \frac{B\cos\frac{\omega T_2}{12}}{\sqrt{1+B^2\cos^2\frac{\omega T_2}{12}}} - \omega^2\left(-B\cos\frac{T_1}{12} + \frac{B\cos\frac{T_1}{12}}{\sqrt{1+B^2\cos^2\frac{T_1}{12}}}\right)}{-B\omega^2\cos\frac{\omega T_2}{12} + \frac{B\cos\frac{\omega T_2}{12}}{\sqrt{1+B^2\cos^2\frac{\omega T_2}{12}}} - \left(-B\cos\frac{T_1}{12} + \frac{B\cos\frac{T_1}{12}}{\sqrt{1+B^2\cos^2\frac{T_1}{12}}}\right)} = \frac{1}{\sqrt{1+\frac{3}{4}B^2}} \tag{4.9.5}$$

The frequency–amplitude formulation (FAF) provides the analogous results with the Harmonic balance method (HBM) and the Hamiltonian approach (HA). From the results presented in Chap. 3, it has been concluded that the EBM has the premier accuracy for this particular nonlinear problem.

4.10 Autonomous Conservative Oscillatory System

The governing equation of an autonomous conservative nonlinear oscillator with the static and inertial cubic nonlinearities can be expressed by the following equation based on Ref. [31]:

$$\ddot{u} + \lambda u + \varepsilon_1 u^2 \ddot{u} + \varepsilon_1 u \dot{u}^2 + \varepsilon_2 u^3 = 0; \\ \text{with initial conditions of}: u(0) = A,\ \dot{u}(0) = 0 \tag{4.10.1}$$

in which λ is an integer that can take the values of either zero, +1 or −1, and also, ε_1 and ε_2 are positive values. In order to solve Eq. (4.10.1), the frequency–amplitude formulation is implemented.

Accordingly, $u_1 = A\cos t$ and $u_2 = A\cos\omega t$ are implemented as the trail functions and the following residual expressions are obtained:

$$R_1(t) = -A\cos t + \lambda A\cos t + \varepsilon_1 (A\cos t)^2(-A\cos t) \\ + \varepsilon_1 (A\cos t)(A\sin t)^2 + \varepsilon_2 (A\cos t)^3 \tag{4.10.2}$$

and

$$R_2(t) = -A\omega^2\cos\omega t + \lambda A\cos\omega t + \varepsilon_1(-A\omega^2\cos\omega t) \times (A\cos\omega t)^2 \\ + \varepsilon_1 (A\cos\omega t)(A\omega\sin\omega t)^2 + \varepsilon_2 (A\cos t)^3 \tag{4.10.3}$$

The weighted residuals of Eqs. (4.10.2) and (4.10.3) are obtained

$$\tilde{R}_1(t) = \frac{2}{\pi}\int_0^{\frac{\pi}{2}} R_1(t)\cos t dt = \frac{1}{2}(A(\lambda - 1) + \varepsilon_1 A^3) \\ + \frac{3}{8}A^3(\varepsilon_2 - 2\varepsilon_1) \tag{4.10.4}$$

and

$$\tilde{R}_2(t) = \frac{2\omega}{\pi}\int_0^{\frac{\pi}{2\omega}} R_2(t)\cos\omega t dt = \frac{1}{2}(A(\lambda - \omega^2) + \varepsilon_1 A^2\omega^2) \\ + \frac{3}{8}(A^3(\varepsilon_2 - 2\varepsilon_1\omega^2)) \tag{4.10.5}$$

The above-weighted residuals are substituted in Eq. (4.2.10), and subsequently, the following formula is achieved for the natural frequency of Eq. (4.10.1) [31]:

$$\omega = \sqrt{\frac{\left[\lambda + \frac{3}{4}\varepsilon_2 A^2\right]}{\left[1 + \frac{1}{2}\varepsilon_1 A^2\right]}} \tag{4.10.6}$$

4.11 Nonlinear Oscillation of a Mass Attached to a Stretched Elastic Wire

In this section, the vibration of a mass attached to a stretched elastic wire is investigated. This system is schematically illustrated in Fig. 4.2 [32, 33].

Dynamic behavior of the mechanical system is presented by the following equation:

$$\ddot{u} + u - \frac{\lambda u}{\sqrt{1+u^2}} = 0; \quad u(0) = A, \dot{u}(0) = 0 \tag{4.11.1}$$

Equation (4.11.1) is solved by the modified FAF, and then, it is analyzed using max-min approach. According to the FAF, the two following trial functions are selected:

$$u_1(t) = A\cos t, \quad u_2(t) = A\cos\omega t \tag{4.11.2}$$

Based on the FAF procedure, two following residuals are obtained as [33]:

$$R_1(t) = -\frac{\lambda A\cos t}{\sqrt{1 + A^2\cos^2 t}} \tag{4.11.3}$$

$$R_2(t) = -A\omega^2\cos\omega t + A\cos\omega t - \frac{\lambda A\cos\omega t}{\sqrt{1 + A^2\cos^2\omega t}} \tag{4.11.4}$$

$\tilde{R}_1$and $\tilde{R}_2$ should be obtained in the form of

$$\tilde{R}_1 = \frac{4}{T_1}\int_0^{\frac{T_1}{4}} R_1(t)\cos t\mathrm{d}t = -\frac{2\lambda}{\pi A}\left[E\left(-A^2\right) - K\left(-A^2\right)\right] \tag{4.11.5}$$

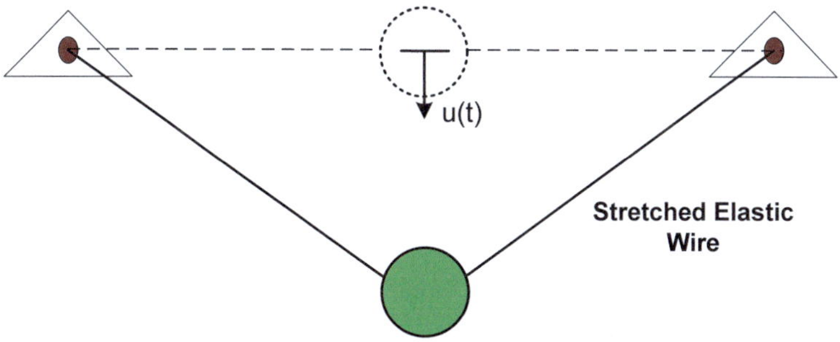

Fig. 4.2 Mass attached to a stretched elastic wire [34]

$$\tilde{R}_2 = \frac{4}{T_2}\int_0^{\frac{T_1}{4}} R_2(t)\cos\omega t\mathrm{d}t = \frac{A}{2} - \frac{2\lambda}{\pi A}\left[E\left(-A^2\right) - K\left(-A^2\right)\right] \tag{4.11.6}$$

Using Eq. (4.2.10), the following frequency–amplitude relationship is obtained:

$$\omega^2 = 1 - \frac{4\lambda}{\pi A^2}\left[E\left(-A^2\right) - K\left(-A^2\right)\right] \tag{4.11.7}$$

To solve Eq. (4.11.1) by the max-min approach, this equation must be altered to the following form:

$$\ddot{u} + uh(u) = 0 \tag{4.11.8}$$

where $h(u) = 1 - \frac{\lambda}{\sqrt{1+u^2}}$.

According to the MMA, $A\cos\omega t$ is used as the trial function. The minimum and maximum values for ω must be found. As a result, we have [33]

$$1 - \lambda < \omega^2 < 1 - \frac{\lambda}{\sqrt{1+A^2}} \tag{4.11.9}$$

Employing He Chengtain's inequality and introducing a new variable $k = \frac{n}{m+n}$, we have

$$\omega^2 = 1 + \lambda(k-1) - \frac{\lambda k}{\sqrt{1+A^2}} \tag{4.11.10}$$

Therefore, solution of Eq. (4.11.1) is obtained to be

$$x = A\cos\left(\sqrt{1 + \lambda(k-1) - \frac{\lambda k}{\sqrt{1+A^2}}}t\right) \tag{4.11.11}$$

Equation (4.11.1) can be altered in the following form by substituting the above approximate solution:

$$\begin{aligned} &\ddot{u} + \left(1 + \lambda(k-1) - \frac{\lambda k}{\sqrt{1+A^2}}\right)f(u) \\ &= \lambda(k-1)u - \frac{\lambda k}{\sqrt{1+A^2}}u + \frac{\lambda}{\sqrt{1+u^2}}u \end{aligned} \tag{4.11.12}$$

If Eq. (4.11.11) coincides with the exact solution of Eq. (4.11.1), then the right-hand side of Eq. (4.11.12) vanishes. Otherwise the right-hand side of Eq. (4.11.12) can be expanded into the Fourier series [33]

$$\begin{aligned}&\lambda(k-1)u-\frac{\lambda k}{\sqrt{1+A^2}}u+\frac{\lambda}{\sqrt{1+u^2}}u\\&=\sum_{j=0}^{\infty}b_{2j+1}\cos[(2j+1)\omega t]=b_1\cos(\omega t)\end{aligned}\tag{4.11.13}$$

After a few mathematical simplifications, one could obtain

$$k=\frac{\sqrt{1+A^2}}{\sqrt{1+A^2}-1}\left[1-\frac{4E(-A^2)-4K(-A^2)}{\pi A^2}\right]\tag{4.11.14}$$

and consequently we have

$$\omega^2=1-\frac{4\lambda}{\pi a^2}\left[E\left(-A^2\right)-K\left(-A^2\right)\right]\tag{4.11.15}$$

It is seen that both the MMA and FAF give the same results for this problem. The solution is fully matched with the one obtained by the Hamiltonian approach and also harmonic balance method. Interested reader can find more information about the above example in Ref. [33].

4.12 Nonlinear Schrödinger Equation

The nonlinear Schrödinger equation is one of the most well-known equations in quantum theory. It mathematically characterizes different phenomena in condensed matter and plasma physics and also nonlinear optics. In this section, the frequency–amplitude formulation is exerted as a tool to solve a kind of nonlinear Schrödinger equation based on the provided solution procedure in Ref. [35]. Zhang et al. studied the following Schrödinger equation using FAF:

$$iu_t+u_{xx}+2|u|^2u=0,\, u(x,0)=\mathrm{e}^{ipx}\tag{4.12.1}$$

According to the rule of FAF, we must have two trial functions for our system. As a result, the following equations are chosen for the objective equations:

$$u_1(x,t)=\mathrm{e}^{i(px+t)};\; u_2\,(x,t)=\mathrm{e}^{i(px+\omega t)}\tag{4.12.2}$$

in which both are the solutions of following linear Schrödinger equation:

$$iu_t + \omega^2 u_{xx} = 0 \tag{4.12.3}$$

The two initial solutions are substituted into Eq. (4.12.1) resulting in the residuals:

$$R_1(x,t) = \left(1 - p^2\right)e^{i(px+t)} \tag{4.12.4}$$

and

$$R_2(x,t) = \left(2 - \omega - p^2\right)e^{i(px+\omega t)} \tag{4.12.5}$$

According to Eq. (4.2.7), we obtain

$$\omega^2 = -\left(\omega + \omega p^2 - 2\omega + p^2 - 2\right) \tag{4.12.6}$$

Simplifying the above relation, one could arrive at

$$\omega = 2 - p^2 \tag{4.12.7}$$

$$u(x,t) = e^{i\left[px + \left(2-p^2\right)t\right]} \tag{4.12.8}$$

It is possible to obtain the exact solution with setting $p = 1$.

As the second example, the following nonlinear Schrödinger equation is considered [35]:

$$iu_t + u_{xx} + 2|u|^2 u = 0, u(x,0) = e^{ipx} \tag{4.12.9}$$

Similar to the example one, after a number of mathematical operations, one could have

$$R_1(x,t) = \left(-3 - p^2\right)e^{i(px+t)} \tag{4.12.10}$$

and

$$R_2(x,t) = \left(-2 + \omega + p^2\right)e^{i(px+\omega t)} \tag{4.12.11}$$

By using Eq. (4.2.7), one may arrive at [35]

$$\omega^2 = -\left(\omega + \omega p^2 + 2\omega + p^2 + 2\right) \tag{4.12.12}$$

Solving the above equation for ω results in Eq. (4.12.13):

$$\omega = -2 - p^2 \tag{4.12.13}$$

and the final solution comes out in the form of Eq. (4.12.14):

$$u(x,t) = e^{i\left[px - (2+p^2)t\right]} \tag{4.12.14}$$

Accordingly, the exact solution can be obtained by setting $p = 1$ [35].

4.13 Rigid Frame Rotates at a Fixed Rate Ω

The motion equation of a rigid frame, which is being rotated at the fixed rate Ω, is as Eq. (4.13.1) [36]:

$$\ddot{u} + (1 - \cos(u))\sin(u) = 0 \tag{4.13.1}$$

Assume the following useful trial functions as

$$u_1 = A\cos(t) \quad \text{and} \quad u_2 = A\cos(2t) \tag{4.13.2}$$

Substituting u_1 and u_2 into Eq. (4.13.1) results in the residual functions as

$$R_1(t) = d^2\frac{(A\cos t)}{dt^2} + (1 - \Delta\cos(A\cos t))\sin(A\cos t) \tag{4.13.3}$$

and

$$R_2(t) = d^2\frac{(A\cos 2t)}{dt^2} + (1 - \Delta\cos(A\cos 2t))\sin(A\cos 2t) \tag{4.13.4}$$

Based on the frequency–amplitude formulation method, we can have

$$\omega^2 = \frac{\begin{array}{r}-4A\cos 2t + (1 - \Delta\cos(A\cos(2t))\sin(A\cos(2t)) + 4A\cos(t)\\ -4(1 - \Delta\cos(A\cos(t))\sin(A\cos(t))\end{array}}{\begin{array}{r}-4A\cos 2t + (1 - \Delta\cos(A\cos(2t))\sin(A\cos(2t)) + A\cos(t)\\ -(1 - \Delta\cos(A\cos(t))\sin(A\cos(t))\end{array}} \tag{4.13.5}$$

Considering $\cos(t) = \cos(2t) = \beta$, we can reformulate Eq. (4.3.15) as follows:

$$\omega = \sqrt{-\frac{(-1 + \Delta\cos(A\beta))\sin(A\beta)}{A\beta}} \tag{4.13.6}$$

Applying the modified FAF to the residual function, one could obtain the value β. Thus, the natural frequency of the system is obtained [36].

4.14 Conservative Lienard Type Equation

Considering the conservative Lienard type equation in the following form [37–39] yields:

$$\ddot{u} + (1 + \dot{u}^2)u = 0 \quad u(0) = A, \dot{u}(0) = 0 \tag{4.14.1}$$

Assuming $u_1 = A\cos\omega_1 t$ as the initial guess, we reach the following form as the residual function:

$$R_1 = \left(1 + \omega_1^2 A^2 \left(\frac{1 - \cos 2\omega_1 t}{2}\right)\right) A\cos\omega_1 t - \omega_1^2 A\cos\omega_1 t \tag{4.14.2}$$

and $u_2 = A\cos\omega_2 t$ is substituted in Eq. (4.14.1), accordingly, the following residual is obtained:

$$R_2 = \left(1 + \omega_2^2 A^2 (\frac{1 - \cos 2\omega_2 t}{2})\right) A\cos\omega_2 t - \omega_2^2 A\cos\omega_2 t \tag{4.14.3}$$

supposing $\omega_1 = 1$ and $\omega_2 = \omega$ using the frequency–amplitude formulation, the natural frequency of the system is obtained as follows:

$$\omega = \frac{2}{\sqrt{4 - A^2}} \tag{4.14.4}$$

Part III: Practice Problems

P.4.1. Use frequency–amplitude formulation to solve the following differential equation considering the first approximation:

P.4.1.a. $\ddot{x} + x^{\frac{1}{3}} = 0 \quad x(0) = A, \dot{x}(0) = 0$
P.4.1.b. $\ddot{x} + \mathrm{sgn}(x) = 0 \quad x(0) = A,\ \dot{x}(0) = 0$

P.4.1.c. $\ddot{x} + \sin(x) = 0 \quad x(0) = A, \dot{x}(0) = 0$
P.4.1.d. $\ddot{x} + \alpha x + \beta x^3 + \gamma x^5 = 0 \quad x(0) = A, \ \dot{x}(0) = 0$
P.4.1.e. $\ddot{x} + \frac{x}{x^2+1} = 0 \quad x(0) = A, \ \dot{x}(0) = 0$

P.4.2. Solve all of the above nonlinear differential equations using the max-min approach and then compare the obtained results with those obtained from Problem 4.1.

P.4.3. Solve the following equations using the frequency–amplitude formulations and consider the first and second approximations for the solution. Assume that $x(0) = A, \ \dot{x}(0) = 0$

P.4.3.a. $\ddot{x} + x^{1/3} = 0$

and

P.4.3.b.$\ddot{x} + x^{-1} = 0$

P.4.4. Use Java program to develop a code, which would be able to solve any type of conservative oscillators using the frequency–amplitude formulations.

P.4.5. Prove that frequency–amplitude formulation is not capable of solving the nonlinear Duffing oscillator with the inclusion of the damping term.

P.4.6. Use the Ren-Gui modification to solve the following nonlinear differential equations [40]:

$$\ddot{x} + \lambda_1 x + \lambda_2 \frac{\text{sgn}(x)|x|^{\alpha_1}}{(1+\alpha_3 x^2)^{\alpha_2}} = 0 \ x(0) = A, \ \dot{x}(0) = 0$$

P.4.7. Compare the results obtained from the max-min approach with those obtained from any numerical methods, such as the Runge–Kutta, for the following nonlinear differential equations [41]:

$$\ddot{x} + \frac{x^3}{x^2+1} = 0 \quad x(0) = A, \ \text{and} \ \dot{x}(0) = 0$$

P.4.8. Consider the following partial differential equation, which mathematically shows the vibration of a curved beam based on the Euler–Bernoulli beam theory [42].

$$EI\frac{d^4w}{dx^4} + \mu\frac{dw}{dt} + \rho A\frac{d^2w}{dt^2} + k_1 w - k_p\frac{\partial^2 w}{\partial x^2}$$
$$= \frac{EA}{l}\int_0^l \left(\frac{\partial y}{\partial x}\frac{\partial w}{\partial x} + \frac{1}{2}(\frac{\partial w}{\partial x})^2\right)dx \left(\frac{\partial^2 w}{\partial x^2} + \frac{\partial^2 y}{\partial x^2}\right)$$

where $E, I, \mu, \rho,$ and A are the modulus of elasticity, moment of inertia, damping coefficient, density, and the cross-sectional area of the beam, respectively. Parameters k_1, k_p, y and l are the respective Winkler and Pasternak coefficients, the function of curvature, and the length of the beam.

P.4.8.a. Obtain the differential equation of the primary mode of the curved beam vibration by using the Galerkin method.
P.4.8.b. Use the frequency–amplitude formulation to obtain the closed form expression for the frequency of the beam's vibration.
P.4.8.c. Compare the obtained results with those found from using the max-min approach.
P.4.8.d. Find the relative error of the results by comparing with the exact numerical simulation.

References

1. Ji-huan H (2002) Ancient Chinese algorithm: the Ying Buzu Shu (method of surplus and deficiency) vs Newton iteration method. Appl Math Mech 23(12):1407–1412
2. Dauben JW (1998) Ancient Chinese mathematics: the (Jiu Zhang Suan Shu) vs Euclid's Elements. Aspects of proof and the linguistic limits of knowledge. Int J Eng Sci 36(12–14):1339–1359
3. He J-H (2004) Solution of nonlinear equations by an ancient Chinese algorithm. Appl Math Comput 151(1):293–297
4. He J-H (2008) Comment on 'He's frequency formulation for nonlinear oscillators'. Eur J Phys 29(4):L19–L22
5. Geng L, Cai X-C (2007) He's frequency formulation for nonlinear oscillators. Eur J Phys 28 (5):923–931
6. Khan Y, Kalami-Yazdi M, Askari H, Saadatnia Z (2011) Dynamic analysis of generalized conservative nonlinear oscillators via frequency amplitude formulation. Arab J Sci Eng 38 (1):175–179
7. He JH (2008) An improved amplitude-frequency formulation for nonlinear oscillators. Int J Nonlinear Sci Numer Simul 9(2):211–212
8. Zhang HL (2008) Application of He's frequency-amplitude formulation to an x1/3 force nonlinear oscillator. Int J Nonlinear Sci Numer Simul 9(3):297–300
9. Davodi AG, Ganji DD, Azami R, Babazadeh H (2009) Application of improved amplitude-frequency formulation to nonlinear differential equation of motion. Mod Phys Lett B 23(28):3427–3436
10. Fan J (2009) He's frequency–amplitude formulation for the Duffing harmonic oscillator. Comput Math Appl 58(11–12):2473–2476
11. Zhang H-L, Qin L-J (2011) An ancient Chinese mathematical algorithm and its application to nonlinear oscillators. Comput Math Appl 61(8):2071–2075
12. Ren Z-F, Gui W-K (2011) He's frequency formulation for nonlinear oscillators using a golden mean location. Comput Math Appl 61(8):1987–1990
13. Kline M (1972) Mathematical thought from Ancient to modern times. Oxford University Press, New Yor
14. He J-H (2004) He Chengtian's inequality and its applications. Appl Math Comput 151 (3):887–891
15. He JH (2008) Max-min approach to nonlinear oscillators. Int J Nonlinear Sci Numer Simul 9 (2):207–210
16. Zeng D-Q (2009) Nonlinear oscillator with discontinuity by the max–min approach. Chaos, Solitons Fractals 42(5):2885–2889

17. Azami R, Ganji DD, Babazadeh H, Davodi AG, Ganji SS (2009) He's Max-Min method for the relativistic oscillator and high-order Duffing equation. Int J Mod Phys B 23(32):5915–5927
18. Demirbağ SA, Kaya MO (2010) Application of He's max-min approach to a generalized nonlinear discontinuity equation. Int J Nonlinear Sci Numer Simul 11(4):269–272
19. Cai X-C, Wu W-Y (2009) He's frequency formulation for the relativistic harmonic oscillator. Comput Math Appl 58(11–12):2358–2359
20. Younesian D, Askari H, Saadatnia Z, KalamiYazdi M (2010) Frequency analysis of strongly nonlinear generalized Duffing oscillators using He's frequency–amplitude formulation and He's energy balance method. Comput Math Appl 59(9):3222–3228
21. Chan G, Alan KL, Clancey Ed (2002) Historical perspectives on East Asian science, technology and medicine. World Scientific Publishing Co. Pte Ltd, London
22. Katz VJ (ed) (2007) The mathematics of Egypt, Mesopotamia, China, India, and Islam, Princeton University press
23. Swetz FJ, Tian A (1984) Se, "A brief chronological and bibliographic guide to the history of Chinese mathematics". Hist Math 11(1):39–56
24. Thomas Hockey MB, Trimble V, Williams TR, Bracher K, Jarrell RA, Marché II JD, Ragep FJ, Palmeri JA (eds) (2007) The biographical encyclopedia of astronomers. Springer
25. Boyd L (2004) Stephen and Vandenberghe. Cambridge University Press, Convex Optimization
26. Ebaid AE (2010) Analytical periodic solution to a generalized nonlinear oscillator: application of He's frequency-amplitude formulation. Mech Res Commun 37(1):111–112
27. Xu L (2011) A Hamiltonian approach for a plasma physics problem. Comput Math Appl 61 (8):1909–1911
28. Cveticanin L, Kalami-Yazdi M, Askari H (2012) Analytical approximations to the solutions for a generalized oscillator with strong nonlinear terms. J Eng Math 77(1):211–223
29. Askari H, Saadatnia Z, Younesian D, Yildirim A, Kalami-Yazdi M (2011) Approximate periodic solutions for the Helmholtz-Duffing equation. Comput Math Appl 62(10):3894–3901
30. Chen G, Tao Z (2012) Amplitude-frequency relationship for the relativistic oscillator. AASRI Procedia 1:400–403
31. Yazdi MK, Khan Y, Madani M, Askari H, Saadatnia Z, Yildirim A (2010) Analytical solutions for autonomous conservative nonlinear oscillator. Int J Nonlinear Sci Numer Simul 11(11):979–984
32. Sun WP, Wu BS, Lim CW (2007) Approximate analytical solutions for oscillation of a mass attached to a stretched elastic wire. J Sound Vib 300(3–5):1042–1047
33. Durmaz S, Altay Demirbağ S, Kaya MO (2011) Approximate solutions for nonlinear oscillation of a mass attached to a stretched elastic wire. Comput Math Appl 61(3):578–585
34. Younesian D, Askari H, Saadatnia Z, KalamiYazdi M (2012) Analytical approximate solutions for the generalized nonlinear oscillator. Appl Anal 91(5):965–977
35. Zhang Y-N, Xu F, Deng L (2009) Exact solution for nonlinear Schrödinger equation by He's frequency formulation. Comput Math Appl 58(11–12):2449–2451
36. Barari A, Kimiaeifar A, Nejad MG, Motevalli M, Sfahani MG (2012) A closed form solution for nonlinear oscillators frequencies using amplitude-frequency formulation. Shock Vib 19 (6):1415–1426
37. Chen G, Tao Z (2011) Notes on a Lienard type equation. In: International conference on information science and technology, pp 5–6
38. Sabatini M (2004) On the period function of $x'' + f(x)x'2 + g(x) = 0$. J Differ Equ 196 (1):151–168
39. Sabatini M (1999) On the period function of Liénard systems. J Differ Equ 152(2):467–487

40. Beléndez A, Arribas E, Francés J, Pascual I (2011) Notes on 'Application of the Hamiltonian approach to nonlinear oscillators with rational and irrational elastic terms.' Math Comput Model 54:11–12
41. Saadatnia Z, Askari H, KalamiYazdi M (2010) Frequency analysis of nonlinear oscillators with rational restoring force via He's energy balance method and He's variational approach. Nonlinear Sci Lett A 1(4):425–430
42. Askari H, Saadatnia Z, Esmailzadeh E (2014) Nonlinear vibration of nanobeam with Quadratic Rational Bezier Arc curvature. In: Volume 4A: dynamics, vibration, and control, vol. 4A, p V04AT04A044

Chapter 5
Semi-inverse and Variational Methods

Abstract Semi-inverse method and variational principle are explained in this chapter, and the Hamiltonian and variational approaches are discussed in detail. The direct relationship between these two techniques is mathematically described. The second part of this chapter is focused on the potential applications of the above-mentioned techniques for studying different nonlinear oscillators. They include the rational and irrational oscillators, nonlinear oscillators with either discontinuity or non-integer fractional characteristic equation, or the rational elastic characteristic equation. Further analyses are done on Duffing equation, nonlinear Schrodinger's equation, Thomas–Fermi equation, heat conduction equation, Lane–Emden equation, and Duffing–Harmonic equation.

Keywords Semi-inverse method · Hamiltonian approach · Variational approach
Duffing equation · Nonlinear Schrodinger's equation · Thomas–Fermi equation
Lane–Emden equation

Part I: Approach

Semi-inverse method is a potent approach, which can be exerted for analyzing different convoluted nonlinear differential equations. In fact, this method is the extension and the modification of the Chein's weighted residual approach, Liu's systematic method [1], and Hu's trial-and-error technique, and it has been broached by J. H. He [2]. The semi-inverse method can readily be utilized for constructing the variational principles without any variational crisis phenomenon. Accordingly, this technique can be implemented as a strong path to obtain various variational principles directly from the field equations and boundary conditions. It is known that variational principle plays an important role in analyzing partial differential equations with variable coefficients [2]. Thus, the semi-inverse method can be very expedient for scrutinizing the nonlinear partial differential equations.

This chapter is focused on different features of the semi-inverse method and variational principle. It will be shown that the semi-inverse method can be

E. Esmailzadeh et al., *Analytical Methods in Nonlinear Oscillations*, Solid Mechanics and Its Applications 252, https://doi.org/10.1007/978-94-024-1542-1_5

effectively used to find the variational principle for different types of differential equations. The variational approach, which has been constructed based on the semi-inverse method, is subsequently presented in detail [3]. Furthermore, the Hamiltonian approach [4] and its potential for solving nonlinear differential equations will be discussed. Afterwards, we will show the direct relationship between the Hamiltonian and variational approach. In the last part of chapter, a number of practical examples are provided, which are based on recently published articles in this area using those approaches. The potency and accuracy of the illustrated methods are examined by solving few complicated nonlinear problems.

5.1 Variational Principle

Variational principle is employed to characterize a wide range of engineering problems. It in fact provides a concise view from the dynamic equations and provides a detailed perception from the influence of different parameters [2]. In addition, it is a potent tool to approximate the solutions of different problems. There are several approaches to find the variational principle of partial differential equations, namely, the Lagrange multiplier method, higher order Lagrange multiplier method, pre-conditioned Lagrange method, and the semi-inverse method [2]. It has been shown that the Lagrange multiplier approach may result in an invalid variation principle for few partial differential equations. In fact, some of the proposed methods cannot lead to a valid variational form for the targeted system. This unreliability is called the variational crisis. There are two main categories for the variational crisis:

Type (I) First kind of variational form: The Lagrange multipliers vanish when the variational constraints are eliminated.
Type (II) Second kind of variational form: A complete form cannot be obtained for the Euler's equations when the modified form of functional stationary is constructed [2].

Having the intention to briefly explain both the Liu's and Chien's methods, which they are the modified approaches to overcome the variational crisis, the following governing equation is considered [2]:

$$\frac{\partial \alpha}{\partial x} + \frac{\partial \beta}{\partial y} = 0 \tag{5.1.1}$$

$$\frac{\partial \Gamma}{\partial x} - \frac{\partial \chi}{\partial y} = 0 \tag{5.1.2}$$

where parameters $\alpha, \beta, \Gamma,$ and χ are functions of pressure, density, and velocity. According to the relationships in Eqs. (5.1.1) and (5.1.2), the following functions are defined:

$$\frac{\partial \psi}{\partial x} = \beta; \quad \frac{\partial \psi}{\partial y} = -\alpha; \quad \frac{\partial \phi}{\partial x} = \chi \quad \text{and} \quad \frac{\partial \phi}{\partial y} = \Gamma \tag{5.1.3}$$

In order to use the Liu's systematic method [1] in establishing the variational principle for the developed model, one could have

$$V(\phi, \psi) = \iint \{\phi, \psi\} \, \mathrm{d}x\mathrm{d}y \tag{5.1.4}$$

in which

$$\{\phi, \psi\} = \frac{\partial \phi}{\partial x}\frac{\partial \psi}{\partial y} - \frac{\partial \phi}{\partial y}\frac{\partial \psi}{\partial x} \tag{5.1.5}$$

$\{\phi, \psi\}$ in the above equation is called the Poisson bracket. Considering F to be an unknown function and by replacing the derivatives of function ψ with α and β, the following trial function can be obtained:

$$V(\phi, \alpha, \beta) = \iint \left(\alpha \frac{\partial \phi}{\partial x} + \beta \frac{\partial \phi}{\partial y} + F \right) \mathrm{d}x\mathrm{d}y \tag{5.1.6}$$

Equation (5.1.6) has the form of energy in a physical plane and is often known as the energy trial function [2]. The Chein's and Liu's methods can also be utilized to develop the energy trial function. It is assumed that there is a function V which satisfies the following equation:

$$\delta V(\phi) = \iint \left(\frac{\partial \alpha}{\partial x} + \frac{\partial \beta}{\partial y} \right) \delta\phi \, \mathrm{d}x\mathrm{d}y \tag{5.1.7}$$

in which ϕ is an independent variable. Appling the integration by part one could arrive at

$$V(\phi) = -\iint \left(\alpha \frac{\partial}{\partial x}(\delta\phi) + \beta \frac{\partial}{\partial y}(\delta\phi) \right) \mathrm{d}x\mathrm{d}y + \delta I \tag{5.1.8}$$

where δI is referred to the constant of integration.

Accordingly, the following functional term with the restricted variable can be obtained:

$$V(\phi) = \iint \left(\tilde{\alpha} \frac{\partial \phi}{\partial x} + \tilde{\beta} \frac{\partial \phi}{\partial y} \right) \mathrm{d}x\mathrm{d}y \tag{5.1.9}$$

in which $\tilde{\alpha}$ and $\tilde{\beta}$ are both referred to as the restricted variables [2]. This method has been used then for analyzing different types of PDE.

5.2 Semi-inverse Method

In this part, the main theme to point out is to discuss the semi-inverse method and its applications in solving different partial differential equations. Accordingly, one could consider an incompressible irrational flow based on the proposed illustration in Ref. [2] as

$$\frac{\partial u}{\partial x} + \frac{\partial v}{\partial y} = 0 \tag{5.2.1}$$

$$\frac{\partial \phi}{\partial x} = u, \quad \text{and} \quad \frac{\partial \phi}{\partial y} = v \tag{5.2.2}$$

with the following set of boundary conditions:

$$\Lambda_{\text{in}} : \xi \cdot n = \xi_0 \tag{5.2.3}$$

$$\Lambda_{\text{out}} : \xi \cdot n = \xi_1 \tag{5.2.4}$$

in which $\xi = ui + vj$.

In order to develop a generalized variational principle with three independent variables ($\phi, u,$ and v), the following function can be formulated by multiplying ϕ to Eq. (5.2.1) with further manipulation. Then, evaluating the integral and adding an unknown function A to the integrate leads to

$$V(\phi, u, v) = \iint \left\{ u\frac{\partial \phi}{\partial x} + v\frac{\partial \phi}{\partial y} + A \right\} \mathrm{d}x\mathrm{d}y + \int\limits_{\Lambda_{\text{in}}} B\,\mathrm{d}S + \int\limits_{\Lambda_{\text{out}}} B\,\mathrm{d}S \tag{5.2.5}$$

where A, B and C are unknown functions. Implementing $\delta J = 0$ would give the following Euler's equation:

$$\delta\phi : -\frac{\partial u}{\partial x} - \frac{\partial v}{\partial y} + \frac{\partial A}{\partial \phi} = 0 \tag{5.2.6}$$

$$\delta u : \frac{\partial \phi}{\partial x} + \frac{\partial A}{\partial u} = 0 \tag{5.2.7}$$

and

$$\partial v : \frac{\partial \phi}{\partial y} + \frac{\partial A}{\partial v} = 0 \tag{5.2.8}$$

Substituting Eqs. (5.2.1) and (5.2.2) in Eq. (5.2.6) through Eq. (5.2.8), one will arrive at

$$A_\phi = 0 \tag{5.2.9}$$

$$A_u = -u \tag{5.2.10}$$

and

$$A_v = -v \tag{5.2.11}$$

Using the above three equations, one could evaluate the unknown function A [2]:

$$A = -\frac{1}{2}(u^2 + v^2) \tag{5.2.12}$$

Using the Green's function approach, then B and C are also obtained as

$$B = -\xi_0 \phi \tag{5.2.13}$$

$$C = -\xi_1 \phi \tag{5.2.14}$$

Accordingly, the following generalized variational principle with three independent parts is constructed [2]:

$$V(\phi, u, v) = \iint \left\{ u\frac{\partial \phi}{\partial x} + v\frac{\partial \phi}{\partial y} - \frac{1}{2}q^2 \right\} \mathrm{d}x\mathrm{d}y - \int_{\Lambda_{\mathrm{in}}} \xi_0 \phi \, \mathrm{d}S - \int_{\Lambda_{\mathrm{out}}} \xi_1 \phi \, \mathrm{d}S \tag{5.2.15}$$

As for the second scheme of the semi-inverse method, He [2] proposed a type of derivation of the generalized multi-variable variational principle from a known variational form.

As an example, one may assume that there is a variational principle with a single variable as

$$V(\psi) = \iint \left(\psi_x^2 + \psi_y^2\right) \mathrm{d}x\mathrm{d}y \tag{5.2.16}$$

The Euler's equations can be readily obtained as

$$\nabla^2 \psi = 0 \tag{5.2.17}$$

The following transformation has now been implemented:

$$\psi_x = h, \quad \psi_y = g \tag{5.2.18}$$

where h and g are the two independent variables. By applying the semi-inverse method, the following trial function is constructed [2]:

$$V(\psi, h, g) = \iint \left(h\psi_x + g\psi_y + A\right) \mathrm{d}x\mathrm{d}y \tag{5.2.19}$$

Considering Euler's equations of the above trial function, and setting $\delta J = 0$, one obtains

$$\delta\psi : A_\psi = -\frac{\partial h}{\partial x} - \frac{\partial g}{\partial y} = 0 \tag{5.2.20}$$

$$\delta h : A_h = -\psi_x = -h \tag{5.2.21}$$

and

$$\delta g : A_g = -\psi_y = -\xi \tag{5.2.22}$$

where A is consequently obtained as

$$A = -\frac{1}{2}(h^2 + g^2) \tag{5.2.23}$$

Thus, the generalized variational principle with three independent variables is found as

$$V(\psi, h, g) = \iint \left\{h\psi_x + g\psi_y - \frac{1}{2}(h^2 + g^2)\right\} \mathrm{d}x\mathrm{d}y \tag{5.2.24}$$

For the third type of semi-inverse method, the variational principle could be developed from the energy integration.

Accordingly, for the incompressible irrational flow, the following trial function is proposed:

$$V(\psi, u, v) = \iint \{u\psi_x + A(\psi, u, v)\} \mathrm{d}x\mathrm{d}y \tag{5.2.25}$$

Employing $\delta J = 0$ and the Euler's equations, the following expressions are developed:

$$\delta\psi : A_\psi = u_x = -v_y \tag{5.2.26}$$

and

$$\delta u : A_u = -\psi_x = -u \tag{5.2.27}$$

The unknown function A can then be readily obtained as

$$A = -\frac{1}{2}u^2 - \psi v_y + a(v) \tag{5.2.28}$$

in which a is an unknown function of v. Substituting Eq. (5.2.28) in (5.2.25) and then integrating it by parts will give the following equation [2]:

$$V(\psi, u, v) = \iint \left\{ -\frac{1}{2}u^2 + u\psi_x + v\psi_y + a(v) \right\} \mathrm{d}x\mathrm{d}y \tag{5.2.29}$$

The Euler's equation will then give

$$\delta v : a_v = -\psi_y = -v \tag{5.2.30}$$

The following expression for a is obtained by integrating Eq. (5.2.30):

$$a = -\frac{1}{2}v^2 \tag{5.2.31}$$

Equation (5.2.31) would then be substituted in Eq. (5.2.29), and finally, the generalized variational principle is obtained. Interested readers could read more information given in Ref. [2].

5.3 Variational Approach

The variational approach has been proved to be a novel and efficient method, which is capable of solving varieties of nonlinear differential equations. To discuss this novel method, the following nonlinear differential equation is examined as done in Ref. [3]:

$$\ddot{x} + \lambda(x) = 0 \tag{5.3.1}$$

According to the semi-inverse method, the variation principle of Eq. (5.3.1) is given as

$$V(x) = \int_0^{T/4} \left\{ -\frac{1}{2}x^2 + \Gamma(x) \right\} \mathrm{d}t \tag{5.3.2}$$

in which T is the period of vibration of the nonlinear oscillator and $\frac{\partial \Gamma}{\partial x} = \lambda$.

The solution of Eq. (5.3.3) is assumed as

$$x = A\cos\omega t \tag{5.3.3}$$

where A and ω are the amplitude and the frequency of oscillation, respectively.

Substituting Eq. (5.3.3) in Eq. (5.3.2) would give the following:

$$\begin{aligned} V(A,\omega) &= \int_0^{T/4}\left\{-\frac{1}{2}A^2\omega^2\sin^2\omega t + \Gamma(A\cos\omega t)\right\}\mathrm{d}t \\ &= \frac{1}{\omega}\int_0^{\pi/2}\left\{-\frac{1}{2}A^2\omega^2\sin^2 t + \Gamma(A\cos t)\right\}\mathrm{d}t \\ &= -\frac{1}{2}A^2\omega\int_0^{\pi/2}\sin^2 t\,\mathrm{d}t + \frac{1}{\omega}\int_0^{\pi/2}\Gamma(A\cos t)\,\mathrm{d}t \end{aligned} \tag{5.3.4}$$

Using the Ritz method, the following formulations have been developed [3]:

$$\frac{\partial V}{\partial A} = 0, \quad \frac{\partial V}{\partial \omega} = 0 \tag{5.3.5}$$

It is quite apparent that for most cases one would have $\frac{\partial V}{\partial \omega} < 0$:

$$\frac{\partial V}{\partial \omega} = \left\{-\frac{1}{2}A^2\int_0^{\pi/2}\sin^2 t\,\mathrm{d}t - \frac{1}{\omega^2}\int_0^{\pi/2}\Gamma(A\cos t)\,\mathrm{d}t\right\} < 0 \tag{5.3.6}$$

Accordingly, Eq. (5.3.6) is implemented to evaluate the natural frequency of the nonlinear differential equation. This method has been modified by using higher order approximations, and hence, the following generalized solution is considered to develop the higher order variational formulation:

$$x = \sum_i^n A_{2i-1}\cos((2i-1)\omega)t \tag{5.3.7}$$

Substituting Eq. (5.3.7) in Eq. (5.3.2) would give [4]

$$
\begin{aligned}
V(A_1, A_3, A_5, \ldots, A_{2n-1}, \omega) &= \int_0^{T/4} \left\{ -\frac{1}{2}\sum_i^n (2i-1)^2 A_{2i-1}^2 \omega^2 \sin^2((2i-1)\omega)t \right. \\
&\qquad \left. + \Gamma\left(\sum_i^n A_{2i-1} \cos((2i-1)\omega)t \right) \right\} \mathrm{d}t \\
&= \frac{1}{\omega} \int_0^{\pi/2} \left\{ -\frac{1}{2}\sum_i^n (2i-1)^2 A_{2i-1}^2 \omega^2 \sin^2((2i-1)\omega)t \right. \\
&\qquad \left. + \Gamma\left(\sum_i^n A_{2i-1} \cos((2i-1)\omega)t \right) \right\} \mathrm{d}t \\
&= -\frac{1}{2}\sum_i^n (2i-1)^2 A_{2i-1}^2 \omega^2 \int_0^{\pi/2} \sin^2((2i-1)\omega)t \, \mathrm{d}t \\
&\qquad + \int_0^{\pi/2} \Gamma\left(\sum_i^n A_{2i-1} \cos((2i-1)\omega)t \right) \mathrm{d}t = 0
\end{aligned}
\tag{5.3.8}
$$

Finally, the following generalized form is defined to find the higher order solution [3]:

$$
\frac{\partial V}{\partial A_1} = 0, \; \frac{\partial V}{\partial A_2} = 0, \; \ldots, \; \frac{\partial V}{\partial A_n} = 0 \tag{5.3.9}
$$

5.4 Hamiltonian Approach—Comparison with Variational Approach

Using the concept of energy and also the variational principle, the corresponding Hamiltonian form can be developed as [4]

$$
H = \frac{1}{2}\dot{x}^2 + \Gamma(x) = H_0 \tag{5.4.1}
$$

or it can be written in another form as

$$H = \frac{1}{2}\dot{x}^2 + \Gamma(x) - H_0 = 0 \tag{5.4.2}$$

It shows that the energy of a conservative oscillatory system would remain unchanged during a period of vibration. In accordance with Eq. (5.4.1), one could write

$$\frac{\partial H}{\partial A} = 0 \tag{5.4.3}$$

A new error function has then been defined as

$$\tilde{H}(x) = \int_0^{T/4} \left\{ \frac{1}{2}\dot{x}^2 + \Gamma(x) \right\} \mathrm{d}t \tag{5.4.4}$$

In accordance with Eq. (5.4.3), the following condition should be fulfilled [4]:

$$\frac{\partial}{\partial A}\left(\frac{\partial \tilde{H}}{\partial T}\right) = 0 \tag{5.4.5}$$

or in another form

$$\frac{\partial}{\partial A}\left(\frac{\partial \tilde{H}}{\partial \left(\frac{1}{\omega}\right)}\right) = 0 \tag{5.4.6}$$

Substituting Eq. (5.3.3) in Eq. (5.4.4) would result in

$$\begin{aligned} \tilde{H}(x) &= \int_0^{T/4} \left\{ \frac{1}{2}A^2\omega^2 \sin^2 \omega t + \Gamma(A\cos\omega t) \right\} \mathrm{d}t \\ &= \int_0^{\pi/2} \left\{ \frac{1}{2}A^2\omega \sin^2 t + \frac{1}{\omega}\Gamma(A\cos t) \right\} \mathrm{d}t \end{aligned} \tag{5.4.7}$$

Substituting Eq. (5.4.7) in Eq. (5.4.6) yields

$$-A\omega^2 \frac{\pi}{4} + \int_0^{\pi/2} \Gamma(A\cos\omega t)\,\mathrm{d}t = 0 \tag{5.4.8}$$

The natural frequency of the system could accordingly be obtained using the above formula. To find higher order solutions for the nonlinear differential equations, the following method, based on the Hamiltonian approach, is proposed.

Initially, Eq. (5.3.7) is substituted in Eq. (5.4.4) in order to find:

$$\begin{aligned}
\tilde{H}(A_1, A_3, A_5, \ldots, A_{2n-1}, \omega) &= \int_0^{T/4} \left\{ \frac{1}{2} \sum_i^n (2i-1)^2 A_{2i-1}^2 \omega^2 \sin^2((2i-1)\omega)t \right. \\
&\quad \left. + \Gamma\left(\sum_i^n A_{2i-1} \cos((2i-1)\omega)t \right) \right\} \mathrm{d}t \\
&= \int_0^{\pi/2} \left\{ \frac{1}{2} \sum_i^n (2i-1)^2 A_{2i-1}^2 \omega \sin^2((2i-1)\omega)t \right. \\
&\quad \left. + \frac{1}{\omega} \Gamma\left(\sum_i^n A_{2i-1} \cos((2i-1)\omega)t \right) \right\} \mathrm{d}t \\
&= \frac{1}{2} \sum_i^n (2i-1)^2 A_{2i-1}^2 \omega \int_0^{\pi/2} \sin^2((2i-1)\omega)t \, \mathrm{d}t \\
&\quad + \frac{1}{\omega} \int_0^{\pi/2} \Gamma\left(\sum_i^n A_{2i-1} \cos((2i-1)\omega)t \right) \mathrm{d}t = 0
\end{aligned} \tag{5.4.9}$$

To find the natural frequency and the unknown amplitudes, the following conditions should simultaneously be satisfied.

$$\begin{aligned}
&\frac{\partial}{\partial A_1} \left(\frac{\partial \tilde{H}(A_1, A_3, A_5, \ldots, A_{2n-1}, \omega)}{\partial \left(\frac{1}{\omega}\right)} \right) = 0; \\
&\frac{\partial}{\partial A_2} \left(\frac{\partial \tilde{H}(A_1, A_3, A_5, \ldots, A_{2n-1}, \omega)}{\partial \left(\frac{1}{\omega}\right)} \right) = 0; \\
&\ldots; \\
&\frac{\partial}{\partial A_n} \left(\frac{\partial \tilde{H}(A_1, A_3, A_5, \ldots, A_{2n-1}, \omega)}{\partial \left(\frac{1}{\omega}\right)} \right) = 0.
\end{aligned} \tag{5.4.10}$$

5.5 Relationship Between Hamiltonian and Variational Approaches

Recently, it has been proven that the Hamiltonian and variational approaches reach analogous results even for the case of higher order approximation. It can be seen from Eq. (5.3.4) that, by considering the first approximation, the variational form of Eq. (5.3.1) can easily be converted to its Hamiltonian form [5]. Based on that discussion, one writes

$$\begin{aligned} V(A,\omega) &= \int_0^{T/4} \left\{ -\frac{1}{2}A^2\omega^2 \sin^2 \omega t + \Gamma(A\cos \omega t) \right\} \mathrm{d}t \\ &= \frac{1}{\omega}\int_0^{T/4} \left\{ \frac{1}{2}A^2\omega \sin^2 t + \frac{1}{\omega}\Gamma(A\cos t) \right\} \mathrm{d}t \end{aligned} \tag{5.5.1}$$

Applying the Ritz method to both sides of Eq. (5.5.1), one may arrive at [5]

$$\begin{aligned} \frac{\partial V(A,\omega)}{\partial A} &= \frac{\partial}{\partial A}\int_0^{T/4} \left\{ -\frac{1}{2}A^2\omega^2 \sin^2 \omega t + \Gamma(A\cos \omega t) \right\} \mathrm{d}t \\ &= \frac{\partial}{\partial A}\left[\frac{\partial}{\partial\left(\frac{1}{\omega}\right)} \left\{ \int_0^{T/4} \frac{1}{2}A^2\omega \sin^2 \omega t + \frac{1}{\omega}\Gamma(A\cos \omega t) \right\} \right] \\ &= H(A,\omega) = 0 \end{aligned} \tag{5.5.2}$$

To show the similarity of results in the second-order approximation, the following solution is considered:

$$x = A_1 \cos \omega t + A_2 \cos 3\omega t \tag{5.5.3}$$

Substituting Eq. (5.5.3) in Eq. (5.5.1), one could obtain

$$V(A_1, A_2, \omega) = \int_0^{T/4} \left\{ \begin{array}{c} -\frac{1}{2}\omega^2(A_1 \sin \omega t + 3A_2 \sin 3\omega t)^2 \\ +\Gamma(A_1 \cos \omega t + A_2 \cos 3\omega t) \end{array} \right\} \mathrm{d}t \tag{5.5.4}$$

in which

$$A_1 + A_2 = A \tag{5.5.5}$$

Based on the Hamiltonian approach, one could convert Eq. (5.5.4) into a Hamiltonian function:

$$V(A_1, A_2, \omega) = \int_0^{T/4} \left\{ -\frac{1}{2}\omega^2 (A_1 \sin \omega t + 3A_2 \sin 3\omega t)^2 + \Gamma(A_1 \cos \omega t + A_2 \cos 3\omega t) \right\} dt$$

$$\frac{\partial}{\partial\left(\frac{1}{\omega}\right)} \int_0^{T/4} \left\{ \frac{1}{2}\omega (A_1 \sin \omega t + 3A_2 \sin 3\omega t)^2 + \frac{1}{\omega}\Gamma(A_1 \cos \omega t + A_2 \cos 3\omega t) \right\} dt \tag{5.5.6}$$

By applying the Ritz method to Eq. (5.5.6), one will arrive at [5]

$$\begin{aligned} \frac{\partial V(A_1, A_2, \omega)}{\partial A_1} &= \frac{\partial}{\partial A_1} \int_0^{T/4} \left\{ -\frac{1}{2}\omega^2 (A_1 \sin \omega t + 3A_2 \sin 3\omega t)^2 + \Gamma(A_1 \cos \omega t + A_2 \cos 3\omega t) \right\} dt \\ &= \frac{\partial}{\partial A_1} \frac{\partial}{\partial\left(\frac{1}{\omega}\right)} \int_0^{T/4} \left\{ \frac{1}{2}\omega (A_1 \sin \omega t + 3A_2 \sin 3\omega t)^2 \right. \\ &\qquad \left. + \frac{1}{\omega}\Gamma(A_1 \cos \omega t + A_2 \cos 3\omega t) \right\} dt \\ &= H(A_1, A_2, \omega) = 0 \end{aligned} \tag{5.5.7}$$

and similarly one could write

$$\begin{aligned} \frac{\partial V(A_1, A_2, \omega)}{\partial A_2} &= \frac{\partial}{\partial A_2} \int_0^{T/4} \left\{ -\frac{1}{2}\omega^2 (A_1 \sin \omega t + 3A_2 \sin 3\omega t)^2 \right. \\ &\qquad \left. + \Gamma(A_1 \cos \omega t + A_2 \cos 3\omega t) \right\} dt \\ &= \frac{\partial}{\partial A_2} \frac{\partial}{\partial\left(\frac{1}{\omega}\right)} \int_0^{T/4} \left\{ \frac{1}{2}\omega (A_1 \sin \omega t + 3A_2 \sin 3\omega t)^2 \right. \\ &\qquad \left. + \frac{1}{\omega}\Gamma(A_1 \cos \omega t + A_2 \cos 3\omega t) \right\} dt \\ &= H(A_1, A_2, \omega) = 0 \end{aligned} \tag{5.5.8}$$

According to Eqs. (5.5.7) and (5.5.8), the Hamiltonian approach (HA) and the variational approach (VA) do indeed present similar results in the second-order approximation. Furthermore, it can be shown that HA and VA would accede to the same result for any order of approximation. In order to prove the similarity of results for both approaches, the following general solution has been considered:

$$\sum_{n=0}^{n} A_n \cos(2n+1)\omega t \tag{5.5.9}$$

By substituting Eq. (5.5.9) in Eq. (5.3.2), one would obtain

$$\begin{aligned} V(A_1,\ldots,A_n,\omega) &= \int_0^{T/4} \left\{ -\frac{1}{2}\omega^2 \left(\sum_{n=0}^{n} A_n(2n+1)\sin^2(2n+1)\omega t \right)^2 \right. \\ &\quad \left. + \Gamma\left(\sum_{n=0}^{n} A_n \cos\omega t \right) \right\} \mathrm{d}t \\ &= \frac{\partial}{\partial\left(\frac{1}{\omega}\right)} \left\{ \int_0^{T/4} \frac{1}{2}\omega \left(\sum_{n=0}^{n} A_n(2n+1)\sin^2(2n+1)\omega t \right)^2 \right. \\ &\quad \left. + \frac{1}{\omega}\Gamma\left(\sum_{n=0}^{n} A_n \cos(2n+1)\omega t \right) \right\} \end{aligned} \tag{5.5.10}$$

The following expression is developed using the Ritz method:

$$\begin{aligned} \frac{V(A_1,\ldots,A_n,\omega)}{\partial A_1} &= \frac{\partial}{\partial A_1} \int_0^{T/4} \left\{ -\frac{1}{2}\omega^2 \left(\sum_{n=0}^{n} A_n(2n+1)\sin^2(2n+1)\omega t \right)^2 \right. \\ &\quad \left. + \Gamma\left(\sum_{n=0}^{n} A_n \cos\omega t \right) \right\} \mathrm{d}t \\ &= \frac{\partial}{\partial A_1}\frac{\partial}{\partial\left(\frac{1}{\omega}\right)} \left\{ \int_0^{T/4} \frac{1}{2}\omega \left(\sum_{n=0}^{n} A_n(2n+1)\sin^2(2n+1)\omega t \right)^2 \right. \\ &\quad \left. + \frac{1}{\omega}\Gamma\left(\sum_{n=0}^{n} A_n \cos(2n+1)\omega t \right) \right\} \\ &= H(A_1,\ldots,A_n,\omega) = 0 \end{aligned} \tag{5.5.11}$$

where

$$A = A_1 + A_2 + \cdots + A_n$$

Applying the Ritz method, one could develop

$$\frac{V(A_1,\ldots,A_n,\omega)}{\partial A_1} = \tilde{H}(A_1,\ldots,A_n,\omega) = 0 \tag{5.5.12}$$

$$\frac{V(A_1,\ldots,A_n,\omega)}{\partial A_2}=\tilde{H}(A_1,\ldots,A_n,\omega)=0 \tag{5.5.13}$$

and eventually one could find

$$\frac{V(A_1,\ldots,A_n,\omega)}{\partial A_n}=\tilde{H}(A_1,\ldots,A_n,\omega)=0 \tag{5.5.14}$$

It can be seen that the results found by employing the variational and Hamiltonian approaches would result in analogous for the nonlinear ordinary differential equations.

Part II: Applications

In the course of better understanding the illustrated procedures, few well-known and practical nonlinear differential equations have been considered and the semi-inverse method, based on recently published articles in this area, is implemented.

5.6 Generalized Duffing Equation

The generalized Duffing equation in its general form of Eq. (5.6.1) is presented here. This has recently been solved by Younesian et al. [6] by using the variational approach. They have considered the following differential equation and the variational approach is applied to obtain the approximate solution:

$$\ddot{x}+x+\alpha_3x^3+\alpha_5x^5+\alpha_7x^7+\cdots+\alpha_{2n-1}x^{2n-1}=0 \tag{5.6.1}$$

Accordingly, the variational function is developed as

$$V(x)=\int_0^{T/4}\left\{-\frac{1}{2}\dot{x}^2+\frac{1}{2}x^2+\frac{1}{4}\alpha_3x^4+\frac{1}{6}\alpha_5x^6+\cdots+\frac{1}{2n}\alpha_{2n-1}x^{2n}\right\}\mathrm{d}x \tag{5.6.2}$$

Substituting Eq. (5.3.3) in Eq. (5.6.2), one obtains

$$\begin{aligned}V(A,\omega)=\int_0^{T/4}&\left\{-\frac{1}{2}A^2\omega^2\sin^2(\omega t)+\frac{1}{2}A^2\cos^2(\omega t)+\frac{1}{4}\alpha_3A^4\cos^4\omega t+\frac{1}{6}\alpha_5A^6\cos^6(\omega t)\right.\\&\left.+\frac{1}{8}\alpha_7A^8\cos^8(\omega t)+\cdots+\frac{1}{2n}\alpha_{2n-1}A^{2n}\cos^{2n}(\omega t)\right\}\mathrm{d}t\end{aligned} \tag{5.6.3}$$

The Ritz method is applied to develop

$$\frac{\partial V(A)}{dA} = \int_0^{T/4} \{-A\omega^2 \sin^2 \omega t + A\cos^2 \omega t + \alpha_3 A^3 \cos^4(\omega t) + \alpha_5 A^5 \cos^6(\omega t) + \cdots + \alpha_{2n-1} A^{2n-1} \cos^{2n}(\omega t)\} dt = 0 \tag{5.6.4}$$

After a simple solution procedure for Eq. (5.6.4), the frequency–amplitude relationship for the system is obtained as [6]

$$\omega = \sqrt{1 + \sum_2^n \alpha_{2n-1} A^{2n-1} \left(\frac{3 \times 5 \times \cdots \times (2n-1)}{4 \times 6 \times 8 \times \cdots \times (2n)} \right)} \tag{5.6.5}$$

5.7 Elastic Force with Rational Characteristic Equation

A generalized nonlinear oscillator having the fractional elastic force is described by the following differential equation [7]:

$$\ddot{x} + x^{\frac{2(m-n)+1}{2n+1}} = 0 \quad \text{for} \quad m \geq n, \text{ with } m \text{ and } n = 0, 1, 2, \ldots \tag{5.7.1}$$

The variational form of Eq. (5.7.1), using the variational approach, is constructed as

$$V(x) = \int_0^t \left(-\frac{1}{2}\dot{x}^2 + \frac{2n+1}{2m+2} x^{\frac{2m+2}{2n+1}} \right) dt \tag{5.7.2}$$

Equation (5.3.3) is substituted in Eq. (5.7.2) to obtain

$$V(x) = \int_0^{T/4} \left\{ -\frac{1}{2} A^2 \omega^2 \sin^2 \omega t + A^{\frac{2m+2}{2n+1}} \cos^{\frac{2m+2}{2n+1}} \omega t \right\} dt \tag{5.7.3}$$

Taking the derivative with respect to A would give

$$\frac{dV}{dA} = \int_0^{T/4} \left\{ -\frac{1}{2} A\omega^2 \sin^2 \omega t + A^{\frac{2(m-n)+1}{2n+1}} \cos^{\frac{2m+2}{2n+1}} \omega t \right\} dt = 0 \tag{5.7.4}$$

Upon implementing few mathematical operations, one could arrive at

$$\omega^2 = \frac{4}{\pi A^{\frac{2(2n-m)}{2n+1}}} \int_0^{\pi/2} \cos^{\frac{2m+2}{2n+1}} t \, dt \tag{5.7.5}$$

Eventually, the nonlinear frequency–amplitude relationship of the system, given by Eq. (5.7.6), is obtained [7]:

$$\omega_{VA} = \frac{1}{A^{\frac{2n-m}{2n+1}}} \left(\frac{2}{\pi} \frac{\Gamma\left(\frac{2(m+n)+3}{2(2n+1)}\right)\Gamma\left(\frac{1}{2}\right)}{\Gamma\left(\frac{m+2n+2}{2n+1}\right)} \right) \tag{5.7.6}$$

5.8 Elastic Force with Non-integer Fractional Characteristic Equation

Recently, a generalized nonlinear oscillator with the non-integer fractional elastic force has been solved by Cveticanin et al. [8]:

$$\ddot{x} + \sum_i d_i^2 x|x|^{\alpha_i - 1} = 0, \quad i = 1, 2 \tag{5.8.1}$$

where d_i is the rigidity coefficient constant and α_i is either a positive fraction or a positive number. The initial conditions are assumed to be

$$x(0) = A, \quad \text{and} \quad \dot{x}(0) = 0 \tag{5.8.2}$$

The Hamiltonian expression of the system is constructed as

$$\tilde{H} = \frac{1}{2}\dot{x}^2 + \sum_i \frac{d_i^2 x^{\alpha_i + 1}}{\alpha_i + 1} \tag{5.8.3}$$

Evaluating the integral of the Hamiltonian function of Eq. (5.8.3) will give

$$\tilde{H} = \int_0^{T/4} \left(\frac{1}{2}\dot{x}^2 + \sum_i \frac{d_i^2 x^{\alpha_i + 1}}{\alpha_i + 1} \right) dt \tag{5.8.4}$$

Substituting Eq. (5.3.3) into Eq. (5.8.3) would reach

$$\tilde{H} = \frac{A^2 \omega a_0}{2} + \frac{1}{\omega} \sum_i \frac{d_i^2 A^{\alpha_i + 1} a_i}{\alpha_i + 1} \tag{5.8.5}$$

in which

$$a_0 = \int_0^{\pi/2} \sin^2 \theta \, \mathrm{d}\theta = \frac{\pi}{4}, \tag{5.8.6}$$

and

$$a_i = \int_0^{\pi/2} \cos^{\alpha_i + 1} \theta \, \mathrm{d}\theta = \frac{\pi}{2^{\alpha_i + 1} (\alpha_i + 1) B\left(\frac{\alpha_i + 3}{2}, \frac{\alpha_i + 1}{2}\right)} \tag{5.8.7}$$

where B denotes the Beta function. Therefore, one may use the algebraic relationship between the beta and the gamma functions.

$$B\left(\frac{\alpha_i + 3}{2}, \frac{\alpha_i + 1}{2}\right) \equiv \frac{\Gamma\left(\frac{\alpha_i + 3}{2}\right) \Gamma\left(\frac{\alpha_i + 1}{2}\right)}{\Gamma(\alpha_i + 2)} \tag{5.8.8}$$

Take the derivative of Eq. (5.8.5) would lead to

$$\frac{\partial \bar{H}}{\partial (1/\omega)} = -\frac{A^2 \omega^2 a_0}{2} + \sum_i \frac{d_i^2 A^{\alpha_i + 1} a_i}{\alpha_i + 1} \tag{5.8.9}$$

Defining the following residual function, one could obtain

$$R(A) \equiv \left(-\frac{A^2 \omega^2 a_0}{2} + \sum_i \frac{d_i^2 A^{\alpha_i + 1} a_i}{\alpha_i + 1} \right) - H_0 \tag{5.8.10}$$

In order to minimize the residual function of $R(A)$, one should have

$$A \omega^2 a_0 - \sum_i d_i^2 A^{\alpha_i} a_i = 0 \tag{5.8.11}$$

Implementing few mathematical operations, the following solution as the natural frequency of the oscillator is found [8]:

$$\omega = \sqrt{\frac{\sum_i d_i^2 \alpha_i A^{\alpha_i - 1}}{a_0}} \tag{5.8.12}$$

As a special case, if one considers $\alpha = 1/3$, the following nonlinear differential equation is then obtained [8]:

$$\ddot{x} + x^{1/3} = 0 \tag{5.8.13}$$

By solving Eq. (5.8.13), the natural frequency of Eq. (5.8.12) is found as

$$\omega = \frac{1}{A^{\frac{1}{3}}}\left(\frac{3}{2^{4/3}}\frac{\Gamma\left(\frac{7}{3}\right)}{\Gamma\left(\frac{5}{3}\right)\Gamma\left(\frac{2}{3}\right)}\right)^{\frac{1}{2}} = \frac{1.076845}{A^{\frac{1}{3}}} \tag{5.8.14}$$

The exact value of the frequency for Eq. (5.8.13) is

$$\omega_{\mathrm{ex}} = \sqrt{\frac{2\pi}{3}}\frac{\Gamma\left(\frac{5}{4}\right)}{A^{\frac{1}{3}}\Gamma\left(\frac{3}{4}\right)} = \frac{1.070451}{A^{\frac{1}{3}}} \tag{5.8.15}$$

As can be seen from Fig. 5.1, the relative error of the approximate solution is found to be as low as only 0.59% [8].

5.9 Higher Order Hamiltonian Approach to Duffing Equations

Consider the standard Duffing equation as shown below [9]:

$$\ddot{x} + \alpha_1 x + \alpha_3 x^3 = 0 \quad x(0) = A, \quad \text{and} \quad \dot{x}(0) = 0 \tag{5.9.1}$$

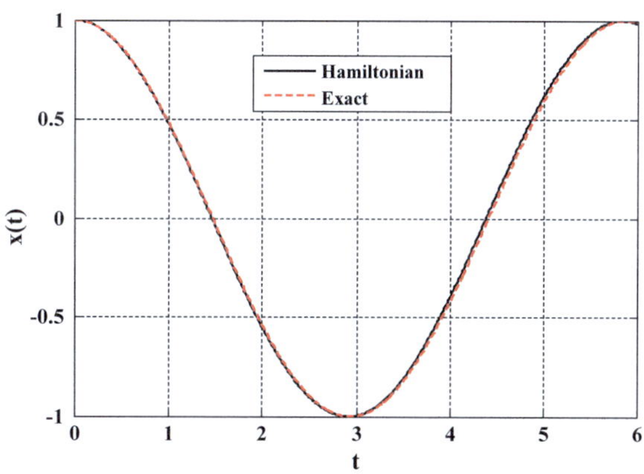

Fig. 5.1 Comparison of the solutions by the HA and exact methods for α = 1/3 [8]

The Hamiltonian of Eq. (5.9.1) can be constructed as

$$\tilde{H}(x) = \int_0^{T/4} \left(\frac{1}{2}\dot{x}^2 + \frac{\alpha_1}{2}x^2 + \frac{\alpha_3}{2}x^4 \right) \mathrm{d}t \tag{5.9.2}$$

Assume the following expression as the initial solution:

$$x = A_1 \cos \omega t + A_2 \cos 3\omega t \tag{5.9.3}$$

in which

$$A = A_1 + A_2 \tag{5.9.4}$$

The following equation is subsequently obtained by substituting Eq. (5.9.3) in (5.9.2):

$$\begin{aligned}\tilde{H}(x) &= \int_0^{T/4} \left[\frac{1}{2}(A_1\omega \sin \omega t + 3A_2\omega \sin 3\omega t)^2 + \frac{1}{2}\alpha_1 (A_1 \cos \omega t + A_2 \cos 3\omega t)^2 \right.\\ &\quad \left. + \frac{1}{4}\alpha_3 (A_1 \cos \omega t + A_2 \cos 3\omega t)^4 \right] \mathrm{d}t \\ &= \int_0^{T/4} \left[\frac{1}{2}\omega(A_1 \sin t + 3A_2 \sin 3t)^2 + \frac{1}{2\omega}\alpha_1 (A_1 \cos t + A_2 \cos 3t)^2 \right.\\ &\quad \left. + \frac{1}{4\omega}\alpha_3 (A_1 \cos t + A_2 \cos 3t)^4 \right] \mathrm{d}t \\ &= \frac{\pi}{8}\omega(A_1^2 + A_2^2) + \frac{\pi}{8\omega}\alpha_1 (A_1^2 + A_2^2) \\ &\quad + \frac{\pi}{64\omega}\alpha_3 (3A_1^4 + 4A_1^3 A_2 + 12A_1^2 A_2^2 + 3A_2^4)\end{aligned} \tag{5.9.5}$$

Equation (5.4.6) would then give the following set of algebraic equations:

$$\frac{\partial}{\partial A_1}\left(\frac{\partial \tilde{H}}{\partial \frac{1}{\omega}}\right) = -\frac{\pi}{4}A_1\omega^2 + \frac{\pi}{4}\alpha_1 A_1 + \frac{\pi}{64}\alpha_3(12A_1^3 + 12A_1^2A_2 + 24A_1A_2^2) = 0 \tag{5.9.6}$$

and

$$\frac{\partial}{\partial A_2}\left(\frac{\partial \tilde{H}}{\partial \frac{1}{\omega}}\right) = -\frac{18\pi A_2}{8}\omega^2 + \frac{\pi}{4}\alpha_1 A_2 + \frac{\pi}{64}\alpha_3(4A_1^3 + 24A_1^2A_2 + 24A_1A_2^2) = 0 \tag{5.9.7}$$

Finally, the following values are obtained for the two parameters A_1 and A_2:

$$A_1 = 0.95714A \tag{5.9.8}$$

and

$$A_2 = 0.04289A \tag{5.9.9}$$

Accordingly, the following natural frequency for the system is obtained [9]:

$$\omega_{\mathrm{SHA}} = \sqrt{\alpha_1 + 0.7205\alpha_3 A^2} \tag{5.9.10}$$

In order to find the third-order approximation, the following solution is considered:

$$x = A_1\cos\omega t + A_2\cos 3\omega t + A_3\cos 5\omega t \tag{5.9.11}$$

where

$$A = A_1 + A_2 + A_3 \tag{5.9.12}$$

Substituting Eq. (5.9.11) into Eq. (5.9.2) will give

$$\begin{aligned}
\tilde{H}(x) &= \int_0^{T/4} \left[\frac{1}{2}(A_1\omega \sin \omega t + 3A_2\omega \sin 3\omega t + 5A_2\omega \sin 5\omega t)^2 \right.\\
&\quad + \frac{1}{2}\alpha_1(A_1 \cos \omega t + A_2 \cos 3\omega t + A_3 \cos 5\omega t)^2\\
&\quad \left. + \frac{1}{4}\alpha_3(A_1 \cos \omega t + A_2 \cos 3\omega t + A_3 \cos 5\omega t)^4 \right] \mathrm{d}t\\
&= \int_0^{T/4} \left[\frac{1}{2}\omega(A_1 \sin t + 3A_2 \sin 3t + 5A_3 \sin 5t)^2 \right.\\
&\quad + \frac{1}{2\omega}\alpha_1(A_1 \cos t + A_2 \cos 3t + A_3 \cos 5t)^2\\
&\quad \left. + \frac{1}{4\omega}\alpha_3(A_1 \cos t + A_2 \cos 3t + A_3 \cos 5t)^4 \right] \mathrm{d}t\\
&= \frac{\pi}{8}\omega(A_1^2 + 9A_2^2 + 25A_3^2) + \frac{\pi}{8\omega}\alpha_1(A_1^2 + A_2^2 + A_3^2)\\
&\quad + \frac{\pi}{64\omega}\alpha_3\big(3A_1^4 + 4A_1^3A_2 + 12A_1^2A_2^2 + 3A_2^4 + 12A_2^2A_3^2\\
&\qquad + 12A_1A_2^2A_3 + 12A_1^2A_2A_3 + 12A_1^2A_3^2 + 12A_1^2A_2^2\big)
\end{aligned} \tag{5.9.13}$$

Manipulating Eq. (5.4.5) through Eq. (5.9.13), one could find the following equation:

$$\begin{aligned}
\frac{\partial}{\partial A_1}\left(\frac{\partial \tilde{H}}{\partial \frac{1}{\omega}}\right) &= -\frac{\pi}{4}A_1\omega^2 + \frac{\pi}{4}\alpha_1 A_1 + \frac{\pi}{64}\alpha_3\big(12A_1^3 + 12A_1^2A_2 + 24A_1A_1^2 + 12A_1^2A_2\\
&\quad + 24A_1A_2^2 + 24A_1A_2A_3 + 24A_1A_3^2\big) = 0
\end{aligned} \tag{5.9.14}$$

$$\begin{aligned}
\frac{\partial}{\partial A_2}\left(\frac{\partial \tilde{H}}{\partial \frac{1}{\omega}}\right) &= -\frac{18\pi A_2}{8}\omega^2 + \frac{\pi}{4}\alpha_1 A_2 + \frac{\pi}{64}\alpha_3\big(4A_1^3 + 24A_1^2A_2 + 24A_1A_2^2\\
&\quad + 4A_2^3 + 24A_1A_2A_3 + 24A_1A_3^2\big) = 0
\end{aligned} \tag{5.9.15}$$

and

$$\begin{aligned}
\frac{\partial}{\partial A_3}\left(\frac{\partial \tilde{H}}{\partial \frac{1}{\omega}}\right) &= -\frac{50\pi A_1}{8}\omega^2 + \frac{\pi}{4}\alpha_1 A_3 + \frac{\pi}{64}\alpha_3\big(12A_3^3 + 24A_2^2A_3 + 12A_1A_2^2\\
&\quad + 12A_1^2A_2 + 24A_1^2A_3\big) = 0
\end{aligned} \tag{5.9.16}$$

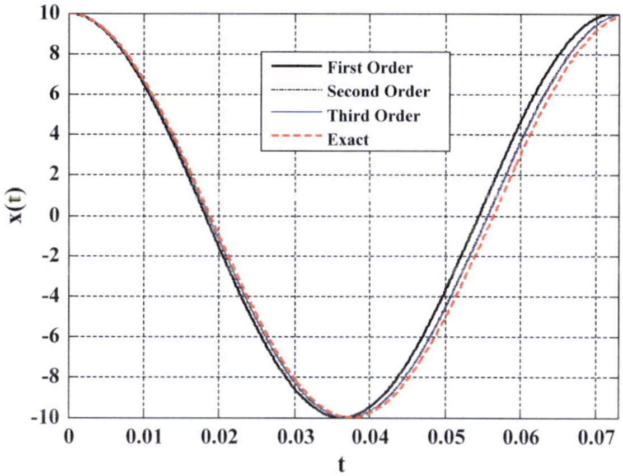

Fig. 5.2 Comparison of the HA higher order solutions with the exact one [9]

Consequently, one could arrive at

$$A_1 = 0.955091A; \quad \text{and} \quad A_2 = 0.0430519A; \quad \text{and} \quad A_3 = 0.0018569A \tag{5.9.17}$$

The third-order approximation for the natural frequency of the Duffing equation is [9]

$$\omega = \sqrt{\alpha_1 + 0.7178\alpha_3 A^2} \tag{5.9.18}$$

Figure 5.2 illustrates a comparison between the higher order approximate solutions of the Hamiltonian approach and the exact numerical simulation. The higher order approximate solution exhibits more accurate results as expected.

5.10 Hamiltonian Approach to Rational and Irrational Oscillator

The following differential equation with the rational expression is considered by Yildirim et al. [10]. They have employed the Hamiltonian approach to solve this equation.

$$\ddot{x} + \frac{x}{x^2+1} = 0 \quad x(0) = A; \quad \text{and} \quad \dot{x}(0) = 0 \tag{5.10.1}$$

It is apparent that the nonlinear equation can be altered to Eq. (5.10.2) for the two cases of the large and the small oscillations:

$$\begin{cases} \text{Small value of } x & \ddot{x} + x = 0 \\ \text{Large value of } x & \ddot{x} + \frac{1}{x} = 0 \end{cases} \tag{5.10.2}$$

The Hamiltonian form of Eq. (5.10.1) is the obtained as

$$H = \frac{1}{2}\dot{x}^2 + \frac{1}{2}\log(x^2 + 1) \tag{5.10.3}$$

Evaluating the integration of Eq. (5.10.3) with respect to time t between the two limits of $t = 0$ and $t = \frac{T}{4}$, the following Hamiltonian function is then obtained:

$$\begin{aligned} \tilde{H}(x) &= \int_0^{T/4} \left\{ \frac{1}{2}\dot{x}^2 + \frac{1}{2}\log(x^2 + 1) \right\} \mathrm{d}t = \int_0^{T/4} \left\{ \frac{1}{2}A^2\omega^2 \sin^2 \omega t + \frac{1}{2}\log(A^2 \cos^2 \omega t + 1) \right\} \mathrm{d}t \\ &= \int_0^{\pi/2} \left\{ \frac{1}{2}A^2\omega \sin^2 \omega t + \frac{1}{2\omega}\log(A^2 \cos^2 \omega t + 1) \right\} \mathrm{d}t \end{aligned} \tag{5.10.4}$$

Using Eq. (5.4.5), one could have

$$\omega = \sqrt{\frac{\int_0^{\pi/2} \left\{ \frac{\cos^2 t}{1 + A^2 \cos^2 t} \right\} \mathrm{d}t}{\int_0^{\pi/2} \sin^2 t \, \mathrm{d}t}} \tag{5.10.5}$$

The exact value of the frequency of Eq. (5.10.1) is obtained [10].

$$\omega_{\mathrm{ex}} = \frac{2\pi}{4\int_0^A \frac{\mathrm{d}x}{\sqrt{[\log(A^2 + 1) - \log(x^2 + 1)]}}} \tag{5.10.6}$$

The effectiveness and the accurate performance of the results obtained in the time domain are confirmed as shown in Figs. 5.3 and 5.4.

The nonlinear oscillator with the irrational elastic force, given by the nonlinear differential equation of Eq. (5.10.7), is solved using Hamiltonian approach [10]:

$$\ddot{x} + x - \frac{\gamma x}{\sqrt{x^2 + 1}} = 0; \; x(0) = A; \; \dot{x}(0) = 0 \quad 0 < \gamma < 1 \tag{5.10.7}$$

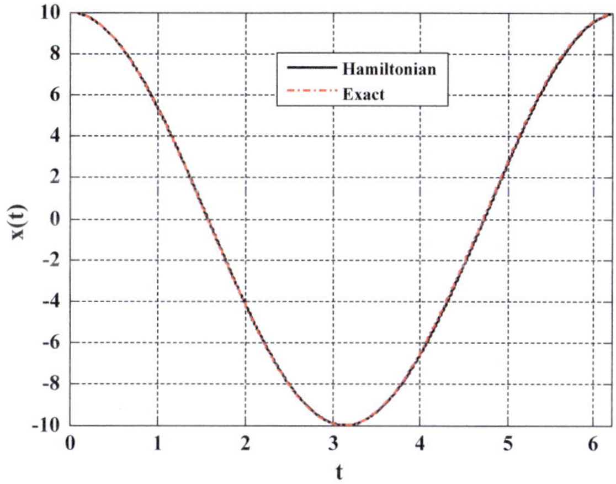

Fig. 5.3 Comparison of the solution by HA with the exact solution for $A = 0.1$

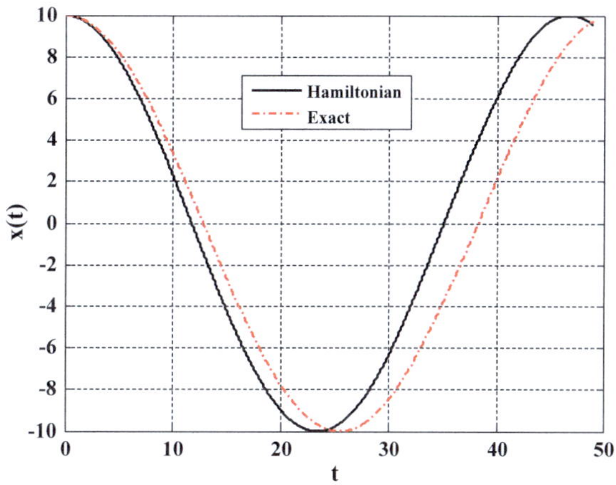

Fig. 5.4 Comparison of the solution by HA with the exact solution for $A = 10$

For either the small or large values of x, Eq. (5.10.7) has been converted to

$$\begin{cases} \text{Largevalueof x}: & \ddot{x} + x \approx 0 \\ \text{Smallvalueof x}: & \ddot{x} + (1-\gamma)x \approx 0 \end{cases} \tag{5.10.8}$$

Accordingly, the following Hamiltonian function is obtained:

$$H = \frac{1}{2}\dot{x}^2 + \frac{1}{2}x^2 - \gamma\sqrt{(1+x^2)} \tag{5.10.9}$$

The following equation is then obtained by integrating Eq. (5.10.9):

$$\begin{aligned}\tilde{H}(x) &= \int_0^{T/4} \left\{\frac{1}{2}\dot{x}^2 + \frac{1}{2}x^2 - \gamma\sqrt{(1+x^2)}\right\}\mathrm{d}t \\ &= \int_0^{T/4} \left\{\frac{1}{2}A^2\omega^2 \sin^2 \omega t + \frac{1}{2}A^2 \cos^2 \omega t - \gamma\sqrt{1+A^2 \cos^2 \omega t}\right\}\mathrm{d}t \\ &= \int_0^{\pi/2} \left\{\frac{1}{2}A^2\omega \sin^2 \omega t + \frac{1}{2\omega}A^2 \cos^2 \omega t - \frac{\gamma}{\omega}\sqrt{1+A^2 \cos^2 t}\right\}\mathrm{d}t\end{aligned} \tag{5.10.10}$$

Using the Hamiltonian operator, the natural frequency of the oscillator is found as

$$\omega_{\mathrm{HA}} = \sqrt{\frac{\int_0^{\pi/2}\left[\cos^2 t - \frac{\gamma \cos^2 t}{\sqrt{1+A^2\cos^2 t}}\right]\mathrm{d}t}{\int_0^{\pi/2} \sin^2 t\,\mathrm{d}t}} \tag{5.10.11}$$

The exact natural frequency is given by [10]

$$\omega_{\mathrm{ex}} = \frac{2\pi}{4\int_0^A \frac{\mathrm{d}x}{\sqrt{(A^2-x^2)+2\gamma\left(\sqrt{1+x^2}-\sqrt{1+A^2}\right)}}} \tag{5.10.12}$$

In order to verify the correctness of the analytical solution, the exact and the approximate solutions are compared and shown in Figs. 5.5 and 5.6.

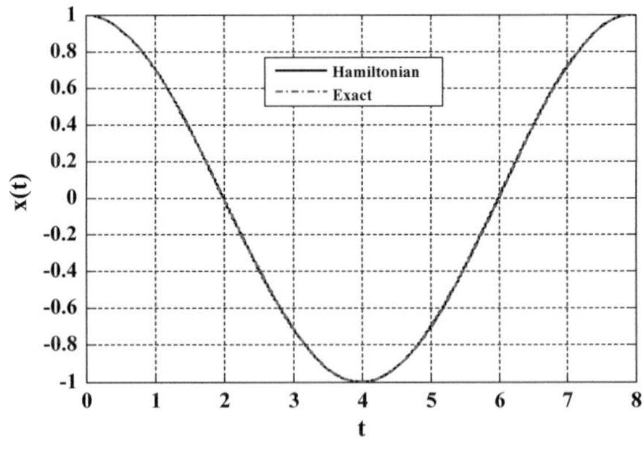

Fig. 5.5 Comparison of the solution by HA with the exact solution for A = 10; $\gamma = 0.5$ and A = 1; and $\gamma = 0.5$

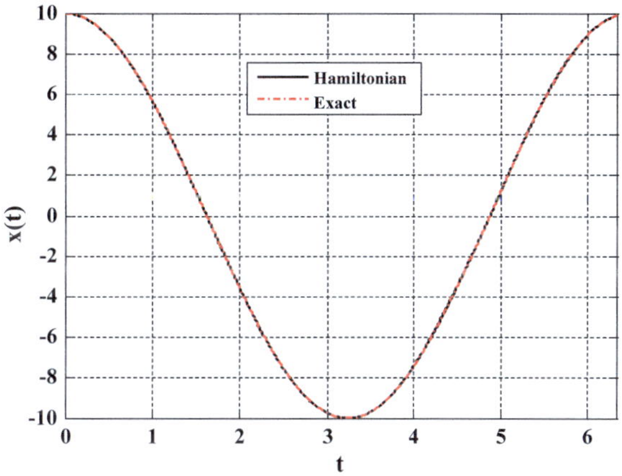

Fig. 5.6 Comparison of the solution by HA with the exact solution for $A = 10$; and $\gamma = 0.5$

5.11 Hamiltonian Approach to Nonlinear Oscillator with Discontinuity

An oscillator with the discontinuity is mathematically described by the following equation:

$$\ddot{x} + x|x| = 0, \;\; x(0) = A \quad \text{and} \quad \dot{x}(0) = 0 \tag{5.11.1}$$

The above equation, using the Hamiltonian approach, is solved by Yildirim et al. [11]. They have first constructed the Hamiltonian as

$$H = \frac{1}{2}\dot{x}^2 + \frac{1}{3}x^2|x| \tag{5.11.2}$$

Integrating Eq. (5.11.2) with respect to time leads to

$$\tilde{H}(x) = \int_0^{T/4} \left(\frac{1}{2}\dot{x}^2 + \frac{1}{3}x^3\right)\mathrm{d}t + \int_{T/4}^{T/2} \left(\frac{1}{2}\dot{x}^2 - \frac{1}{3}x^3\right)\mathrm{d}t \tag{5.11.3}$$

Considering Eq. (5.3.3) in the above calculation would result in

$$\begin{aligned}
\tilde{H}(x) &= \int_0^{T/4} \left(\frac{1}{2}A^2\omega^2 \sin^2 \omega t + \frac{1}{3}A^3 \cos^3 \omega t\right) \mathrm{d}t + \int_{T/4}^{T/2} \left(\frac{1}{2}A^2\omega^2 \sin^2 \omega t - \frac{1}{3}A^3 \cos^3 \omega t\right) \mathrm{d}t \\
&= \int_0^{\pi/2} \left(\frac{1}{2}A^2\omega \sin^2 t + \frac{1}{3\omega}A^3 \cos^3 t\right) \mathrm{d}t + \int_{\pi/2}^{\pi} \left(\frac{1}{2}A^2\omega \sin^2 t - \frac{1}{3\omega}A^3 \cos^3 t\right) \mathrm{d}t \\
&= \frac{\pi}{4}\omega A^2 + \frac{4}{9\omega}A^3
\end{aligned} \tag{5.11.4}$$

The following expression, as the closed-form frequency expression of the considered oscillator with the discontinuity, using the Hamiltonian operator, is obtained:

$$\omega_{\mathrm{FAH}} = \sqrt{\frac{8}{3\pi}A} \tag{5.11.5}$$

The relative error for the closed-form solution is obtained to be about 0.72% [11].

In order to find the higher order solution for Eq. (5.11.1), the following function is considered:

$$x = A_1 \cos \omega t + A_2 \cos 3\omega t, \quad A_1 + A_2 = A \tag{5.11.6}$$

Substituting Eq. (5.11.6) in Eq. (5.11.3) and applying the Hamiltonian operator yields

$$\frac{\partial}{\partial A_1}\left(\frac{\partial \tilde{H}}{\partial\left(\frac{1}{\omega}\right)}\right) = -\frac{\pi}{2}A_1\omega^2 + \frac{4}{3}A_1^2 + \frac{8}{15}A_1A_2 + \frac{36}{35}A_2^2 = 0 \tag{5.11.7}$$

and

$$\frac{\partial}{\partial A_2}\left(\frac{\partial \tilde{H}}{\partial\left(\frac{1}{\omega}\right)}\right) = -\frac{9\pi}{2}A_2\omega^2 + \frac{4}{15}A_2^2 + \frac{75}{35}A_1A_2 - \frac{4}{9}b^2 = 0 \tag{5.11.8}$$

Employing a straightforward mathematical procedure, the values of A_1 and A_2 are obtained:

$$A_1 = 0.97424661242708A \tag{5.11.9}$$

and

$$A_2 = 0.002575338757292A \tag{5.11.10}$$

The second-order approximation for the natural frequency is found as [11]

$$\omega_{\mathrm{SOH}} = 0.91441568459242\sqrt{A} \tag{5.11.11}$$

Apparently, more accuracy can be obtained for the solution of Eq. (5.11.1) when using the second-order Hamiltonian approach. For the second-order Hamiltonian approach, the relative error is almost equal to 0.029% which is a good indication for the high accuracy and substantial potency of the second-order Hamiltonian approach.

To obtain a higher order approximation, one should utilize the third-order Hamiltonian approach. After applying the Hamiltonian operator, the following set of equation is obtained:

$$\begin{aligned}\frac{\partial}{\partial A_1}\left(\frac{\partial \tilde{H}}{\partial\left(\frac{1}{\omega}\right)}\right) = &-\frac{\pi}{4}A_1\omega^2 + \frac{2}{3}A_1^2 + \frac{4}{15}A_1A_2 + \frac{18}{35}A_1^2 \\ &-\frac{4}{105}A_1A_3 + \frac{20}{63}A_2A_3 + \frac{50}{99}A_3^2 = 0\end{aligned} \tag{5.11.12}$$

$$\begin{aligned}\frac{\partial}{\partial A_2}\left(\frac{\partial \tilde{H}}{\partial\left(\frac{1}{\omega}\right)}\right) = &-\frac{9\pi}{4}A_2\omega^2 + \frac{2}{15}A_1^2 + \frac{36}{35}A_1A_2 - \frac{2}{9}A_2^2 \\ &+\frac{36}{55}A_2A_3 + \frac{20}{63}A_1A_3 - \frac{50}{273}A_3^2 = 0\end{aligned} \tag{5.11.13}$$

and

$$\begin{aligned}\frac{\partial}{\partial A_3}\left(\frac{\partial \tilde{H}}{\partial\left(\frac{1}{\omega}\right)}\right) = &-\frac{25\pi}{4}A_3\omega^2 - \frac{2}{105}A_2^2 + \frac{20}{63}A_1A_2 + \frac{18}{55}A_2^2 \\ &-\frac{100}{273}A_2A_3 + \frac{100}{99}A_1A_3 + \frac{2}{15}A_3^2 = 0\end{aligned} \tag{5.11.14}$$

Solving the above set of equations will give

$$A_1 = 0.97491406508A \tag{5.11.15}$$

$$A_2 = 0.0257281322963A \tag{5.11.16}$$

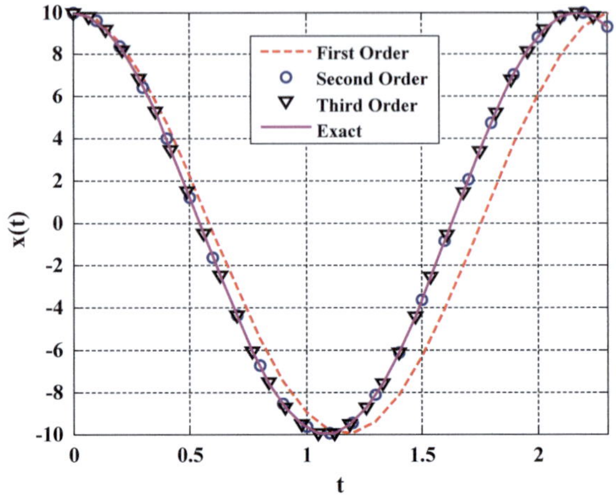

Fig. 5.7 Comparison of the approximate and exact solutions for $A = 100$ [9]

and

$$A_3 = -0.0006421973766A \quad (5.11.17)$$

Moreover, the natural frequency of Eq. (5.11.1) is obtained as [11]

$$\omega = 0.9147335203935\sqrt{A} \quad (5.11.18)$$

Remarkably, the application of the third-order Hamiltonian approach is quite satisfactory and approximately has 0.00571%. The accuracy of the employed higher order solution is illustrated in Fig. 5.7.

5.12 Nonlinear Oscillator with Quintic Nonlinearity

The study of an oscillator with the quintic nonlinearity is conducted by Yildirim, et al. [11]:

$$\ddot{x}+x+x^5=0 \quad \text{with initial conditions of} \quad x(0)=A, \text{ and } \dot{x}(0)=0 \quad (5.12.1)$$

The Hamiltonian function for Eq. (5.12.1) is constructed as

$$H=\frac{1}{2}\dot{x}^2+\frac{1}{2}x^2+\frac{1}{6}x^6 \quad (5.12.2)$$

Integrating Eq. (5.12.2) with respect to time will result in

$$\tilde{H}(x) = \int_{0}^{T/4} \left(\frac{1}{2}\dot{x}^2 + \frac{1}{2}x^2 + \frac{1}{6}x^6 \right) \mathrm{d}t \tag{5.12.3}$$

Substituting Eq. (5.3.3) into Eq. (5.12.3) and using the Hamiltonian operator yields

$$\frac{\partial}{\partial A}\left(\frac{\partial \tilde{H}}{\partial\left(\frac{1}{\omega}\right)} \right) = -\frac{\pi}{4}A\omega^2 + \frac{\pi}{4}A + \frac{5\pi}{32}A^5 = 0 \tag{5.12.4}$$

Eventually, the natural frequency is obtained as

$$\omega = \sqrt{1 + \frac{5}{8}A^4} \tag{5.12.5}$$

To obtain the second-order approximation, Eq. (5.11.6) is substituted into Eq. (5.12.1) and then the Hamiltonian operator is applied. Accordingly, the following set of algebraic equation is obtained:

$$\frac{\partial}{\partial A_1}\left(\frac{\partial \tilde{H}}{\partial\left(\frac{1}{\omega}\right)} \right) = -\frac{\pi}{8}2A_1\omega^2 + \frac{\pi}{8}2A_1 + \frac{5\pi}{192\omega}\left(6A_1^5 + 15A_1^4A_2 + 36A_1^3A_2^2 + 18A_1^2A_2^3 + 18A_1A_2^4\right) = 0 \tag{5.12.6}$$

and

$$\frac{\partial}{\partial A_2}\left(\frac{\partial \tilde{H}}{\partial\left(\frac{1}{\omega}\right)} \right) = -\frac{\pi}{8}18A_2\omega^2 + \frac{\pi}{8}2A_2 + \frac{5\pi}{192\omega}\left(3A_2^5 + 18A_1^4A_2 + 18A_1^3A_2^2 + 72A_1A_2^3 + 6A_2^3\right) = 0 \tag{5.12.7}$$

The first- and the second-order approximate natural frequencies are compared with those found from the exact solutions, as shown in Fig. 5.8.

5.13 Nonlinear Schrodinger's Equation

The nonlinear Schrodinger's equation is represented by the following expression [12]:

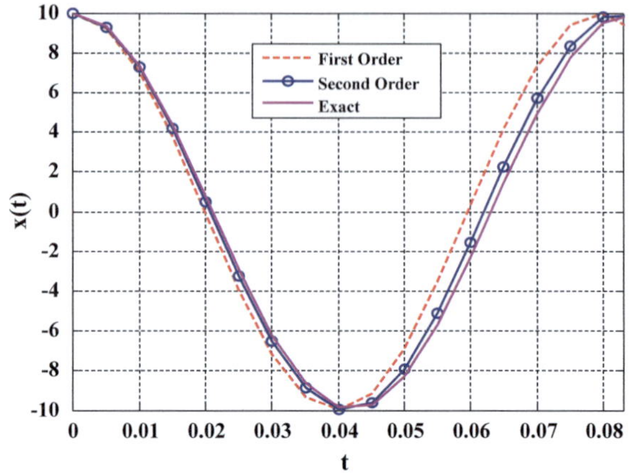

Fig. 5.8 Comparison between the approximate and exact solutions for $A = 10$ [11]

$$i\psi_t + at^n\psi_{xx} + bt^m|\psi|^2\psi = 0 \tag{5.13.1}$$

in which $\psi(x,t)$ is a complex function of variables x and t. Also, m and n are real numbers, while coefficients a and b are both constant parameters [12].

By substituting $\psi(x,t) = u(x,t) + iv(x,t)$ in Eq. (5.13.1), the following equation is obtained [12]:

$$\left[-v_t + at^n u_{xx} + bt^m(u^3 + v^2u)\right] + i\left[u_t + at^n v_{xx} + bt^m(u^2v + v^3)\right] = 0 \tag{5.13.2}$$

where $u(x,t)$ and $v(x,t)$ are both real functions, which are functions of both x and t.

Equation (5.13.2) is converted into a set of two second-order equations as

$$-v_t + at^n u_{xx} + bt^m(u^3 + v^2u) = 0 \tag{5.13.3}$$

and

$$u_t + at^n v_{xx} + bt^m(u^2v + v^3) = 0 \tag{5.13.4}$$

In order to find the corresponding variational principle, a trial function is proposed:

$$V(u,v) = \int_\Omega \left[-v_t u - \frac{1}{2}at^n u_x^2 + \frac{1}{2}bt^m\left(\frac{1}{2}u^4 + v^2u^2\right) + F(v)\right] \mathrm{d}\Omega \tag{5.13.5}$$

which is equivalent to [12]

$$V(u,v) = \int_{\Omega} \left[u_t v - \frac{1}{2} a t^n u_x^2 + \frac{1}{2} b t^m \left(\frac{1}{2} u^4 + v^2 u^2 \right) + F(v) \right] \mathrm{d}\Omega \tag{5.13.6}$$

Accordingly, the variational principle of Eq. (5.13.3) in its general form would be

$$V(u,v) = \int_{\Omega} \left[\lambda u_t v - (1-\lambda) v_t v - \frac{1}{2} a t^n u_x^2 + \frac{1}{2} b t^m \left(\frac{1}{2} u^4 + v^2 u^2 \right) + F(v) \right] \mathrm{d}\Omega \tag{5.13.7}$$

in which λ could be an arbitrary constant and being function of x. The following Euler–Lagrange equation would then be obtained:

$$u_t + b t^m v u^2 + \frac{\delta F}{\delta v} = 0 \tag{5.13.8}$$

In order to find the unknown F, one may rearrange it:

$$\frac{\delta F}{\delta v} = -u_t - b t^m v u^2 = a t^n v_{xx} + b t^m v^3 \tag{5.13.9}$$

Accordingly, the following expression for F is obtained [12]:

$$F = -\frac{1}{2} a t^n v_x^2 + \frac{1}{4} b t^m v^4 \tag{5.13.10}$$

The corresponding variational principle can be achieved by the following equation [12]:

$$V(u,v) = \int_{\Omega} \left[\lambda v u_t - (1-\lambda) v_t u - \frac{1}{2} a t^n u_x^2 + \frac{1}{2} b t^m \left(\frac{1}{2} u^4 + v^2 u^2 \right) - \frac{1}{2} a t^n v_x^2 + \frac{1}{4} b t^m v^4 \right] \mathrm{d}\Omega \tag{5.13.11}$$

Detailed procedure for solving Eq. (5.13.1) and further analysis of nonlinear partial differential equations can be found in Ref. [12].

5.14 Thomas–Fermi Equation

The well-known Thomas–Fermi differential equation is commonly represented in the following general form [13]:

$$z''(x) = x^{-1/2} z^{3/2} \text{ with extreme values of } z(0) = 1, \text{ and } z(\infty) = 0 \tag{5.14.1}$$

The semi-inverse method is utilized by He [13] to obtain the variational form of the functional Thomas–Fermi equation as

$$V = \int_0^\infty \left(\frac{1}{2} z'^2 + \frac{2}{5} x^{-1/2} z^{5/2} \right) \mathrm{d}x \tag{5.14.2}$$

The following trial function for z is considered:

$$z = \mathrm{e}^{-\Psi x} \tag{5.14.3}$$

Equation (5.14.3) has then been substituted in Eq. (5.14.2) to obtain

$$V(\Psi) = \int_0^\infty \left(\frac{1}{2} \Psi^2 \mathrm{e}^{-2\Psi x} + \frac{2}{5} x^{-1/2} \mathrm{e}^{-5\Psi x/2} \right) \mathrm{d}x = \frac{1}{4} + \Gamma(1/2) \sqrt{\frac{2}{5\Psi}} \tag{5.14.4}$$

The stationary condition of the function must be satisfied when

$$\frac{\partial V}{\partial \Psi} = \frac{1}{4} - \sqrt{\frac{\pi}{10}} \Psi^{-3/2} = 0 \tag{5.14.5}$$

The following value for Ψ is obtained by solving Eq. (5.14.5):

$$\Psi = \left(\frac{8\pi}{5} \right)^{1/3} \tag{5.14.6}$$

Results found indicate that the accuracy of the solutions obtained, using the semi-inverse method, is 7.8% [13]. In order to find a higher order solution, the following trial function could be considered:

$$z = \mathrm{e}^{-\Psi x} (1 + c_1 x + c_2 x^2 + \cdots) \tag{5.14.7}$$

Further discussions on the Thomas–Fermi equation and the procedure for solving these types of problem can be found in the referenced article [13].

5.15 Heat Conduction Equation

In the general area of heat transfer and energy engineering, the heat conduction equation is conventionally expressed as [14]

$$\frac{\partial^2 \theta}{\partial x^2} = k^2 \frac{\partial \theta}{\partial t} \tag{5.15.1}$$

in which k and θ are the thermal conductivity constant and temperature, respectively.

Using the method of separation of variables will lead to

$$\theta(x,t) = [\phi(x)] \cdot [\psi(t)] \tag{5.15.2}$$

The following boundary conditions have been assumed for this case:

$$\theta(\alpha,t) = \lambda(t) \quad \text{and} \quad \theta(\beta,t) = \eta(t) \tag{5.15.3}$$

also, with the arbitrary initial condition of

$$\theta(x,0) = q(x) \tag{5.15.4}$$

Accordingly, the following set of differential equations will be obtained:

$$\frac{\mathrm{d}^2\phi}{\mathrm{d}x^2} + \lambda k^2 \phi = 0 \tag{5.15.5}$$

and

$$\frac{\mathrm{d}\psi}{\mathrm{d}t} + \lambda \psi = 0 \tag{5.15.6}$$

in which λ is a constant coefficient. The trial function of Eq. (5.15.5) is defined as

$$V(\phi,\psi) = \int_0^{t_0} \int_\alpha^\beta \left[\left(\frac{\mathrm{d}^2\phi}{\mathrm{d}x^2} + \lambda k^2 \phi \right) \psi + Q \right] \mathrm{d}x\mathrm{d}t \tag{5.15.7}$$

indicating that

$$x \in [\alpha, \beta]; \quad \text{and} \quad t \in [0, t_0] \tag{5.15.8}$$

Note that Q in Eq. (5.15.7) is an unknown function.

The following Euler–Lagrange equations are developed to determine Q:

$$\frac{\mathrm{d}^2\phi}{\mathrm{d}x^2} + \lambda k^2 \phi - k^2 \frac{\mathrm{d}\phi}{\mathrm{d}t} = 0 \tag{5.15.9}$$

and

$$\frac{\mathrm{d}^2\phi}{\mathrm{d}x^2} + \lambda k^2 \phi + k^2 \frac{\mathrm{d}\phi}{\mathrm{d}t} = 0 \tag{5.15.10}$$

where ϕ and ψ are functions of x and t, respectively.

Equation (5.15.7) will now be converted to Eq. (5.15.12) in terms of only temperature by using the following new definition for the temperature function:

$$\theta_t = \phi\psi_t \quad \text{and} \quad \theta_{xx} = \phi_{xx}\psi \tag{5.15.11}$$

By utilizing Eq. (5.15.11), the following constrained function is obtained.

$$V(\phi, \psi) = \int_0^{t_0} \int_\alpha^\beta \left(\frac{\partial^2\theta}{k^2 \partial x^2} + \frac{\partial\theta}{\partial x} \right) \mathrm{d}x\mathrm{d}t \tag{5.15.12}$$

Consider the following constraint:

$$\int_0^{t_0} \int_a^b \theta \, \mathrm{d}x\mathrm{d}t = \text{Constant} \tag{5.15.13}$$

The variational principle of Eq. (5.15.1) can now be found as [14]

$$V(\phi, \psi) = \int_0^{t_0} \int_\alpha^\beta \left[\left(\frac{\mathrm{d}^2\phi}{\mathrm{d}x^2} + \lambda k^2 \phi \right) \psi + k^2 \phi \frac{\mathrm{d}\psi}{\mathrm{d}t} \right] \mathrm{d}x\mathrm{d}t \tag{5.15.14}$$

The application of semi-inverse method in the heat conduction equation can further be examined in the referenced paper [14].

5.16 Lane–Emden-Type Equation

The well-known Lane–Emden-type equation is generally described in the following form [15]:

$$u'' + \frac{2}{x} u' + g(u) = 0 \tag{5.16.1}$$

with the two initial conditions of $u(0) = u_0$ and $u'(0) = 0$.

The Lane–Emden-type equation was first analyzed by He [15]. Using the semi-inverse method, one could construct the variational principle as

$$V(u) = \int_0^a \left\{ -\frac{1}{2} x^2 u'^2 + x^2 G(u) \right\} \mathrm{d}x \tag{5.16.2}$$

The unknown function $G(u)$ will be satisfied if

$$\frac{\partial G}{\partial u} = g \tag{5.16.3}$$

Consequently, considering $g = u^p$ and using Eq. (5.16.3), one could arrive at [15]

$$V(u) = \int_0^a \left\{ -\frac{1}{2} x^2 u^2 + \frac{1}{m+1} x^2 u^{p+1} \right\} \mathrm{d}x \tag{5.16.4}$$

5.17 Dynamic Analysis of Centrifugal Governor System

Dynamical behavior of a centrifugal governor system is mathematically modeled by the following nonlinear differential equation [16]:

$$\ddot{\theta} + \omega_0^2 (1 - \Delta \cos\theta) \sin\theta = 0 \tag{5.17.1}$$

in which

$$\Delta = (\Omega^2 \times l)/g \tag{5.17.2}$$

and

$$\omega_0^2 = \frac{g}{l} \tag{5.17.3}$$

Using the Taylor expansion, Eq. (5.17.1) could be converted into

$$\ddot{\theta} + \omega_0^2 (1 - \Delta)\theta + \omega_0^2 \left(-\frac{1}{6}(1 - \Delta) + \frac{1}{2}\Delta \right)\theta^3 + \omega_0^2 \left(\frac{1}{120}(1 - \Delta) - \frac{1}{8}\Delta \right)\theta^5 + \omega_0^2 \left(-\frac{1}{5040}(1 - \Delta) + \frac{\Delta}{80} \right)\theta^7 = 0 \tag{5.17.4}$$

Accordingly, the following new variables may be defined to simplify Eq. (5.17.4):

$$\alpha_1 = \omega_0^2(1 - \Delta) \tag{5.17.5}$$

$$\alpha_2 = \omega_0^2\left(-\frac{1}{6}(1 - \Delta) + \frac{1}{2}\Delta\right) \tag{5.17.6}$$

$$\alpha_3 = \omega_0^2\left(\frac{1}{120}(1 - \Delta) - \frac{1}{8}\Delta\right) \tag{5.17.7}$$

and

$$\alpha_4 = \omega_0^2\left(-\frac{1}{5040}(1 - \Delta) - \frac{\Delta}{80}\right) \tag{5.17.8}$$

The nonlinear differential Eq. (5.17.1) will be formed as

$$\ddot{\theta} + \alpha_1\theta + \alpha_2\theta^3 + \alpha_3\theta^5 + \alpha_4\theta^7 = 0 \tag{5.17.9}$$

The variational formulation is constructed as

$$V(\theta) = \int_0^{T/4} \left(-\frac{1}{2}\dot{\theta}^2 + \alpha_1\frac{\theta^2}{2} + \alpha_2\frac{\theta^4}{4} + \alpha_3\frac{\theta^6}{6} + \alpha_4\frac{\theta^8}{8}\right)\mathrm{d}t \tag{5.17.10}$$

The initial approximation is assumed to be $A\cos\omega t$, and hence

$$V(A, \omega) = \int_0^{T/4} \left(-\frac{1}{2}A^2\omega^2\sin^2\omega t + \frac{\alpha_1}{2}A^2\cos^2\omega t + \frac{\alpha_2 A^4}{4}\cos^4\omega t + \frac{\alpha_3 A^6}{6}\cos^6\omega t + \frac{\alpha_4 A^8}{8}\cos^8\omega t\right)\mathrm{d}t \tag{5.17.11}$$

After performing the integration and using the limits, one may arrive at

$$V(A, \omega) = -\frac{\pi}{8}A^2\omega^2 + \frac{\pi}{8}\alpha_1 A^2 + \frac{3\pi}{64}\alpha_2 A^4 + \frac{5\pi}{192}\alpha_3 A^6 + \frac{3\pi}{2048}\alpha_4 A^8 \tag{5.17.12}$$

Applying the Ritz method to Eq. (5.17.12) yields

$$\frac{\partial V(A, \omega)}{\partial A} = -\frac{\pi}{4}A\omega^2 + \frac{\pi}{4}\alpha A + \frac{3\pi}{16}\beta A^3 + \frac{5\pi}{32}\gamma A^5 + \frac{3\pi}{256}\Omega A^7 = 0 \tag{5.17.13}$$

Finally, the expression for the frequency–amplitude would be obtained as [16]

$$\omega = \sqrt{\alpha_1 + \frac{3}{4}\alpha_2 A^2 + \frac{5}{8}\alpha_3 A^4 + \frac{35}{64}\alpha_4 A^6} \tag{5.17.14}$$

5.18 Duffing Harmonic Equation

The Duffing harmonic equation is represented by the following differential equation based on the referenced article [17]:

$$\ddot{x} + \frac{x^3}{x^2+1} = 0 \text{ with the initial conditions of } x(0) = A, \quad \text{and} \quad \dot{x}(0) = 0 \tag{5.18.1}$$

For choosing either the small or large quantities of x, Eq. (5.18.1) will then become [17]

$$\begin{cases} \text{For small values of } x: & \rightarrow \quad \ddot{x} + x \approx 0 \\ \text{For large values of } x: & \rightarrow \quad \ddot{x} + x^3 \approx 0 \end{cases} \tag{5.18.2}$$

The variational principle of Eq. (5.18.1) is given by

$$V(x) = \int_0^{T/4} \left\{ -\frac{1}{2}\dot{x}^2 + \frac{1}{2}x^2 - \frac{1}{2}\log(x^2+1) \right\} \mathrm{d}t \tag{5.18.3}$$

Substituting Eq. (5.3.3) in Eq. (5.18.3) yields

$$V(A) = \int_0^{T/4} \left\{ -\frac{1}{2}A^2\omega^2 \sin^2 \omega t + \frac{1}{2}A^2 \cos^2 \omega t - \frac{1}{2}\log(A^2 \cos^2 \omega t + 1) \right\} \mathrm{d}t \tag{5.18.4}$$

By equating to zero, the derivative of Eq. (5.18.4) leads to

$$\frac{\partial V}{\partial A} = 0 \tag{5.18.5}$$

Subsequently, one could determine the final form of Eq. (5.18.5) as

$$\frac{\partial V}{\partial A} = \int_0^{\pi/2} \left\{ -A\omega^2 \sin^2 \omega t + A\cos^2 \omega t - \frac{A\cos^2 \omega t}{A^2\cos^2 \omega t + 1} \right\} dt \\ = 0 \tag{5.18.6}$$

The closed-form solution to express the frequency–amplitude is then obtained as [17]

$$\omega = \sqrt{\frac{\int_0^{\pi/2} \left\{ \cos^2 \omega t - \frac{\cos^2 \omega t}{A^2\cos^2 \omega t + 1} \right\} dt}{\int_0^{\pi/2} \sin^2 \omega t \, dt}} \tag{5.18.7}$$

The approximate solution obtained from Eq. (5.18.7) could be matched with the exact value of the frequency obtained from the following equation:

$$\omega_{ex} = \frac{2\pi}{4\int_0^A \frac{dx}{\sqrt{(A^2 - x^2) + [\log(x^2+1) - \log(A^2+1)]}}} \tag{5.18.8}$$

Part III: Practice Problems

P.5.1. Based on a recent article, published in Applied Mathematics Letters, develop the generalized variational principle for a shell using the variational principle [18].
P.5.2. The heat balance in a finned tube heat exchangers can mathematically be modeled by the following equation [19]:

$$\frac{d^2\theta}{dt^2} + \frac{1}{r}\frac{d\theta}{dr} - m^2\theta = 0$$

where $\theta = T_f - T$ and $m^2 = \frac{4\lambda_{eff}}{\lambda_f \delta X}$.
Use the variational approach to find the approximate solution of the above equation.
P.5.3. Consider the following differential equation for a foam drainage system [20]:

$$u_t + \left(u^2 - \frac{1}{2}u\sqrt{u}\right)_x = 0$$

Use the semi-inverse approach to develop its variational function.
P.5.4. Use the variational approach to solve the following differential equations considering their first approximation:

P.5.4.a. $\ddot{x} + x^{\frac{1}{3}} = 0$ with the initial conditions of $x(0) = A,\ \dot{x}(0) = 0$
P.5.4.b. $\ddot{x} + \mathrm{sgn}(x) = 0$ with the initial conditions of $x(0) = A,\ \dot{x}(0) = 0$

P.5.4.c. $\ddot{x} + \sin(x) = 0$ with the initial conditions of $x(0) = A,\ \dot{x}(0) = 0$
P.5.4.d. $\ddot{x} + \alpha x + \beta x^3 + \gamma x^5 = 0$ with the initial conditions of $x(0) = A,\ \dot{x}(0) = 0$
P.5.4.e. $\ddot{x} + \frac{x}{x^2+1} = 0$ with the initial conditions of $x(0) = A,\ \dot{x}(0) = 0$

P.5.5. Using the Hamiltonian approach, solve all the nonlinear differential equations outlined in problem **P.5.4.** Compare their results with those obtained from problem **P.4.1**.
P.5.6. Prove the relationship between the Harmonic balance method and the variational approach for the conservative nonlinear oscillators. Show that both of them will reach a similar solution for the first approximation when they are used for solving the nonlinear conservative oscillators.
P.5.7. The vibration of a curved nanobeam, based on the nonlocal Euler–Bernoulli beam theory, is described by the following partial differential equation:

$$EI\frac{\mathrm{d}^4 w}{\mathrm{d}x^4} + \left(1 - (e_0 a)\frac{\partial^2 w}{\partial x^2}\right)\left(\mu\frac{\mathrm{d}w}{\mathrm{d}t} + \rho A\frac{\mathrm{d}^2 w}{\mathrm{d}t^2} + k_1 w - k_p\frac{\partial^2 w}{\partial x^2} - \frac{EA}{l}\int_0^l\left(\frac{\partial y}{\partial x}\frac{\partial w}{\partial x} + \frac{1}{2}\left(\frac{\partial w}{\partial x}\right)^2\right)\mathrm{d}x\left(\frac{\partial^2 w}{\partial x^2} + \frac{\partial^2 y}{\partial x^2}\right)\right) = 0$$

in which $E, e_0 a, I, \mu, \rho, A, k_1, k_p, y$ and l are the respective elasticity modulus, scale effect, moment of inertia, damping coefficient, density, cross-sectional area of beam, Winkler and Pasternak foundation coefficients, curvature function, and the beam length.

P.5.7.a. Using Galerkin method, obtain the equation of the primary mode of vibration of the curved beam.
P.5.7.b. Use Hamiltonian approach to find the natural frequency of vibration of the beam.
P.5.7.c. Compare the results obtained with those found from using the variational approach.
P.5.7.d. Find the relative error of results by comparing with the exact numerical simulation.

P.5.8. Breaking soliton equation is mathematically described by the following equation [21]:

$$u_{xt} - u_{xxxy} - 2u_{xx}u_y - 4u_x u_{xy} = 0$$

Use the semi-inverse method to obtain the variational function of the equation.
P.5.9. The Benjamin Ono equation is mathematically presented by the following nonlinear differential equation [22]:

$$u_{tt} + \beta\left(u^2\right)_{xx} + \gamma u_{4x} = 0$$

Develop the variational function of the Benjamin Ono equation using the semi-inverse method.

P.5.10. The following partial differential equation describes the flow of one-dimensional inviscid compressible fluid as presented in the published paper [23]:

$$u_t + \left(uu_x + \omega^2 x\right)_x = 0$$

Develop the variational function for the above equation using the semi-inverse method.

P.5.11. Use both variational and Hamiltonian approaches to solve the following nonlinear differential equation [24]:

$$\ddot{u} + u + \alpha u^2 \ddot{u} + \alpha u \left(\frac{\mathrm{d}u}{\mathrm{d}t}\right)^2 + \beta u^3 = 0$$

with the two initial conditions $u(0) = A$, and $\dot{u}(0) = 0$.

Compare the results obtained with those found from the numerical simulation.

P.5.12. The analytical model for the free vibration of an axially functionally graded beam is developed by Simsek [25]. The following partial differential equation describes the size-dependent nonlinear free vibration of an axially functionally graded microbeam [25].

$$\frac{\partial^2}{\partial x^2}\left[\left(E(x)I + \mu(x)Al^2\right)\frac{\partial^2 w}{\partial x^2}\right] - \left[\frac{A}{2K}\int_0^L \left(\frac{\partial w}{\partial x}\right)^2 \mathrm{d}x\right]\frac{\partial^2 w}{\partial x^2} + \rho(x)A\frac{\partial^2 w}{\partial t^2} = q(x)$$

where $q(x)$ is referred to the distributed transverse load. Both the elasticity modulus and density vary continuously in the axial direction and are defined by the following formulas:

$$E(x) = (E_L - E_R)(1 - x/L)^k + E_R$$

$$\rho(x) = (\rho_L - \rho_R)(1 - x/L)^k + \rho_R$$

- P.5.12.a. Use Galerkin method to obtain the first principle mode of the above equation.
- P.5.12.b. Use variational approach to obtain the nonlinear natural frequency of the model.
- P.5.12.c. Compare the results obtained with those found from the numerical simulation.
- P.5.12.d. Perform a parametric study for the vibration of system, based on the solution obtained. Definition of the parameters can be found in published paper [25].

P.5.13. The nonlocal model for the vibration of functionally graded beams, resting on nonlinear elastic foundation, is mathematically described by the following differential equation [26]:

$$\left(D_{11} - \frac{B_{11}^2}{A_{11}} + \mu N\right)\frac{\partial^4 w}{\partial x^4} - N\frac{\partial^2 w}{\partial x^2} - \mu I_0 \frac{\partial^4 w}{\partial x^2 \partial t^2} + I_0 \frac{\partial^4 w}{\partial t^2} + \mu \frac{\partial^2 f}{\partial x^2} - f(x,t) = 0$$

where $f(x,t) = -k_L W - k_{NL} w^3 + k_s \frac{\partial^2 w}{\partial x^2}$.

P.5.13.a. Use Galerkin method to obtain the first principle mode of the above equation.
P.5.13.b. Use variational approach to obtain the nonlinear frequency of the model.
P.5.13.c. Compare the results obtained with those found from the numerical simulation.
P.5.13.d. Perform parametric study for the vibration of system based on the solution obtained. Definition of the parameters can be found in published paper [26].

P.5.14. The nanoscale vibration of protein microtubules is mathematically modeled using the modified strain gradient theory. The governing equation of the model is [27]

$$V\frac{\partial^6 w}{\partial x^6} - B\frac{\partial^4 w}{\partial x^4} - \rho A\frac{\partial^2 w}{\partial x^2} + \left(-EA\alpha\Delta T + \frac{EA}{2L}\int_0^L \left(\frac{\partial w}{\partial x}\right)^2 \mathrm{d}x\right)\frac{\partial^2 w}{\partial x^2} - 2b\left(k_1 w - k_2 \frac{\partial^2 w}{\partial x}\right) = 0$$

P.5.14.a. Consider the simply–simply and the clamped–clamped boundary conditions for the protein microtubules and use the Galerkin method to obtain the primary mode of vibration.
P.5.14.b. Utilize both the variational and Hamiltonian approaches to obtain the nonlinear natural frequency of the considered model.
P.5.14.c. Compare the results obtained with those found from the numerical simulation.
P.5.14.d. Perform parametric study for the vibration of system based on the solution obtained. Definition of the parameters can be found in the published paper [27].

P.5.15. Develop the variational formulation for the vibration of a trammel pendulum, which is mathematically described by the following differential equation:

$$\left(a\cos^2\theta + b\sin^2\theta\right)\ddot{\theta} + 0.5\left(b^2 - a^2\right)\dot{\theta}^2 + gb\sin\theta = 0$$

Definition of parameters in the above equation can be found in Refs. [28, 29].

References

1. Liu GL (2000) Derivation and transformation of variational principles with emphasis on inverse and hybrid problems in fluid mechanics: a systematic approach. Acta Mech 140 (1):73–89
2. Jihuan H (2000) Semi-inverse method and generalized variational principles with multi-variables in elasticity. Appl Math Mech 21(7):797–808
3. He JH (2007) Variational approach for nonlinear oscillators. Chaos, Solitons Fractals 34 (5):1430–1439
4. He JH (2010) Hamiltonian approach to nonlinear oscillators. Phys Lett A 374(23):2312–2314
5. Yildirim A, Askari H, Yazdi MK, Khan Y (2012) A relationship between three analytical approaches to nonlinear problems. Appl Math Lett 25(11):1729–1733
6. Younesian D, Askari H, Saadatnia Z, Yazdi MK (2011) Free vibration analysis of strongly nonlinear generalized Duffing oscillators using He's variational approach & homotopy perturbation method. Nonlinear Sci Lett A 2(1):11–16
7. Younesian D, Askari H, Saadatnia Z, Yildirim A (2010) Periodic solutions for the generalized nonlinear oscillators containing fraction order elastic force. Int J Nonlinear Sci Numer Simul 11(12):1027–1032
8. Cveticanin L, Kalami-Yazdi M, Saadatnia Z, Askari H (2010) Application of Hamiltonian approach to the generalized nonlinear oscillator with fractional power. Int J Nonlinear Sci Numer Simul 11(12):997–1002
9. Askari H, Saadatnia Z, Yildirim A, Yazdi MK, Khan Y (2013) Application of higher order Hamiltonian approach to nonlinear vibrating systems. J Theor Appl Mech 51(2):287–296
10. Yildirim A, Saadatnia Z, Askari H (2011) Application of the Hamiltonian approach to nonlinear oscillators with rational and irrational elastic terms. Math Comput Model 54 (1):697–703
11. Yildirim A, Saadatnia Z, Askari H, Khan Y, Kalami-Yazdi M (2011) Higher order approximate periodic solutions for nonlinear oscillators with the Hamiltonian approach. Appl Math Lett 24(12):2042–2051
12. He JH (2004) Variational principles for some nonlinear partial differential equations with variable coefficients. Chaos, Solitons Fractals 19(4):847–851
13. He JH (2003) Variational approach to the Thomas-Fermi equation. Appl Math Comput 143 (2):533–535
14. He JH, Lee EWM (2009) A constrained variational principle for heat conduction. Phys Lett A 373(31):2614–2615
15. He JH (2003) Variational approach to the Lane-Emden equation. Appl Math Comput 143 (2):539–541
16. Younesian D, Askari H, Saadatnia Z, Yazdi MK (2011) Periodic solutions for nonlinear oscillation of a centrifugal governor system using the He's frequency-amplitude formulation and He's energy balance method. Nonlinear Sci Lett A 2(3):143–148
17. Saadatnia Z, Askari H, Kalami-Yazdi M (2010) Frequency analysis of nonlinear oscillators with rational restoring force via He's energy balance method and He's variational approach. Nonlinear Sci Lett A 1(4):425–430
18. He J-H (2017) Generalized equilibrium equations for shell derived from a generalized variational principle. Appl Math Lett 64:94–100

19. He JH, Mo LF (2013) Variational approach to the finned tube heat exchanger used in hydride hydrogen storage system. Int J Hydrogen Energy 38(36):16177–16178
20. He JH (2011) Variational approach to foam drainage equation. Meccanica 46(6):1265–1266
21. Tao ZL (2009) Solving the breaking soliton equation by He's variational method. Comput Math Appl 58(11):2395–2397
22. Tao ZL (2009) Variational approach to the Benjamin Ono equation. Nonlinear Anal Real World Appl 10:1939–1941
23. Tao ZL (2008) Variational approach to the inviscid compressible fluid. Acta Applicandae Mathematicae 100(3):291–294
24. Razzak MA, Alam MS (2016) An analytical coupled technique for solving nonlinear large-amplitude oscillation of a conservative system with inertia and static non-linearity. Springerplus 5(1):1–9
25. Şimşek M (2015) Size dependent nonlinear free vibration of an axially functionally graded (AFG) microbeam using He's variational method. Compos Struct 131:207–214
26. Niknam H, Aghdam MM (2015) A semi analytical approach for large amplitude free vibration and buckling of nonlocal FG beams resting on elastic foundation. Compos Struct 119:452–462
27. Zeverdejani MK, Beni YT (2013) The nano scale vibration of protein microtubules based on modified strain gradient theory. Curr Appl Phys 13(8):1566–1576
28. Askari H, Younesian D, Saadatnia Z (2015) Nonlinear oscillations analysis of the elevator cable in a drum drive elevator system. Adv Appl Math Mech 7(1):43–57
29. Younesian D, Askari H, Saadatnia Z, Esmailzadeh E (2011) Analytical solutions for nonlinear lateral sloshing in partially-filled elliptical tankers. In: Proceedings of the ASME design engineering technical conference, vol 1, no Parts A & B

Chapter 6
Integral Based Methods

Abstract Three well-known semi-analytical techniques are introduced in this chapter. The background and mathematical procedure of these techniques, namely, Adomian Decomposition method (ADM), Variational Iteration method (VIM) and Homotopy Analysis method (HAM) are presented in detail. Eleven examples based on recently published works, using ADM, VIM, and HAM, are fully analyzed. The applications of these methods for solving four well-known differential equations are discussed. The historical development and applications of Volterra Integro-differential equation, nonlinear Schrödinger equation, the van der Pol and Korteweg-de Vries equations are presented. Thirteen examples, which could be solved by utilizing the integral based methods, are given as practice problems for the readers, in the last section of this chapter.

Keywords Semi-analytical method · Adomian decomposition method
Variational iteration method · Homotopy analysis method · van der Pol equation
Korteweg-de Vries equation · Volterra integro-differential equation

Part I: Approaches

The main focus of this chapter is on three well-known semi analytical methods, namely, the Adomian decomposition [1], variational iteration [2], and the Homotopy analysis methods [3], which are very vigorous for analyzing diverse types of nonlinear differential equations.

The Adomian decomposition method (ADM) is a very powerful and useful approach for scrutinizing both the nonlinear partial differential and the ordinary differential equations, which investigated by G. Adomian during the late 90s [1, 4]. Furthermore, ADM can provide analytical approximation to many nonlinear and even stochastic equations without any demand for linearization and perturbation, which can engender a gigantic numerical or mathematical procedure [5–10]. Finally, ADM has been adopted by many researchers to enhance further knowledge in the field of nonlinear differential equations [11–14].

E. Esmailzadeh et al., *Analytical Methods in Nonlinear Oscillations*, Solid Mechanics and Its Applications 252, https://doi.org/10.1007/978-94-024-1542-1_6

Variational iteration method (VIM) is another strong tool for solving the nonlinear partial and ordinary differential equations. This approach has been proposed based on a paper by Inokuti et al. [2, 15–17] in which, Lagrange multiplier method was developed for solving nonlinear problems in quantum mechanics. The fundamental concept of the above-mentioned work was then implemented for introducing the Variational iteration method [2, 16]. It has been delineated that the method is very strong for solving plenty of nonlinear problems with high accuracy. This method can be also utilized for analyzing differential difference equations [17]. The effectiveness and reliability of the method have prompted many researchers to use and develop VIM for probing different kinds of nonlinear systems [18–21].

Homotopy analysis method (HAM) is presented as the last analytical method of this chapter. This method implements the concept of Homotopy from topology to provide convergent series solution for nonlinear systems [22–25]. This method was originally illustrated in a Ph.D. dissertation by Liao in 1992 [3]. He developed a novel method in which, there is no need to small parameters, and for that reason it can be exploited for solving nonlinear problems without either small or large parameters. The interesting capability of HAM is its potency for transforming any nonlinear problems into an infinite number of linear problems without any demand for assuming small parameters [22, 26–28]. As a consequence, it is competent and valid for dissecting the nonlinear problems even with very strong nonlinearity. The pioneering work of Liao engrossed many researchers in applied science. His interesting and strong method has been used for analyzing many convoluted nonlinear problems [29–31].

It should be noted that there are no small parameters in many nonlinear problems. Accordingly, perturbation methods are not too dexterous for analyzing many nonlinear problems. In view of that, ADM, VIM and HAM are fully described in this chapter. Furthermore, recent developments of these methods are presented.

In the last section, different types of nonlinear problems based on a few recently published articles are selected to show the potency and accuracy of the considered approaches for solving highly nonlinear problems.

6.1 Adomian Decomposition Method

6.1.1 Background

Adomian decomposition method (ADM) was used since 70s to date, to develop and analyze a large number of differential equations. A simple search in Scopus database for the word "Adomian decomposition method" results in more than 1500 papers, which is a key emblem for the remarkable impact of this method on the world of applied mathematics. A detailed survey of literature reveals the significant impact of ADM. The idea of ADM originally emerged in a pioneering paper by Adomian [4]. Thus, many researchers focused their attention on the modification and utilizing of the proposed approach for different types of nonlinear ordinary and partial differential equations.

Researchers who made most significant contributions in the applications and developments of ADM are Wazwaz [13, 32], Rach [14], Abbaoui [5], Bellomo [9], Abbasbandi [7] Repaci [33], among others. Adomian decomposition method was implemented in a number of linear and nonlinear differential problems, namely, Lane Edmen type equation [34], Voltera population model [35], fractional differential type equation [36], Thomas Fermi equation [37], generalized Korteweg-de Vries (KdV) and regularized long-wave (RLW) equations [38], quadratic Riccati differential equations [39], Bratu type equations [40], coupled Burger equation [8], Fisher equation [12], Klein Gorden equation [41], nonlinear Schrödinger equation [42], and Fredholm integral equations [43]. These are just few examples of applications of ADM. The easiness and accuracy of this method may explain as why so many researchers have become intrigued by these ideas.

Furthermore, the convergence of ADM is investigated by a number of researchers. The proof of the convergence of ADM was first proposed by Cherruault and Adomian [44]. It was revealed that the speed of the convergence of ADM presents the capability of solving nonlinear functional equations. Cut from the same cloth, a sufficient condition of the convergence of ADM was provided by Abbaoui and Cherruault [5, 6]. The order of the convergence of ADM was obtained by Babolian and Biazar. It was shown that any improvement of the convergence of ADM leads to a more accurate results [10].

In the next part, the Adomian decomposition method (ADM) is fully discussed, and different types of its modification are fully illustrated.

6.1.2 Description of Adomian Decomposition Method

The differential equation of motion in its general form is presented as [1, 45]:

$$Lu + Nu = g(t) \tag{6.1.1}$$

where L, N and $g(t)$ are the linear operator, nonlinear operator and the source inhomogeneous term, respectively. In accordance with the ADM, the following infinite series for the unknown function, $u(x)$ is defined:

$$u(x) = \sum_{n=0}^{\infty} u_n(x) \tag{6.1.2}$$

where the components of $u_n(x)$ could easily be obtained. In accordance with the standard ADM, the operator L can be written in the following form [1, 15]:

$$L = \tilde{L} + R \tag{6.1.3}$$

The highest order of the derivative is represented by $\tilde{L}$, which is invertible. The parameter R shows a linear operator of one order less than $\tilde{L}$. According to ADM, one could have [15]:

$$u = h - \tilde{L}^{-1}Ru - \tilde{L}^{-1}Nu \tag{6.1.4}$$

in which h shows the terms resulting from the integration of the term g with the given initial conditions. In accordance with ADM, the nonlinear term Nu and the solution u can be expressed in the respective forms of [1, 15]:

$$Nu = \sum_{n=0}^{\infty} A_n, \quad \text{and } u = \sum_{n=0}^{\infty} u_n \tag{6.1.5}$$

where A_n is called the Adomian polynomials. The algorithm of the standard ADM can be expressed as [1, 45]:

$$u_0 = h \quad \text{and} \quad u_{n+1} = -\tilde{L}^{-1}Ru_n - \tilde{L}^{-1}A_n, \;\; n > 0 \tag{6.1.6}$$

in which, A_n are called the Adomian polynomials of $u_0, u_1, \ldots, u_n$ defined by [1, 45]:

$$A_n = \frac{1}{n!}\frac{d^n}{d\lambda^n}\left[H\left(\sum_{i=0}^{n} \lambda^i u_i\right)\right]_{\lambda=0}; \quad n = 0, 1, 2, \ldots \tag{6.1.7}$$

where $H(u) = \sum_{n=0}^{\infty} A_n$. According to the referenced papers [1, 45], one could write the following list of formulas:

$$A_1 = u_1 H'(u_0) \tag{6.1.8}$$

$$A_2 = u_2 H'(u_0) + \frac{1}{2}u_1^2 H''(u_0) \tag{6.1.9}$$

$$A_3 = u_3 H'(u_0) + u_1 u_2 H''(u_0) + \frac{1}{3}u_1^3 H'''(u_0) \tag{6.1.10}$$

and

$$\begin{aligned} A_4 = {} & u_4 H'(u_0) + \left(u_1 u_3 + \frac{1}{2}u_2^2\right) H''(u_0) \\ & + (1/2)u_1^2 u_2 H'''(u_0) + \frac{1}{24}u_1^4 H^{(iv)}(u_0) \end{aligned} \tag{6.1.11}$$

6.1.2.1 Wazwaz's Modification

In order to reduce the computational work, and also to accelerate the convergence of the method, the following assumption is made by Wazwaz [32]:

$$h = h_1 + h_2 \tag{6.1.12}$$

where h_1 is assigned to the zeroth component u_0 and h_2 has been used to define u_1. Consequently, the following modified recursive algorithm is implemented:

$$\begin{aligned} &u_0 = h_1 \\ &u_1 = h_2 - L^{-1}(Ru_0) - L^{-1}(Nu_0) \\ &\text{and} \\ &u_{k+2} = h_2 - L^{-1}(Ru_{k+1}) - L^{-1}(Nu_{k+1}), \quad k \geq 0 \end{aligned} \tag{6.1.13}$$

6.1.2.2 Laplace Adomian Decomposition Method

The following form of the differential equation can be considered as

$$L_t u + R(u) + N(u) = g \tag{6.1.14}$$

with the corresponding initial condition assumed as:

$$u(x, 0) = h(x) \tag{6.1.15}$$

in which L_t, R, N and g represent a first-order differential operator, a linear operator, a nonlinear operator and the source term, respectively.

Implementing the Laplace transform to both sides of Eq. (6.1.14) and considering the given initial condition, one could obtain:

$$\ell[L_t u] + \ell[R(u)] + \ell[N(u)] = \ell[g] \tag{6.1.16}$$

$$\{s\ell[u] - h(x)\} = -\ell[R(u)] - \ell[N(u)] + \ell[g] \tag{6.1.17}$$

and

$$\ell[u] = \frac{h(x)}{s} + \frac{\ell[g]}{s} - \frac{\ell[R(u)]}{s} - \frac{\ell[N(u)]}{s} \tag{6.1.18}$$

Substituting Eq. (6.1.5) into Eq. (6.1.16) will give:

$$\ell\left[\sum_{n=0}^{\infty} u_n\right] = \frac{h(x)}{s} + \frac{\ell[g]}{s} - \frac{\ell[R(u)]}{s} - \frac{\ell\left[\sum_{n=0}^{\infty} A_n\right]}{s} \tag{6.1.19}$$

Accordingly, the following recursive formula can be defined:

$$\ell[u_0] = \frac{h(x)}{s} + \frac{\ell[g]}{s} \tag{6.1.20}$$

and

$$\ell[u_{n+1}] = -\frac{\ell[R(u)]}{s} - \frac{\ell[A_n]}{s} \tag{6.1.21}$$

For the sake of obtaining u_n, the inverse Laplace transform is applied to both sides of Eqs. (6.1.20) and (6.1.21).

6.2 Variational Iteration Method

6.2.1 Background

The Variational iteration method (VIM) is a very effective approach for analyzing many types of nonlinear differential equations. The phrase "Variational iteration method" was originally mentioned in 1949 in a paper by Kohn [46], where a simple and rapidly convergent method was proposed for solving homogeneous linear equations. About 50 years later, He [47] used the same phrase for referring to a novel method, which is very powerful for solving linear and nonlinear differential equations.

It was shown that the results obtained using the VIM are the same or even better than those found from ADM [47]. This claim was then proved by many researchers who used VIM for solving many complicated nonlinear problems. Indeed, over the past 25 years, many researchers have shed light on different features of VIM and also, compared this method with other powerful analytical, semi-analytical and the numerical approaches for solving differential equations. Indeed, this tantalizing method led researchers to crucial developments in the field of differential equations.

Searching the Scopus database from 1997 to 2014, the Variational iteration method was used or cited in more than 1100 papers. In a somewhat surprising view of its accuracy and simplicity, the method was used as a powerful and efficient way for scrutinizing many differential equations. In order to show the significant effect of VIM in the area of differential equations, one could refer to its application in solving the Blasius equation [48], Fokker Plank equation [49], Burger's and the coupled Burger's equations [50], nonlinear Schrödinger equation [51], KdV Burger's equation [21], Laplace equation [52], and the Boussinesq equation [53].

The Variational iteration method was used to find the analytical solution of the seepage flow with the fractional derivatives in porous media [54], extracting the resonant condition in the nonlinear vibration of rotor bearing system [55], analyzing the nonlinear fractional partial differential equations arising in fluid mechanics [56], probing nonlinear viscoelastic model with the fractional derivatives [57], epidemic and prey-predator model [18], studying the heat and wave like equations [20], vibration analysis of nanowires [58] and a plate cavity system [59], extracting the natural frequency of nonlinear vibration of beams [60], and the analysis of wave propagation in shallow media [61].

A series of valuable results was reported by Ramos [62] to confirm that the VIM is in fact the Picard-Lindelof technique for first-order ordinary differential equations. Furthermore, he mentioned that it will lead to the same results as the well-known fixed-point theorem for the first-order (in time) nonlinear partial differential equations of the evolution type [62]. Citation from his published paper ([63], p. 1) reads:

> …He's variational iteration method for second-order ordinary differential equations can be derived by means of integration by parts, Green's function, adjoint operators, and the method of weighted residuals without the need for introducing variational principles and constrained variations such as the ones introduced by He.

Based on the above quotation, there is a direct relationship between VIM and different well-known iterative methods. It was also confirmed that VIM results are exactly the same sequence of iteration as Picard's method if the initial guess/iterate is the same as the initial condition [62]. The mathematical procedure of VIM is developed in Sect. 6.2.2 and its capability in solving fractional differential equation, differential and difference equations and also diverse types of modifications of VIM are presented.

6.2.2 Description of Method

There are three main algorithms for the variational iteration method (VIM) [17]. Initially, the general form of the differential equations can be considered as:

$$L[u(t)] + N[u(t)] = 0 \tag{6.2.1}$$

where the linear and the nonlinear operators are represented by L and N, respectively.

6.2.2.1 Algorithm I

In accordance with the fundamental concept of the method, the corrected form of Eq. (6.2.1) could be developed as [17]:

$$u_{n+1}(t) = u_n(t) + \int_{t_0}^{t} \lambda[Lu_n(s) + N\tilde{u}_n(s)]\mathrm{d}s \tag{6.2.2}$$

where λ represents a general Lagrange multiplier, obtained by utilizing the variational principle, u_n is the nth approximate solution, and $\tilde{u}_n$ shows a limited variation, which is $\delta\tilde{u}_n = 0$. An interesting article by He [17] explains the approach in detail for using the Lagrange multiplier in VIM.

6.2.2.2 Algorithm II

When the Lagrange multiplier was considered, the following form can be developed as the variation procedure for Eq. (6.2.1) [17]:

$$u_{n+1}(t) = u_0(t) + \int_{t_0}^{t} \lambda N u_n(s)\mathrm{d}s \tag{6.2.3}$$

More information regarding the proof of this algorithm can be examined in Ref. [16].

The important limitation of this approach is that u_0 must satisfy both the initial and the boundary conditions, simultaneously.

6.2.2.3 Algorithm III

In this case, by using the discussed Variational Algorithm II, one could have [17]:

$$u_{n+2}(t) = u_0(t) + \int_{t_0}^{t} \lambda N u_{n+1}(s)\mathrm{d}s \tag{6.2.4}$$

The following equation is obtained by subtracting Eq. (6.2.4) from Eq. (6.2.3) as presented in Ref. [17].

$$u_{n+2}(t) = u_{n+1}(t) + \int_{t_0}^{t} \lambda[N u_{n+1}(s) - N u_n(s)]\mathrm{d}s \tag{6.2.5}$$

The third algorithm is appropriate for the boundary-value problems having higher orders.

6.2.2.4 Application to the Fractional Differential Equation

Consider the following fractional differential equation [17]:

$$\frac{D^q u}{Dt^q} + h = 0 \tag{6.2.6}$$

where $\frac{D^q u(t)}{Dt^q}$ is known as the Caputo's functional derivative, which being widely implemented, and found by

$$\frac{D^q u(t)}{Dt^q} = \frac{1}{\Gamma(m+1-q)} \int_0^t \frac{u^{(m+1)}(\tau)}{(t-\tau)^{q-m}} d\tau, \quad m<q<m+1 \tag{6.2.7}$$

For some real numbers q is called the order of fractional derivative with the gamma function.

In order to develop the variational iteration for Eq. (6.2.6), it is rewritten as:

$$\frac{du}{dt} + \frac{D^q u}{Dt^q} - \frac{du}{dt} + h = 0 \tag{6.2.8}$$

in which $0<q<1$ and accordingly, the variational iteration algorithms are obtained as [17]

$$\begin{cases} u_{n+1}(t) = u_n(t) - \int_0^t \left(\frac{D^q u_n}{Dt^q} + h_n\right) ds \\ u_{n+1}(t) = u_0(t) - \int_0^t \left(\frac{D^q u_n}{Dt^q} - \frac{du_n}{ds} + h_n\right) ds \\ u_{n+1}(t) = u_0(t) - \int_0^t \left\{\left(\frac{D^q u_n}{Dt^q} - \frac{du_n}{ds} + h_n\right) - \left(\frac{D^q u_{n-1}}{Dt^q} - \frac{du_{n-1}}{ds} + h_{n-1}\right)\right\} ds \end{cases} \tag{6.2.9}$$

For more information regarding the application of VIM for the fractional differential equations, one could examine the article published by Odibat and Momani [64] in which, they modified the VIM for the fractional nonlinear oscillators.

6.2.2.5 Application to the Fractional Differential Difference Equation

One powerful approach for modelling discontinuous problems is the differential difference equations, which in general has the following form [17]:

$$\frac{\mathrm{d}^p u_i}{\mathrm{d}t^p} + h(\ldots, u_{i+1}, u_i, u_{i+1}, \ldots) = 0 \tag{6.2.10}$$

Hence, one could write the following formulation as the variational iteration algorithms for the differential difference equation:

$$u_{i,n+1}(t) = u_{i,n}(t) + (-1)^p \int_0^t \frac{1}{(p-1)!}(s-t)^{p-1}\left(\frac{\mathrm{d}^p u_{i,n}}{\mathrm{d}t^k} + h_n\right)\mathrm{d}s \tag{6.2.11}$$

6.2.2.6 Modified Variational Iteration Method

Optimal Variational Iteration Methods

In this part, two types of modified versions of variational iteration method are presented. Addition of an unknown auxiliary parameter $\hbar$ can be very effective for obtaining more accurate approximate solutions using VIM. Accordingly, the variational algorithm could be changed to the following form using the auxiliary term [65]:

$$\begin{cases} u_1(x,\hbar) = u_0(x) + \hbar \int_0^x \lambda(s) H u_0(s)\mathrm{d}s \\ u_{n+1}(x,\hbar) = u_n(x,\hbar) + \hbar \int_0^x \lambda(s) H u_n \\ n \geq 1 \end{cases} \tag{6.2.12}$$

One may use commercially available software, e.g., Maple or Mathematica, to obtain $u_n(x,\hbar)$ [65, 66].

Turkyilmazoglu [67] proposed the following algorithm for the variational iteration method:

$$u_{k+1}(t) = aL^{-1}\{L(u_k(t)) + N(u_k(t)) - g(t)\}^2 \mathrm{d}t \tag{6.2.13}$$

In his article, the following comment has been written:

> … The presence of the parameter a plays very crucial role that may not be handled via the Lagrange multiplier of the classical variational iteration method, for instance, in the case of failure of convergence of the classical variational iteration method.

Accordingly, the optimal value of the convergence accelerating parameter a or $\hbar$ can be determined on the order of approximation by the residual error:

$$\mathrm{Re}[s(a)] = \int_{\alpha}^{\beta} \{L(u_k + N(u_k(t)) - g(t)\}^2 \mathrm{d}t \tag{6.2.14}$$

One could easily minimize Eq. (6.2.14) by imposing the requirement [66, 67]:

$$\frac{\mathrm{d}\{\mathrm{Re}[s(a)]\}}{\mathrm{d}a} = 0 \tag{6.2.15}$$

Finding the exact residual $\mathrm{Re}[s(a)]$ of Eq. (6.2.15) requires much larger CPU time. Accordingly, the constant a curves can be utilized in finding the optimal value of a [67].

Fractional Variational Iteration Method with Adomian Series

Recently, Ji et al. [11] utilized the Adomian series in order to develop the fractional variational iteration method. In accordance with their article, an initial value problem is considered as:

$$\mathrm{D}^{\mu}u(x) + L[u(x)] + N[u(x)] = g(x), \quad \mu = m\alpha > 0, \ 0 < \alpha \leq 1 \tag{6.2.16}$$

and

$$u^{(k\alpha)}(0) = c_k, \quad k = 0, 1, \ldots, m - 1 \tag{6.2.17}$$

where D^{μ} represents $\mathrm{D}^{\alpha}, \ldots, \mathrm{D}^{\alpha}$, with L and N correspond to the linear and non-linear operators, respectively. In accordance with VIM, one could have [11]:

$$u_{n+1}(x) = u_n(x) + \frac{1}{\Gamma(1+\alpha)} \int_{0}^{x} \lambda \left\{ u_n^{(m\alpha)}(\xi) + L[u_n(\xi)] + N[u_n(\xi)] - g(\xi) \right\} (\mathrm{d}\xi)^{\alpha} \tag{6.2.18}$$

and the generalized multiplier is

$$\lambda = -(1)^m \frac{(\xi - x)^{(m-1)\alpha}}{\Gamma(1 + (m-1)\alpha)} \tag{6.2.19}$$

Therefore, the nonlinear term $N[u(\xi)]$ can then be obtained by

$$N[u(\xi)] = N\left(\sum_{n=0}^{\infty} u_n\right) = \sum_{n=0}^{\infty} A_n(u_0, u_1, \ldots, u_n) \tag{6.2.20}$$

Finally, one could obtain the following expression:

$$A_n = \frac{1}{n!}\left[\frac{\mathrm{d}^n}{\mathrm{d}\lambda^n}\left(\sum_{n=0}^{\infty} \lambda^i u_i\right)\right]_{\lambda=0} \tag{6.2.21}$$

Wu–Lee's Fractional Variational Iteration Algorithm

The following partial differential equation is considered to illustrate the application of the variational iteration method by Wu-Lee [19, 68]:

$$\frac{\partial u}{\partial t} + \frac{\partial^{2\beta} u}{\partial x^{2\beta}} = 0 \tag{6.2.22}$$

Using the formulation performed by Wu-Lee for the variational iteration method, the following equation is obtained:

$$\begin{aligned} u_{n+1}(x,t) = u_n(x,t) + \frac{1}{\Gamma(1+\beta)} \int_0^x \frac{(s-x)^\beta}{\Gamma(1+\beta)} \\ \left\{\frac{\partial^{2\beta} u_n(s,t)}{\partial s^{2\beta}} + \frac{\partial u_n(s,t)}{\partial s}\right\} \mathrm{d}s^\beta \end{aligned} \tag{6.2.23}$$

The fractional variational iteration algorithm developed by Wu-Lee could be employed for many types of fractional differential equations [19, 68].

Laplace Variational Iteration Method

The Laplace variational iteration method is another type of the modified version of VIM. In accordance with this modification, the following algorithm is employed [69]:

$$\ell[u_{m+1}(t)] = \ell[u_m(t)] + \ell \int_0^t \lambda[Lu_m(s) + N\tilde{u}_m(s)]\mathrm{d}s \tag{6.2.24}$$

in which ℓ corresponds to the Laplace transform.

Variational Homotopy Perturbation Method

Variational Homotopy perturbation method (VHPM) is the fusion of two powerful approaches for solving nonlinear problems. In order to introduce VHPM, a brief illustration of the Homotopy perturbation method, based on Liao's Homotopy analysis method, is presented [70]. Considering the following equation

$$F(u) - f(r) = 0, \quad r \in \Omega \tag{6.2.25}$$

with the following boundary conditions

$$BC(u, \partial u/\partial n) = 0, \quad r \in \Gamma, \tag{6.2.26}$$

in which F represents a general form of the differential operator, $f(r)$ is a known as an analytic function, BC indicates a boundary operator, and Γ is the boundary of the domain [70].

To construct the Homotopy expression of $u(r,p) : \Omega \times [0,1] \to \Re$

$$H(u,p) = (1-p)[L(u) - L(u_0)] + p[F(u) - f(r)] = 0, \quad p \in [0,1], \ r \in \Omega. \tag{6.2.27}$$

u_0 is the initial solution of Eq. (6.2.25) and p shows an embedded parameter which varies between the values of zero and one.

In accordance with Eq. (6.2.27), one could write:

$$H(u,0) = L(u) - L(u_0) = 0 \tag{6.2.28}$$

and

$$H(u,1) = F(u) - f(r) = 0 \tag{6.2.29}$$

Equations (6.2.28) and (6.2.29) are obtained by changing the embedded parameter p from zero to unity in Eq. (6.2.27). Hence,

$$u = \sum_{i=0}^{\infty} p^i u_i = u_0 + pu_1 + p^2 u_2 + p^3 u_3 + \cdots \tag{6.2.30}$$

If $p \to 1$, then Eq. (6.2.30) could be simplified to Eq. (6.2.31) and then it would reach to the approximate solution in the form of Eq. (6.2.31) [11]:

$$f = \mathrm{Lim}_{p\to 1} u = \sum_{i=0}^{\infty} u_i \qquad (6.2.30)$$

Substituting Eq. (6.2.30) into Eq. (6.2.2) for constructing a new form of VIM, one could obtain the following formula as the variational Homotopy perturbation method (VHPM) [71]:

$$\sum_{n=0}^{\infty} p^n u_n = u_0 + \int_{t_0}^{t} \lambda \left[\sum_{n=0}^{\infty} p^n L u_n(s) + \sum_{n=0}^{\infty} p^n N \tilde{u}_n(s) \right] \mathrm{d}s \tag{6.2.31}$$

6.2.3 Homotopy Analysis Method

The well-known and powerful Homotopy analysis method (HAM), developed by Liao [3], is based on both Homotopy in topology and the McLaurin series. In one of his pioneering articles, he claimed that the method does not require either small or large parameters comparing with the perturbation techniques [23]. The general concept of this method has been considered by many researchers in their published works [22, 24, 25, 72, 73].

HAM was then utilized by a number of mathematicians to study different linear and nonlinear differential equations. One could review important works such as: (a) finding analytical solution for the flow of an incompressible non-Newtonian liquid over an unsteady stretching surface [74]; (b) study of nonlinear progressive waves in deep water [75]; (c) the Sturm-Liouville problems [76]; (d) nonlinear second-order Boundary Value Problem (BVP) equation [30]; (e) solutions of time-dependent Emden-Fowler [77]; (f) Lane-Emden equations [78]; (g) nonlinear fractional partial differential equations [29]; (h) solitary wave solution of Kuramoto-Sivashinsky equation [27]; (i) Fisher equation [79]; (j) Cauchy reaction-diffusion problems [31]; (k) generalized Zakharov equation [80]; among many other linear and nonlinear problems.

Among most influential researchers in developing the Homotopy analysis method one could review the works of: Abbasbandy [26, 27, 76, 80], Hayat [28, 81, 82], Nadeem [83], Alsaedi [84], van Gorder [78], and Vajravelu [78].

In this section, the basic procedure of HAM is illustrated and then, one of its modified version of the technique, namely, the optimal Homotopy analysis method (OHAM), is discussed.

6.2.4 Description of Method

The basic concept behind the Homotopy analysis method (HAM) is illustrated by using the following nonlinear equation [26]:

$$L[u(\tau)] + N[u(\tau)] = 0 \tag{6.3.1}$$

where N, τ and $u(\tau)$ are a nonlinear operator, an independent variable, and an unknown function, respectively. Accordingly, HAM for a zero-order approximate equation can be constructed as:

$$(1 - p)\ell[\phi(\tau; p) - u_0(\tau)] = p\hbar H(\tau)N[\phi(\tau; p)] \tag{6.3.2}$$

where $p \in [0, 1]$ is an embedding parameter, $\hbar \neq 0$ is a non-zero auxiliary linear, $H(\tau) \neq 0$ is an auxiliary function, ℓ is an auxiliary linear operator, $u_0(\tau)$ is an initial guess of $u(\tau)$, and $\phi(\tau, p)$ is a known function [26]. There is a great freedom for selecting the auxiliary parameters in HAM. By expanding $\phi(\tau, p)$ in the Taylor series with respect to p, one could have:

$$\phi(\tau, p) = u_0(\tau) + \sum_{m=1}^{+\infty} u_m(\tau)p^m \tag{6.3.3}$$

in which

$$u_m(\tau) = \frac{1}{m!} \frac{\partial^m \phi(\tau; p)}{\partial p^m}\bigg|_{p=0} \tag{6.3.4}$$

with the proper choice of auxiliary linear operator, the auxiliary parameter $\hbar$, initial guess, and the auxiliary function, one could see that Eq. (6.3.3) will converge at $p = 1$ as:

$$u(\tau) = u_0(\tau) + \sum_{m=1}^{+\infty} u_m(\tau) \tag{6.3.5}$$

By considering $\hbar = -1$ and $H(t) = 1$, Eq. (6.3.2) would then be converted to:

$$(1 - p)\ell[\phi(\tau; p) - u_0(\tau)] + pN[\phi(\tau; p)] = 0 \tag{6.3.6}$$

Equation (6.3.6) is frequently utilized in the Homotopy perturbation method. Based on Eq. (6.3.6), the zero-order deformation equation is implemented to find the governing equation. Thus, the vector is defined as:

$$\vec{u}_n = \{u_0(\tau), u_1(\tau), \ldots, u_n(\tau)\} \tag{6.3.7}$$

The higher-order equation (mth order) is developed as [26]:

$$\ell[u_m(\tau) - \chi_m u_{m-1}(\tau)] = \hbar H(\tau)R_m(\vec{u}_{m-1}) \tag{6.3.8}$$

where

$$R_m(\vec{u}_{m-1}) = \frac{1}{(m-1)!}\frac{\partial^{m-1}N[\varphi(\tau;p)]}{\partial p^{m-1}}\bigg|_{p=0} \tag{6.3.9}$$

and

$$\chi_m = \begin{cases} 0, & m \leq 1, \\ 1, & m > 1. \end{cases} \tag{6.3.10}$$

6.3 Optimal Homotopy Asymptotic Method (OHAM)

In order to illustrate OHAM based on the article referenced [85], one could consider the following differential equation with the assumed boundary condition as:

$$L(u(\tau)) + g(\tau) + N(u(\tau)) = 0, \quad BC(u) = 0. \tag{6.3.11}$$

in which $L, \tau, u(\tau), g(\tau), N((u(\tau))$, and BC represent a nonlinear operator, independent variable, unknown function, known function, a nonlinear operator, and the boundary condition operator, respectfully.

The Homotopy is then constructed as:

$$\begin{aligned}(1-p)[L\varphi(\tau,p) + g(\tau)] = H(p)[L(\varphi(\tau,p)) + g(\tau) \\ + N(\phi(\tau;p))], \quad \text{with } BC(\varphi(\tau,p)) = 0.\end{aligned} \tag{6.3.12}$$

where $p \in [0,1]$ and $H(p)$ represent an embedding parameter and a non-zero auxiliary function for $p \neq 0$, respectively. Furthermore, $\varphi(\tau,p)$ refers to an unknown function. Thus, $H(p)$ being an auxiliary function defined as the following series:

$$H(p) = pB_1 + p^2B_2 + \cdots \tag{6.3.13}$$

in which $B_1, B_2, \ldots$ are constants, which can easily be determined and illustrated later. Using the well-known Taylor series, one could develop:

$$\phi(\tau,p,B_i) = u_0(\tau) + \sum_{k\geq 1} u_k(\tau,B_i)p^k, \quad i = 1,2,\ldots \tag{6.3.14}$$

Equation (6.3.14) is substituted into Eq. (6.3.12), and then by collecting those terms with the same power of p, and equating each coefficient of p to zero, the following equation is obtained:

$$L(u_1(\tau)) = B_1 N_0(u_0(\tau)), \quad \text{with } B(u_1) = 0$$

and

$$\begin{aligned} L(u_k(\tau) - u_{k-1}(\tau)) &= B_k N_0(u_0(\tau)) \\ &+ \sum_{i=1}^{k-1} B_i [L(u_{k-i}(\tau) + N_{k\ i}(u_0(\tau), u_1(\tau), \ldots, u_{k-i}(\tau)], \\ &\text{for } k = 2, 3, \ldots, \quad \text{with } B(u_k) = 0, \quad \text{and } B_k = k(t). \end{aligned} \tag{6.3.15}$$

in which, N_i, with $i \geq 0$, is the coefficient of p^i appeared in the nonlinear operator N:

$$\begin{aligned} N(u(\tau)) = N_0(u_0(\tau)) + pN_1(u_0(\tau), u_1(\tau)) \\ + p^2 N_2(u_0(\tau), u_1(\tau), u_2(\tau)) + \cdots \end{aligned} \tag{6.3.16}$$

The solution of Eq. (6.3.11) is obtained in the form of Eq. (6.3.17) as:

$$\tilde{u}^m = u_0(\tau) + \sum_{k=1}^{m} u_k(\tau, B_i) \tag{6.3.17}$$

The following residual is achieved by combining Eqs. (6.3.17) and (6.3.11)

$$R(\tau, B_i) = L\tilde{u}^m(\tau, B_i) + g(\tau) + N(\tilde{u}^m(\tau, B_i)). \tag{6.3.18}$$

Therefore, the residual function is formulated as:

$$J(B_1, B_2, \ldots, B_n) = \int_a^b R^2(\tau, B_1, B_2, \ldots, B_m) \mathrm{d}\tau \tag{6.3.19}$$

in which, the integral limits a and b are chosen based on the objective problem. To identify the unknown constants B_i, the following formula is utilized:

$$\frac{\partial J}{\partial B_1} = \frac{\partial J}{\partial B_2} = \cdots = 0. \tag{6.3.20}$$

After determining the unknown B_i, the semi-analytical solution (mth-order) is obtained [85].

Part II: Applications

6.4 Volterra-Integro Differential Equations—(History, Development and Applications)

The Volterra-integro differential equations have first reported by Vito Volterra, a prominent Italian mathematician and physicist, when he studied a population growth model [86]. He is a pioneer in the areas of functional analysis and integral equations and developed the theory of functional. His outstanding works have been briefly illustrated in his prominent book and a search in Scopus database for phrase "Volterra equation or model" can easily indicate the crucial impact of Volterra on scientific world. Interested readers for the Volterra nonlinear models of interacting populations are referred to the article by Goel et al., and in a detail study entitled "Sur les equations integro-differernielles et leurs applications", Volterra fully illustrated the integro-differential equations [87]. It is worth mentioning the quote from the first paragraph of his revolutionary work in which, he has mentioned (in French):

> Dans beaucoup de questions de physics mathematique, de mechanique et d'analyse il est neccessaire de consider des relations analytiques qui ont en meme tems le caractere des equations integrals et celui des equations differentielles. Je les ai appelees equations integro differentials.

In the above quote, Volterra illustrates that in many mechanical and mathematical physics problems it is necessary to describe the system with an analytical relationship, which contain the combination of the character and integral form of the equations. He called these types of equations as integro-differential equations. Then, he reported the following differential equations [87]:

$$\frac{\partial^2 u(x,y,z,t)}{\partial x^2}+\frac{\partial^2 u(x,y,z,t)}{\partial y^2}+\frac{\partial^2 u(x,y,z,t)}{\partial z^2}$$
$$+\int_{t_0}^{t}\left\{\frac{\partial^2 u(x,y,z,t)}{\partial z^2}f(t,\tau)+\frac{\partial^2 u(x,y,z,t)}{\partial y^2}\phi(t,\tau)+\frac{\partial^2 u(x,y,z,\tau)}{\partial z^2}\phi(t,\tau)\right\}\mathrm{d}\tau=0 \tag{6.4.1}$$

The integro type differential equation was then analyzed by many researchers by using different analytical techniques. Brunner [88, 89], Maleknejad [90, 91], Vecchio [92], Wazwaz [93], Dafermos [94], Baker [95], and Chen [96] are the most influential researchers in analyzing the Volterra and integro type equations.

Few examples from literature on the integro-differential equations are presented here to show the capability and effectiveness of the illustrated methods of this chapter.

6.4.1 Example 1

The first example is chosen from an article by Wazwaz [86], in which he used ADM for analyzing the following nonlinear Volterra integro-differential equations as [35]:

$$\frac{\mathrm{d}u}{\mathrm{d}t} = 10u(t) - 10u^2(t) - 10u(t)\int_0^t u(x)\mathrm{d}t, \quad u(0) = 0.1 \tag{6.4.2}$$

Accordingly, the following form was deduced for Eq. (6.4.2):

$$u(t) = 0.1 + L^{-1}\left(10u(t) - 10u^2(t) - 10\int_0^t u(x)u(t)\mathrm{d}x\right) \tag{6.4.3}$$

in which $L^{-1}(.) = \int_0^t (.)\mathrm{d}t$.

Using the ADM technique, the following expression can be written [35]:

$$\begin{aligned}\sum_{n=0}^{\infty} u_n(t) = 0.1 + L^{-1}\Bigg(&10\sum_{n=0}^{\infty} u_n(t) - 10\sum_{n=0}^{\infty} G_n(t) \\ &- 10\int_0^t \sum_{n=0}^{\infty} G_n(t) - 10\int_0^t \sum_{n=0}^{\infty} H_n(x,t)\Bigg)\end{aligned} \tag{6.4.4}$$

The Adomian polynomials was accordingly, been implemented by Wazwaz for representing $G_n(t)$ and $H_n(x,t)$ to define the nonlinear terms $u^2(t)$ and $u(x)u(t)$[86]. Thus

$$u^2(t) = \sum_{n=0}^{\infty} G_n(t) \tag{6.4.5}$$

and

$$u(x)u(t) = \sum_{n=0}^{\infty} H_n(t) \tag{6.4.6}$$

Then, the Adomian polynomials are obtained as [35]:

$$G_0(t) = u_0^2(t); \quad G_1(t) = 2u_0(t)u_1(t) \tag{6.4.7}$$

$$G_2(t) = u_1^2(t) + 2u_0(t)u_2(t) \tag{6.4.8}$$

and

$$G_3(t) = 2u_1(t)u_2(t) + 2u_0(t)u_3(t) \tag{6.4.9}$$

However, for $H_n(x,t)$, one could have:

$$H_0(x,t) = u_0(x)u_0(t) \tag{6.4.10}$$

$$H_1(x,t) = u_0(x)u_1(t) + u_1(x)u_0(t) \tag{6.4.11}$$

$$H_2(x,t) = u_0(x)u_2(t) + u_1(x)u_1(x) + u_2(x)u_0(t) \tag{6.4.12}$$

and

$$H_3(x,t) = u_0(x)u_3(t) + u_1(x)u_2(t) + u_2(x)u_1(t) + u_3(x)u_0(t) \tag{6.4.13}$$

The following recursive relationship was used to determine the components of $u(t)$:

$$u_0(t) = 0.1 \tag{6.4.14}$$

and

$$u(t) = L^{-1}\left(10u_k(t) - 10G_k(t) - 10\int_0^t H_k(x,t)\mathrm{d}x\right), \quad k \geq 0 \tag{6.4.15}$$

Accordingly, the following form is obtained [35]:

$$\begin{aligned} u_1(t) &= L^{-1}\left(10u_0(t) - 10G_0(t) - 10\int_0^t H_0(x,t)\mathrm{d}t\right) \\ &= 0.9t - 0.05t^2 \end{aligned} \tag{6.4.16}$$

$$u_2(t) = L^{-1}(10u_1(t) - 10G_1(t) - 10\int_0^t H_1(x,t)\mathrm{d}t) = 3.6t^2 - \frac{7}{12}t^3 + \frac{1}{60}t^4 \tag{6.4.17}$$

and

$$u_3(t) = L^{-1}\left(10u_2(t) - 10G_2(t) - 10\int_0^t H_2(x,t)\mathrm{d}t\right. \\ = 6.9t^3 - 3.154166668t^3 \\ +0.2425t^4 - 0.00472222223t^6 \tag{6.4.18}$$

After finding the values of $u_4(t), u_5(t), u_6(t), u_7(t)$ and $u_8(t)$, the following approximation is then obtained [35]:

$$u(t) = 0.1 + 0.9t + 3.55t^2 + 6.316666667t^3 \\ - 5.5375t^4 - 63.70916667t^5 - 156.0804167t^6 \\ - 18.47323411t^7 + 1056.288569t^8 + O(t^9) \tag{6.4.19}$$

This confirms that the solution obtained is in close agreement with the series solution method [86].

6.4.2 Example 2

The combined Laplace transform-Adomian decomposition method (LADM) was implemented by Wazwaz for solving the following nonlinear Volterra integro-differential equation [97]:

$$u'(x) = \frac{3}{2}\mathrm{e}^x - \frac{1}{2}\mathrm{e}^{3x} + \int_0^x \mathrm{e}^{x-t}u^3(t)\mathrm{d}t, \quad u(0) = 1 \tag{6.4.20}$$

The expression given below is obtained by using the Laplace transform-Adomian decomposition method (LADM):

$$\ell(u'(x)) = \ell\left(\frac{3}{2}\mathrm{e}^x - \frac{1}{2}\mathrm{e}^{3x}\right) + \ell(\mathrm{e}^{x-t}u^3(t)) \tag{6.4.21}$$

and then,

$$sU(s) - U(0) = \frac{3}{2(s-1)} - \frac{1}{2(s-3)} + \frac{1}{(s-1)}\ell(u^3(x)) \tag{6.4.22}$$

The Adomian polynomials $A_k(x)$ is employed for the nonlinear term $u^3(x)$.

$$A_0 = u_0^3 \tag{6.4.23}$$

$$A_1 = 3u_0^2 u_1 \tag{6.4.24}$$

$$A_2 = 3u_0^2 u_2 + 3u_0 u_1^2 \tag{6.4.25}$$

and

$$A_3 = 3u_0^2 u_3 + 6u_0 u_1 u_2 + u_1^3 \tag{6.4.26}$$

Accordingly, the following recursive relationship has been considered [97]:

$$U_0(s) = \frac{1}{s} + \frac{3}{2s(s-1)} - \frac{1}{2s(s-3)} \tag{6.4.27}$$

and

$$\ell\{u_{k+1}(x)\} = \frac{1}{s(s-1)}\ell\{A_k(x)\} \tag{6.4.28}$$

Taking the inverse Laplace transform of both sides of Eq. (6.4.28) and then utilizing the recursive formula, one could obtain:

$$u_0(x) = 1 + x - \frac{1}{2}x^3 - \frac{1}{2}x^4 - \frac{13}{40}x^5 + \cdots \tag{6.4.29}$$

$$u_1(x) = \frac{1}{2}x^2 + \frac{2}{3}x^3 + \frac{5}{12}x^4 + \frac{7}{120}x^5 + \cdots \tag{6.4.30}$$

and

$$u_2(x) = \frac{1}{8}x^4 + \frac{11}{40}x^5 + \cdots \tag{6.4.31}$$

Finally, the following equation is obtained as the solution, which converges to the exact solution of the expansion of $u(x) = e^x$ [97]:

$$u(x) = 1 + x + \frac{1}{2!}x^2 + \frac{1}{3!}x^3 + \frac{1}{4!}x^4 + \frac{1}{4!}x^5 + \cdots \tag{6.4.32}$$

6.4.3 Example 3

The following form of the nonlinear mixed Volterra-Fredholm integral equation was considered by Yousefi et al. [98]:

$$u(x,y) = xy - e^{y} + y + 1 + \int_0^y \int_0^1 t e^{u(s,t)} \mathrm{d}s\mathrm{d}t, \quad 0 \le y \le 1 \tag{6.4.33}$$

In accordance with VIM and the referenced article, one could develop the following iteration formula for Eq. (6.4.33):

$$u_{n+1}(x,y) = u_n(x,y) - \int_0^y \left[\frac{\partial u_n}{\partial \tau} - \tau \int_0^1 e^{u_n(s,\tau)} \mathrm{d}s + e^{\tau} - x - 1 \right] \mathrm{d}\tau \tag{6.4.34}$$

Using the above iteration formula, one could reach to a very close agreement with the results found from the exact solution.

6.5 Nonlinear Schrödinger Equations—(History, Development and Applications)

Dr. Erwin Schrödinger, an Austrian physicist and Nobel Prize winner, developed many fundamental topics in the field of quantum theory and formulated a partial differential equation to describe the physical system changes with respect to time for the quantum state. The Schrödinger equation [107, 108] was originally reported in a revolutionary article entitled "An undulatory theory of mechanics of atoms and molecules". Apparently, his research opened up a new dimension to the theory of quantum mechanics. The general form of the time-dependent Schrödinger equation is:

$$i\hbar \frac{\partial}{\partial t} \psi = - \left[\frac{\hbar^2}{2m} \nabla^2 + V(r,t) \right] \psi(r,t) \tag{6.5.1}$$

where $i, \hbar, \hat{H}, \psi, V, \nabla^2$ and m are the imaginary unit, Plank constant, Hamiltonian operator, wave function, potential energy, Laplacian, and the reduced particle mass, respectively.

The phrase "Schrödinger equation" has been mentioned in more than 30,000 publications, which indicates the importance of his work in the field of physics. Many scientists focused their attention on the work of Schrödinger and developed some interesting ideas in physics. Many physicists and mathematicians worked on finding the analytical solutions of Schrödinger equation. Simos [99], Biswas [100,

101], Feit [102], Koslof [103], Kevrekidis [104], Tian [105], and Hayashi [106] provided most significant contributions for analyzing Schrödinger equation.

ADM, VIM and HAM were utilized to solve different types of Schrödinger equation. One of the earliest attempts for utilizing ADM to analyze the Schrödinger equation was the work done by Datta [107]. Following his work, the linear and nonlinear Schrödinger equations were studied by employing ADM [108]. Variational iteration method was then implemented to the Schrödinger equation by Sweilam [51]. Similarly, explicit series solutions were obtained for some linear and nonlinear Schrödinger equations via HAM [109]. This section investigates few examples for analyzing the Schrödinger equations by considering ADM, VIM and HAM.

6.5.1 Example 1

Alomari et al. [109] developed explicit series solution for the nonlinear Schrödinger equation using the HAM. They have considered the following nonlinear one-dimensional Schrödinger equation with a zero trapping potential:

$$iu_t = -\frac{1}{2}u_{xx} - |u|^2 u, \quad t \geq 0\, u(x,0) = e^{ix} \tag{6.5.2}$$

Accordingly, the following initial condition is assumed for the system as:

$$u_0(x,t) = e^{ix} \tag{6.5.3}$$

Applying the HAM procedure, they have developed the expression below [109]:

$$R_m(\vec{u}_{m-1}) = i\frac{\partial u_{m-1}}{\partial t} + \frac{1}{2}\frac{\partial^2 u_{m-1}}{\partial x^2} + \sum_{i=0}^{m-1}\sum_{k=0}^{m-i-1} |u_i||u_k|u_{m-k-i-1} \tag{6.5.4}$$

The following successive solution is therefore, obtained:

$$u_1(x,t) = \frac{1}{2}\frac{\hbar e^{ix} t}{i} \tag{6.5.5}$$

$$u_2(x,t) = \frac{1}{2}\frac{\hbar e^{ix} t(\hbar t + 4i + 4i\hbar)}{i^2} \tag{6.5.6}$$

and

$$u_3(x,t) = \frac{1}{48}\frac{\hbar e^{ix} t(\hbar^2 i^4 t^2 + 12\hbar^2 i^3 t + 12i^3 \hbar t - 24\hbar^2 - 48\hbar - 24 + 24i\hbar t + 24\hbar^2 it)}{i^3} \tag{6.5.7}$$

It is assumed that $\hbar = -1$, and the series solution is considered as [109]:

$$u_0(x,t) = e^{ix} \tag{6.5.8}$$

$$u_1(x,t) = \frac{1}{1!}\left(\frac{1}{2}it\right)e^{ix} \tag{6.5.9}$$

$$u_2(x,t) = \frac{1}{2!}\left(\frac{1}{2}it\right)^2 e^{ix} \tag{6.5.10}$$

and

$$u_3(x,t) = \frac{1}{2!}\left(\frac{1}{2}it\right)^3 e^{ix} \tag{6.5.11}$$

The general solution based on the above equations is obtained as:

$$u(x,t) \approx u_0(x,t) + \sum_{m=1}^{\infty}\frac{1}{m!}\left(\frac{1}{2}it\right)^m \tag{6.5.12}$$

The above series solution yields to the exact solution given in the form of [109]:

$$u(x,t) = e^{i\left(x+\frac{1}{2}t\right)} \tag{6.5.13}$$

6.5.2 Example 2

The following form of Eq. (6.5.2) was solved by Sweilam [51] when utilizing VIM,

$$iu_t + u_{xx} + p|u|^2 u = 0 \quad p \geq 0, \quad A \leq x \leq B \tag{6.5.14}$$

Assuming the form of the function to be as:

$$u = u(x,t) = v(x,t) + iw(x,t) \tag{6.5.15}$$

Postulating the boundary and initial conditions yields:

$$u(x,t=0) = g(x) = gQ(x) + ig_1(x) \quad A \le x \le B \tag{6.5.16}$$

and

$$\frac{\partial u(A,t)}{\partial x} = \frac{\partial u(B,t)}{\partial x} = 0; \quad t \ge t_0 \tag{6.5.17}$$

The two Eqs. (6.5.18) and (6.5.19) are obtained by substituting Eq. (6.5.15) into (6.5.14), and separating the real and the imaginary parts:

$$v_t + w_{xx} + p(v^2 + w^2)w = 0 \tag{6.5.18}$$

and

$$w_t - v_{xx} - p(v^2 + w^2)w = 0 \tag{6.5.19}$$

The initial and the boundary conditions are given as:

$$v(x,0) = gQ(x) \quad \text{and} \quad w(x,0) = gS(x) \tag{6.5.20}$$

and

$$\frac{\partial v(A,t)}{\partial x} = \frac{\partial v(B,t)}{\partial x} = 0 \quad \frac{\partial w(A,t)}{\partial x} = \frac{\partial w(B,t)}{\partial x} = 0 \tag{6.5.21}$$

By applying VIM to Eqs. (6.5.15) and (6.5.16), the following formulations are obtained:

$$v_{n+1}(x,t) = v_n(x,t) - \int_0^t \left[v_{nt} + w_{nxx} + p(v_n^2 + w_n^2)w_n\right] \mathrm{d}t \tag{6.5.22}$$

and

$$w_{n+1}(x,t) = w_n(x,t) - \int_0^t \left[w_{nt} - v_{nxx} - p(v_n^2 + w_n^2)w_n\right] \mathrm{d}t \tag{6.5.23}$$

As for the initial approximation, one could consider the form of:

$$v_0(x,t) = \sqrt{\frac{2\alpha}{p}}\cos(0.5cx)\mathrm{sech}(\sqrt{\alpha}x) \tag{6.5.24}$$

and

$$w_0(x,t) = \sqrt{\frac{2\alpha}{p}}\sin(0.5cx)\mathrm{sech}(\sqrt{\alpha}x) \tag{6.5.25}$$

Applying the iteration to Eqs. (6.5.22) and (6.5.23), the results could be obtained as [51]:

$$\begin{aligned} v_1(x,t) = \sqrt{\frac{2\alpha}{p}}(&\cos(0.5x)\mathrm{sech}(\sqrt{\alpha}x) - [t\mathrm{sech}^3(\sqrt{\alpha}x)\sin(0.5x) + t\mathrm{sech}(\sqrt{\alpha}x) \\ &(-0.25\sin(0.5cx) - \cos(0.5cx)\tanh(\sqrt{\alpha}x) + \sin(0.5cx)\tanh^2(x)]) \end{aligned} \tag{6.5.26}$$

and

$$\begin{aligned} w_1(x,t) = \sqrt{\frac{2\alpha}{p}}(&\sin(0.5x)\mathrm{sech}(\sqrt{\alpha}x) - [-t\mathrm{sech}^3(\sqrt{\alpha}x)\cos(0.5x) + t\mathrm{sech}(\sqrt{\alpha}x) \\ &(0.25\cos(0.5cx) - \sin(0.5cx)\tanh(\sqrt{\alpha}x) - \cos(0.5cx)\tanh^2(\sqrt{\alpha}x)]) \end{aligned} \tag{6.5.27}$$

It has been shown that VIM can provide a very accurate solution for Eq. (6.5.14) [51].

6.5.3 Example 3

The generalized nonlinear Schrödinger equation employing ADM was analyzed by Adomian and Meyers [110]. The differential equation of their developed work is:

$$u_t = -\frac{1}{2}\frac{\beta_t}{\beta}u + \frac{i}{2\beta}\nabla^2 u - \frac{i}{2}\beta u - \frac{i\alpha}{\beta}|u|^2 u \tag{6.5.28}$$

in which $\beta(t)$ is a real function and $\alpha = \alpha(x)$, and also, $x \in R^3$.

By applying the inverse operator to Eq. (6.5.28), one could find:

$$u = u(0) - \frac{1}{2}L_t^{-1}\frac{\beta_t}{\beta}u + \frac{i}{2}L_t^{-1}\frac{1}{\beta}\nabla^2 u - \frac{i}{2}L_t^{-1}\beta u - iL_t^{-1}\frac{\alpha}{\beta}|u|^2 u \tag{6.5.29}$$

Using the Adomian polynomials for $|u|^2 u$, the following recursive algorithm could be obtained as:

$$\begin{aligned} &u_0 = u(0) \\ &\text{and} \quad \text{for } n \geq 0 \\ &u_{n+1} = -\frac{1}{2}L_t^{-1}\frac{\beta_t}{\beta}u_n + \frac{i}{2}L_t^{-1}\frac{1}{\beta}\nabla^2 u_n - \frac{i}{2}L_t^{-1}\beta u_n - iL_t^{-1}\frac{\alpha}{\beta}A_n \end{aligned} \tag{6.5.30}$$

The coefficient A_n is a nonlinear term, obtained by means of the vector notation for the complex quantities:

$$A_0 = (u_0 \cdot u_0)u_0 \tag{6.5.31}$$

$$A_1 = (2u_0 \cdot u_1)u_1 \tag{6.5.32}$$

and

$$A_2 = (u_1 \cdot u_1 + 2u_0 \cdot u_2)u_2 \tag{6.5.33}$$

The analytical solution for Eq. (6.5.28) was obtained by Adomian and Meyers [110] in the form of utilizing the above polynomial component.

6.6 Van der Pol Equation (History, Development and Applications)

The van der Pol equation was first reported by the Dutch electrical engineer and physicist, Balthazar van der Pol. In his pioneer article, the concept of relaxation oscillation was introduced to explain the oscillation produced by a self-sustained oscillating system, namely a triode circuit. Actually, van der Pol during the period of 1926–1930 popularized the term "relaxation oscillation" and turned it from the experiments to concept [111]. Mary-Lucy Cartwright, the first woman mathematician being elected a Fellow of the Royal Society of England, wrote in one of her articles that "*his work formed the basis of much of the modern theory of nonlinear oscillations*" [112].

A new type of nonlinear differential equation was reported by van der Pol, which in fact is a representation of the so-called "relaxation oscillations". In his 1925 article, van der Pol stated that there exists a type of oscillation, which diverge considerably from the sine curve, and being more peaked. Moreover, the frequency is no longer determined by the equation $\omega = 1/\sqrt{LC}$ and the time period is approximately the product of a resistance by a capacitor, i.e., a relaxation time. Such oscillations belong to a separate group and are conveniently named the 'relaxation oscillations' [111].

In order to develop the governing differential equation of the considered system, a Taylor-Maclaurin series expansion was implemented to define the characteristic equation of the triode by assuming a cubic function (van der Pol 1920).

He obtained the following differential equation for the considered model.

$$C\frac{d^2u}{dt^2} + \left(\frac{1}{R} - \alpha\right)\frac{du}{dt} + \beta\frac{du^2}{dt} + \gamma\frac{du^3}{dt} = 0 \tag{6.6.1}$$

and the following form of Eq. (6.6.1) was developed as

$$\frac{d^2u}{dt^2} + \frac{d}{dt}\left(\frac{Ru}{L} + \frac{\psi(u)}{C}\right) + \omega_0^2 u = 0 \tag{6.6.2}$$

where

$$\psi(u) = \alpha u + \beta u^2 + \gamma u^3 + \delta u^4 + \varepsilon u^5 + \cdots \tag{6.6.3}$$

Few years later, the dimensionless form of Eq. (6.6.2) was introduced by van der Pol as:

$$\ddot{v} - \varepsilon(1 - v^2)\dot{v} + v = 0 \tag{6.6.4}$$

Equation (6.6.4) brought much attention to many researchers, and based on that a number of different types of nonlinear oscillatory systems were developed. Selected few of recently developed nonlinear equations, based on the work of van der Pol, are listed in Table 6.1.

Few selected examples are considered in this section to demonstrate the potency of the illustrated approaches for solving different types of van der Pol equation.

6.6.1 Example 1

Consider the following differential equation and its solution is sought using VIM [122]:

$$\begin{aligned} &\ddot{u} + u - \varepsilon(1 - u^2)\dot{u} = 0 \\ &u(0) = A, \dot{u}(0) = 0, \quad t \geq 0 \end{aligned} \tag{6.6.5}$$

Table 6.1 Development of the van der Pol equation

Nonlinear system	Equation of motion
Micromechanical resonator embedded with a nano mechanical resonator [113]	$m_\mu x''_\mu + (m_\mu \gamma_u + \eta x_n^2)x'_\mu + \left(k_\mu + \beta_\mu x_\mu^2 + \Gamma x_n^2\right)x_\mu = F_\mu(t)$ $m_n x''_n + \left(m_n \gamma_n + \eta x_\mu^2\right)x'_\mu + \left(k_\mu + \beta_n x_n^2 + \Gamma x_\mu^2\right)x_n = F_n(t)$
Edge emitting semi-conductor laser [114]	$LC\frac{\mathrm{d}^2 v}{\mathrm{d}t^2} - v\left(1 - \frac{I_0 L}{v V_0}\cosh\left(\frac{v}{2V_0}\right)\right)\frac{\mathrm{d}v}{\mathrm{d}t} + \left(1 - r\left(\frac{1}{R_1} - \frac{1}{R_2}\right)\right)v$ $+2I_0 r\sinh\left(\frac{v}{2V_0}\right) = \frac{r}{R_2}E$
Stiffness-nonlinear stochastic dynamics of a tall building to the random wind load [115]	$m\ddot{x} + c(1 - \mu x^2)\dot{x} + kx = \frac{1}{2}\rho \bar{U}_H^2 A_p C_D\left(\frac{2u'(t)}{U_H}\right)$
Deep-water risers under vertex induced vibration [116]	$EI\frac{\partial^4 x(z,t)}{\partial z^4} - \frac{\partial}{\partial z}\left[T(z,t)\frac{\partial x(z,t)}{\partial z}\right] + \bar{m}\frac{\partial^2 x(z,t)}{\partial t^2} = f_x(z,t)$ $F_z = \frac{\rho g H \pi R^2[6\cosh(k(L-d) - kR\sinh(k(L-d))]}{6\cosh(kL)}$
Van der Pol oscillators coupled by fractional derivative [117]	$x'' - \varepsilon(1 - x^2)x' + x = \varepsilon y D^\alpha(y - x)$ $y'' - \varepsilon(1 - x^2)y' + y = \varepsilon y D^\alpha(x - y)$
Using delay to quench undesirable vibrations [118]	$\ddot{x} + x - \varepsilon x + \varepsilon\alpha x^2\dot{x} = \beta\dot{x}(t - T) + \delta x(t - T)$
Modeling cardiac pacemakers [119]	$\ddot{x} + a(x - W_1)(x - W_2)\dot{x} - \frac{x(x+\mathrm{d})(x+\mathrm{e})}{\mathrm{ed}} = \Gamma(t)$
Solar magnetic cycle [120]	$\frac{\mathrm{d}^2 B_\phi}{\mathrm{d}t^2} + \omega^2 B_\phi + \mu(3\xi B_\phi^2 - 1)\frac{\mathrm{d}B_\phi}{\mathrm{d}t} - \lambda B_\phi^3 = 0$
Motion of two ions in a Paul trap [121]	$\ddot{x}^{(k)} = -\gamma\dot{x}^{(k)} - k(a + 2q\cos(2\tau))x^{(k)}$ $K = 1$ and -2 are related to the equation for the r and z coordinates, respectively

Based on VIM, one could obtain:

$$u_{n+1}(t) = u_n(t) + \int_0^t \lambda\left(\frac{\mathrm{d}^2 u_n(t)}{\mathrm{d}t^2} + u_n(t) - \varepsilon(1 - \tilde{u}^2(t))\frac{\mathrm{d}u_n(t)}{\mathrm{d}t}\right)\mathrm{d}t \tag{6.6.6}$$

After identifying the Lagrange-multiplier, one could have:

$$u_{n+1}(t) = u_n(t) + \int_0^t \sin(\tau - t)\left(\frac{d^2 u_n(t)}{dt^2} + u_n(t) - \varepsilon(1 - \tilde{u}^2(t))\frac{du_n(t)}{dt}\right)dt \tag{6.6.7}$$

The following solution has been assumed as a good initial guess:

$$u_0(t) = A\cos t \tag{6.6.8}$$

Accordingly, the following equation is obtained:

$$u_1(t) = 2\cos t + \frac{3\varepsilon}{4}\sin t - \frac{\varepsilon}{4}\sin 3t \tag{6.6.9}$$

and the second approximation is obtained as:

$$u_2 = 2\cos t + \left(\frac{3}{4}\sin t - \frac{1}{4}\sin 3t\right) + \frac{1}{96}(-13\cos t + 18\cos 3t - 5\cos 5t + 12t\sin t)\varepsilon^2 + \left(\frac{(-120t\cos t - 165\sin t + 162\sin 3t - 50\sin 5t + 7\sin 7t)\varepsilon^4}{1536}\right) + O(4) \tag{6.6.10}$$

6.6.2 Example 2

In this example, the Adomian decomposition method is used to analyze the forced van der Pol equation. Very recently, Ramana and Pradad [123] modified the ADM to utilize it in analyzing the forced van der Pol equation.

Accordingly, the governing differential equation is developed [123]:

$$\ddot{u} + \mu(c - bu)\dot{u} + u = F\cos(\omega t) \quad u(0) = a_1, \;\; u_t(0) = a_2, \;\; t \geq 0 \tag{6.6.11}$$

Applying the ADM to both sides of Eq. (6.6.11), one could write [123]:

$$L_t u = -\mu(c - bu^2)L_{t1}u - u + F\cos(\omega t) \tag{6.6.12}$$

in which $L_t = \frac{d^2}{dt^2}$ and $L_{t1} = \frac{d}{dt}$.

Multipling both sides of Eq. (6.6.12) by L_t^{-1} will result in

$$L_t^{-1}[L_t u] = L_t^{-1}[-\mu(c - bu^2)L_{t1}u - u + F\cos(\omega t)] \tag{6.6.13}$$

Hence

$$u = a_1 + ta_2 + L_t^{-1}(F\cos(\omega t) - \mu(c - bu^2)L_{t1}u - u) \tag{6.6.14}$$

Using the Adomian polynomials for the nonlinear term, the following equation for the first approximation is developed:

$$\begin{aligned} u_1 = L_t^{-1}(-\mu(c - bA_0)L_{t1}u_0 - u_0) &= \frac{1}{2}\mu c a_2 t^2 + \left(\frac{1}{6}\mu c f - \frac{1}{6}a_2\right)t^3 \\ &+ \left(-\frac{1}{12}\mu b a_2^3 - \frac{1}{24}f\right)t^4 + \mu\left(-\frac{1}{120}cf\omega^2 - \frac{1}{10}ba_2^2 f\right)t^5 + \cdots \end{aligned} \tag{6.6.15}$$

and

$$\begin{aligned} u_2 = L_t^{-1}(-\mu(c - bA_1)L_{t1}u_1 - u_1) &= \frac{1}{2}\mu^2 c^2 a_2 t^3 + \frac{1}{12}\mu c a_2 t^4 - \frac{1}{24}\mu c a_2 t^4 \\ &- \frac{1}{120}\mu c f + \frac{1}{120}a_2 + \frac{1}{20}\mu c - \frac{1}{3}\mu b a_2^3 - \frac{1}{6}ft^5 + \frac{1}{360}\mu b a_2^3 + \frac{1}{720}f \\ &+ \mu\frac{1}{30}c\mu - \frac{1}{24cf}\omega^2 - \frac{1}{2}ba_2^2 f - \frac{1}{30}ba_2^3\mu^2 c^2 t^6 + \cdots \end{aligned} \tag{6.6.16}$$

and accordingly, u_3 could be obtained as:

$$u_3 = L_t^{-1}(-\mu(c - bA_1)L_{t1}u_2 - u_2) \tag{6.6.17}$$

Finally, the following equation is developed as:

$$u = \sum_{i=0}^{n} u_i \tag{6.6.18}$$

6.7 Korteweg-de Vries Equation (History, Development and Applications)

Joseph Valentin Boussinesq, a prominent French mathematician and physicist, first developed a nonlinear partial differential equation for the water waves, published in his 1872 article known as Boussinesq equation [53]. He developed the water waves mathematically, turbulence and, generally speaking, made momentous contributions in fluid mechanics. 'Boussinesq equation' has been cited in more than 6500 articles, which indicates the importance of his proposed model.

In fact, he was instigated by the experiment of J.S. Russel and M. Bazin, as he mentioned in the first line of his paper:

> Tous les ingenieurs connaissent les belles experiences de J. Scott Russell et de M. Bazin sur la production et la propagation des ondes solitaries.

The nonlinear partial differential equation for the water waves was developed by Boussinesq [53] as:

$$\frac{\partial^2 h}{\partial t^2} = gH\frac{\partial^2 h}{\partial x^2} + gH\frac{\partial^2}{\partial x^2}\left(\frac{3}{2}\frac{h^2}{H} + \frac{H^2}{3}\frac{d^2 h}{dx^2}\right) \tag{6.7.1}$$

A shallow canal with a rectangular cross-section has been considered in his work. He assumed that the fluid is both incompressible and rotation free, and also, friction along the boundaries of the canal has been neglected.

Equation (6.7.1) fascinated many researchers working in areas of coastal engineering and fluid mechanics, and has repeatedly been developed for computer modeling and simulation of water waves in shallow seas. Furthermore, assumptions by Boussinesq has been implemented for testing atmospheric models [124], Tsunami propagation [125], nonlinear model of wave motions induced by a bottom jet [126], a propagating front in the presence of a spatiotemporally chaotic flow field [127], modeling of ground water flow [128] and the hydrogen dispersion behavior [129].

Boussinesq's theoretical study of the water waves and S. Russel's experiments led to the development of Korteweg-de Vries equation (KdV equation) by Diederik Korteweg and Gustav de Vries [130]. The shallow water surface, which was mathematically modeled by Korteweg-de Vries, is one of the most well-known models for shallow water, and a classical example of the propagation of weakly dispersive nonlinear waves. Studying the water waves, during more than half of a century, and pursuing the S. Russel's experiments, Rayleigh and Boussinesq theoretical works and many others results led to the development of KdV equation.

To give a short biography for Diederik and Gustav, one could mention that Diederik Korteweg was a Dutch mathematician (1848–1941) who received his Ph. D. from University of Amsterdam after defending his dissertation entitled "On the Propagation of Waves in Elastic Tubes". His supervisor was Prof. J. D. van der Waals who developed the equation of state and the continuity of the gas and fluid. Gustav de Vries was also a Dutch mathematician (1866–1934) who worked on his Ph.D. dissertation entitled "Bijdrage tot de Kennis der Lange Golven" under the supervision of Prof. Diederik Korteweg. A passage of his Ph.D. thesis was then published in the *Philosophical Magazine* back in 1895.

Collaborations of Kortweg and de Vries result in a hot topic for researchers to work in different fields of engineering and mathematics. The simplified form of KdV equation can be presented as:

Table 6.2 Recent development of KdV equations

Model	Equation
General system of differential-difference KdV equation [131]	$\frac{\mathrm{d}}{\mathrm{d}_t}U_n = U_n^2(E_{\sigma_1}U_{n+1}E_{\sigma 2} - E_{\sigma_2}U_{n-1}E_{\sigma_1})$ where U_n is a diagonal $N \times N$ matrix, $(U_n)_{ij} = \delta_{ij}u_j(n,t), i,j = 1,\ldots,N$, E_σ are permutation matrix
Generalized fifth order KdV equation	$u_t + \alpha u u_{xxx} + \beta u_x u_{xx} + \gamma u^2 u_x + u_{xxxx} = 0$ where α, β and γ are arbitrary parameters
Singularly perturbed generalized higher-order KdV equation [132]	$U_t + \alpha U^{n+1}U_x + \beta U_{xxx} + \varepsilon(U_{xx} + U_{xxxx}) = 0$ where α, β are positive parameters, ε is a small positive parameters
System of coupled nonlinear Schrödinger–KdV equations [133]	$\begin{cases} iu_t + u_{xx} + \alpha uv + \lvert u\rvert^2 u = 0 \\ v_t + v_{xxx} + vv_x + \frac{1}{2}\alpha\left(\lvert u\rvert^2\right)_x = 0 \end{cases}$ where $u = u(x,t) \in C$ while $v = v(x,t) \in R$, and $\alpha < 0$ is the real coupling constant
Nonlinear fractional KdV-Burgers equation [134]	$\frac{\partial^\alpha u(x,t)}{\partial t^\alpha} + \varepsilon u^r(x,t)\frac{\partial^\beta u(x,t)}{\partial x^\beta} + \eta\frac{\partial^2 u(x,t)}{\partial x^2}$ $v\frac{\partial^3 u(x,t)}{\partial x^3} = 0$ where: $0 < \alpha$; $\beta \geq 1$; and $x, t > 0$
Rosenau-KdV-RLW equation [135]	$u_t + au_x + b_1 u_{xxx} + b_2 u_{xxt} + cu_{xxxxt} + k(u^n) = R$ $R = \alpha u + \beta u_{xx} + \gamma u_x u_{xx} + \delta u^m u_x + \lambda u u_{xxx}$ $vuu_x u_{xx} + \sigma u_x^3 + \xi u_x u_{xxxx} + \eta u_{xx} u_{xxx} + \rho u_{xxxx}$ $+ \psi u_{xxxxx} + kuu_{xxxxxx}$
KdV-Caudrey-Dodd-Gibbon equation [136]	$u_t + k(u_{xx} + \frac{1}{5}\alpha u^2) + b\left(\frac{1}{15}\alpha u^3 + \alpha u u_{xx} + u_{xxxx}\right)_x = 0$
Hirota–Satsuma coupled KdV equation [137]	$u_t = \frac{1}{2}u_{xxx} - 3uu_x + 3(uw)_x$ $v_t = -v_{xxx} + 3uv_x$ $w_t = -w_{xxx} + 3uw_x$

$$\frac{\partial u}{\partial t} - 6u\frac{\partial u}{\partial x} + \frac{\partial^3 u}{\partial x^3} = 0 \tag{6.7.2}$$

A number of physical problems in ocean engineering and fluid mechanics were mathematically modeled using KdV equation. Recent developed forms of KdV equation are listed in Table 6.2.

6.7.1 Example 1

The solution procedure of the coupled Schrodinger-KdV equations is presented here as a first example. This equation has been solved by Abdou and Soliman [138] using the VIM as:

$$iu_t - (u_{xx} + uv) = 0 \tag{6.7.3}$$

and

$$v_t + 6uv_x + v_{xxx} - \left(|u|^2\right)_x = 0 \tag{6.7.4}$$

Considering the following initial condition [138]:

$$u(x,0) = 6\sqrt{2}\mathrm{e}^{i\alpha x}k^2\mathrm{sech}^2(kx), v(x,0) = \frac{\alpha + 16k^2}{3} - 16k^2\tanh^2(kx) \tag{6.7.5}$$

where α and k are arbitrary constants. In accordance with the VIM, one could develop the following system of iteration formula for Eqs. (6.7.3) and (6.7.4) [138].

$$u_{n+1}(x,t) = u_n(x,t) - \int_0^t (u_{nt} + i(u_{nxx} + u_n v_n))\mathrm{d}\tau, \tag{6.7.6}$$

and

$$v_{n+1} = v_n(x,t) - \int_0^t (v_{nt} + 6u_n v_{nx} + v_{nxxx} - u^2)\mathrm{d}\tau, \tag{6.7.7}$$

The first and the second approximations, using the above iteration, are developed as:

$$\begin{aligned} u_1(x,t) &= 6\sqrt{2}e^{i\alpha x}k^2\mathrm{sech}^2(kx) \\ &\quad + 6it\sqrt{2}e^{i\alpha x}k^2\mathrm{sech}^2(kx)[\alpha^2 - 2k^2\mathrm{sech}^2(kx) \\ &\quad - 4i\sqrt{2}\alpha k\tanh(kx) + 4\sqrt{2}k^2\tanh^3(kx)] \end{aligned} \tag{6.7.8}$$

and

$$\begin{aligned} v_1(x,t) &= \frac{\alpha + 16k^2}{3} - 16k^2\tanh^2(kx) \\ &\quad - 48tk^5\tanh(kx)\mathrm{sech}^2(kx)[2\mathrm{sech}^2(kx) - \tanh^2(kx)] \end{aligned} \tag{6.7.9}$$

The second approximation of the solution of Eqs. (6.7.3) and (6.7.4) is found as:

$$\begin{aligned} u_2(x,t) = {} & u_1(x,t) - it\sqrt{2}\mathrm{e}^{i\alpha x}\mathrm{sech}^4(kx)[\alpha + 34k^2 + \alpha\cosh(2x) - 2k^2\cosh(2kx) \\ & + \frac{3}{8}t\mathrm{sech}^2(kx)\{-3i\alpha^4 - 72i\alpha^2k^2 - 528ik^4 \\ & - 4i\alpha^4\cosh(2kx) - 48i\alpha^2k^2\cosh(2kx) \\ & + 416ik^4\cosh(2kx) - i\alpha^4\cosh(4kx) \\ & + 24i\alpha^2k^2\cosh(4kx) - 16ik^4\cosh(4kx) \\ & + 16\alpha^3k\sinh(2kx) + 320\alpha k^3\sin(2kx) \\ & + 8\alpha k\sinh(4kx)(\alpha^2 - 4k^2)\}] \end{aligned} \tag{6.7.10}$$

and

$$\begin{aligned} v_2(x,t) = {} & v_1(x,t) - 6tk^4\mathrm{sech}^5(kx)[k^4t\mathrm{sech}^3(kx)(1208 - 1191\cosh(2kx) \\ & + 120\cosh(4kx) - \cosh(6kx)) \\ & + 24\mathrm{e}^{2i\alpha x}(-i\alpha\cosh(kx) + 2k\sinh(kx))] \end{aligned} \tag{6.7.11}$$

6.7.2 *Example 2*

The solution procedure for a type of coupled KdV equation, employing the VIM, is illustrated in this section. Assas [137] used the VIM to solve the following equation.

Accordingly,

$$u_t + 6\alpha uu_x - 2bvv_x + \alpha u_{xxx} = 0 \tag{6.7.12}$$

and

$$v_t + 3\beta uu_x + \beta u_{xxx} = 0 \tag{6.7.13}$$

Assuming the following two initial conditions as

$$u(x,0) = f(x) \quad \text{and} \quad v(x,0) = g(x) \tag{6.7.14}$$

One has the iteration formula for Eqs. (6.7.12) and (6.7.13) in the same way as was given in Example 1:

$$u_{n+1}(x,t) = u_n(x,t) - \int_0^t (u_{nt} + 6\alpha u_n u_{nx} - 2bv_n v_{nx} + \alpha u_{nxxx})\,\mathrm{d}\tau \tag{6.7.15}$$

and

$$v_{n+1} = v_n(x,t) - \int_0^t (v_{nt} + 3\beta u_n v_{nx} + \beta v_{nxxxx})\mathrm{d}\tau \tag{6.7.16}$$

The following initial approximations have now been assumed:

$$u_0 = \frac{\lambda}{\alpha}\operatorname{sech}^2\left[\frac{1}{2}\sqrt{\frac{\lambda}{\alpha}}x\right] \quad v_0 = \frac{\sqrt{\frac{1}{\alpha}}\lambda\operatorname{sech}^2\left[\frac{1}{2}\sqrt{\frac{\lambda}{\alpha}}x\right]}{\sqrt{2}} \tag{6.7.17}$$

Substituting u_0 and v_0 in Eqs. (6.7.15) and (6.7.16), the following approximate solutions are obtained [137]:

$$\begin{aligned} u_1 = & \frac{\lambda}{\alpha}\operatorname{sech}^2\left[\frac{1}{2}\sqrt{\frac{\lambda}{\alpha}}x\right] + \frac{7t\lambda^2\sqrt{\frac{\lambda}{\alpha}}\operatorname{sech}^4\left[\frac{1}{2}\sqrt{\frac{\lambda}{\alpha}}x\right]\tanh\left[\frac{1}{2}\sqrt{\frac{\lambda}{\alpha}}x\right]}{2\alpha} \\ & - \frac{bt\lambda^2\sqrt{\frac{\lambda}{\alpha}}\operatorname{sech}^4\left[\frac{1}{2}\sqrt{\frac{\lambda}{\alpha}}x\right]\tanh\left[\frac{1}{2}\sqrt{\frac{\lambda}{\alpha}}x\right]}{\alpha} \\ & + \frac{t\lambda^2\sqrt{\frac{\lambda}{\alpha}}\cosh\left[\sqrt{\frac{\lambda}{\alpha}}x\right]\operatorname{sech}^4\left[\frac{1}{2}\sqrt{\frac{\lambda}{\alpha}}x\right]\tanh\left[\frac{1}{2}\sqrt{\frac{\lambda}{\alpha}}x\right]}{2\alpha} \end{aligned} \tag{6.7.18}$$

and

$$\begin{aligned} v_1 = & \frac{\sqrt{\frac{1}{\alpha}}\lambda\operatorname{sech}^2\left[\frac{1}{2}\sqrt{\frac{1}{\alpha}}x\right]}{\sqrt{2}} \\ & + \frac{t(\frac{1}{\alpha})^{\frac{3}{2}}\lambda^2\beta\sqrt{\frac{\lambda}{\alpha}}\operatorname{sech}^2\left[\frac{1}{2}\sqrt{\frac{\lambda}{\alpha}}x\right]\tanh\left[\frac{1}{2}\sqrt{\frac{\lambda}{\alpha}}x\right]}{\sqrt{2}} \end{aligned} \tag{6.7.19}$$

6.7.3 Example 3

To show different applications of HPM, the solution procedure of the fractional KdV Burgurs equation is presented in this example [139].

$$\frac{\partial^\alpha u}{\partial t^\alpha} + \varepsilon u \frac{\partial^\beta u}{\partial x^\beta} + \eta \frac{\partial^2 u}{\partial x^2} + v \frac{\partial^3 u}{\partial x} = 0 \quad \text{for } t > 0 \qquad (6.7.20)$$
$$\text{with } 0 < \alpha; \text{ and } \beta \leq 1$$

and incorporating the following initial condition:

$$u(x, 0) = h(x) \qquad (6.7.21)$$

In accordance with the HPM, the following Homotopy formula for Eq. (6.7.20) has been developed:

$$(1 - p)D_t^\alpha u + p\left(\varepsilon u D_x^\beta u + L_x u\right) = 0 \qquad (6.7.22)$$

The approximate solution is then expanded by using the Homotopy parameter p:

$$u = u_0 + pu_1 + p^2 u_2 + p^3 u_3 + \cdots \qquad (6.7.23)$$

Substituting Eq. (6.7.23) into Eq. (6.7.22) and separating the terms with same powers, one could have:

$$\begin{aligned} &D_t^\alpha u_0 = 0 \\ &D_t^\alpha u_1 + \varepsilon u_0 D_x^\beta = 0 \\ &D_t^\alpha u_2 + \varepsilon u_0 D_x^\beta u_1 + \varepsilon u_1 D_x^\beta u_0 + L_x u_1 = 0 \\ &\text{and} \\ &D_t^\alpha u_3 + \varepsilon u_0 D_x^\beta u_2 + \varepsilon u_1 D_x^\beta u_1 + \varepsilon u_2 D_x^\beta u_0 + L_x u_2 = 0 \end{aligned} \qquad (6.7.24)$$

Considering the definition for the fractional operator, the approximate solution of Eq. (6.7.20) is obtained:

$$\begin{aligned} u_0 &= h_0(x) \\ u_1 &= h_1(x) - h^1(x) \frac{t^\alpha}{\Gamma(\alpha + 1)} \\ u_2 &= h_2(x) - h^2(x) \frac{t^\alpha}{\Gamma(\alpha + 1)} + h^3(x) \frac{t^{2\alpha}}{\Gamma(2\alpha + 1)} \end{aligned} \qquad (6.7.25)$$

in which $h_i(x) = u(x, 0)$, and $\sum_{i=0}^{\infty} h_i(x) = h(x)$[139]:

$$h^1 = \varepsilon f_0 D_x^\beta h_0 + \eta h_{0xx} + v h_{0xx} \qquad (6.7.26)$$

$$h^2 = \varepsilon \left[f_0 D_x^\beta h_1 + h_1 D_x^\beta h_0\right] + \eta h_{1xx} + v h_{1xxx} \qquad (6.7.27)$$

and

$$h^3 = \varepsilon\left[h_0 D_x^{\beta} h_1 + h_1 D_x^{\beta} h_0\right] + \eta h_{xx}^1 + \nu h_{xxx}^1 \tag{6.7.28}$$

Part III: Practice Problems

P.6.1. Dynamics of spreading rumor with consideration of forgetting mechanism in social networks were recently modelled by Zhao et al. [140]. Figure 6.1 shows this system schematically.

According to their work, the following model was developed for the considered dynamical system:

$$\begin{aligned}
\frac{dI(t)}{dt} &= -\lambda\bar{k}I(t)S(t) \\
\frac{dI(t)}{dt} &= \lambda\bar{k}I(t)S(t) - \alpha\bar{k}S(t)(S(t)+R(t)) - \delta S(t) \\
\frac{dR(t)}{dt} &= \alpha\bar{k}S(t)(S(t)+R(t)) + \delta S(t)
\end{aligned}$$

with the initial conditions given as:

$$I(0) = \frac{N-1}{N}; \quad S(0) = \frac{1}{N}; \quad \text{and} \quad R(0) = 0.$$

Functions $I(t), S(t)$ and $R(t)$ denote the density of ignorant, spreaders, and the stifles, respectively. Also, $\bar{k}, \lambda, \delta$ and α are the average degree of social networks, spreading, forgetting, and the stifling rates, respectively.

P.6.1.a. Use the VIM, ADM and HAM to find analytical solutions for the above dynamic systems.

P.6.1.b. Identify the most powerful method among the VIM, ADM and HAM to solve the rumor mathematical model.

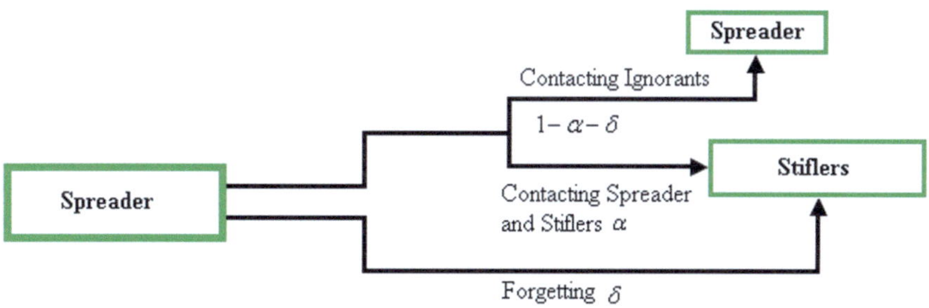

Fig. 6.1 Schematic of rumor spreading procedure

Project:

- Consider a real Facebook group and investigate rumor spreading in it, using the developed mathematical model.

P.6.2. The mathematical model of a food chain eco-epidemic with infection is modelled by the following coupled differential equations, adopted from the referenced article [141]:

$$\begin{aligned}
\frac{dW}{dt} &= -mW + pVW \\
\frac{dV}{dt} &= -lV + eSV - hVW + qIV \\
\frac{dI}{dt} &= \beta IS - nIV - \gamma I - vI \\
\frac{dS}{dt} &= aS\left(1 - \frac{S+I}{K}\right) - cVS - \beta SI + \gamma I
\end{aligned}$$

where W and V represent a top predator and the intermediate population, respectively.

Parameters N, S and I describe the bottom prey affected by an epidemic, and the two subpopulation of the population of the bottom prey, which are called susceptible and infected, respectively.

The population of the bottom prey is subdivided in two subpopulations of susceptible S and infected I. Definition of other parameters can be found in [141].

P.6.2.a. Utilize the VIM, LADM and HAM to obtain the analytical solutions for the assumed mathematical model.
P.6.2.b. Discuss and compare the convergence of the solutions, which have been obtained from VIM, LADM and HAM.

Project:

Use Java to write a code for finding nth iteration of the solutions obtained from VIM, LADM and HAM.

P.6.3. The one-dimensional tsunami run up over a shore has been mathematically modelled by the following coupled partial differential equation [61]:

$$\begin{aligned}
u_t + uu_x + g\eta_x &= 0 \\
\eta_t + (u(\eta + H))_x &= 0
\end{aligned}$$

where, $\eta(x, t)$ and $u(x, t)$ represent the respective surface elevation and the velocity time-dependent profile. $H(x)$ shows the variable depth and g stands for the acceleration due to gravity, all adopted from Ref. [61].

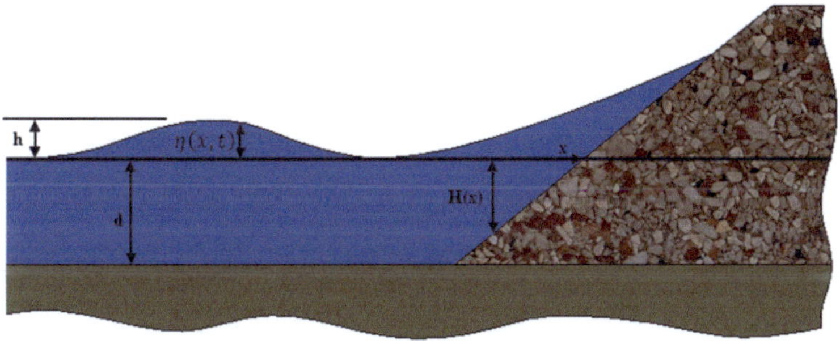

Fig. 6.2 The one-dimensional tsunami run-up on a shore. Adopted with permission from Ref. [61]

Figure 6.2 illustrates the one-dimensional tsunami run-up on a shore.
The initial condition can be considered as:

$$\eta(x,0) = h.\mathrm{sech}^2\sqrt{\frac{3h}{4d^3}}x \quad \text{and} \quad u(x,0) = \eta\sqrt{\frac{g}{d}}$$

where d and h denote the stationary elevation and the initial wave height, respectively.

P.6.3.a. Find the third approximate solution by employing VIM.
P.6.3.b. Investigate the influence of the variable depth on the behavior of the wave.

Project:

- Optimize the variable depth to reduce the velocity and the wave height, based on the obtained solutions from VIM.

P.6.4. Free vibration of a plate cavity system can mathematically be presented by the following partial differential equation:

$$\begin{aligned} \mathrm{D}\nabla^4 W(x,y,t) + \rho\frac{\partial^2 w}{\partial t^2} &= P_E(x,y,t) - P_i(x,y,t) \\ &\quad + h\left(\frac{\partial^2\varphi}{\partial y^2}\frac{\partial^2 W}{\partial x^2} + \frac{\partial^2\varphi}{\partial x^2}\frac{\partial^2 W}{\partial y^2} - 2\frac{\partial^2\varphi}{\partial x\partial y}\frac{\partial^2 W}{\partial x\partial y}\right) \end{aligned}$$

and

$$\nabla^4\phi = E\left[\left(\frac{\partial^2 W}{\partial x\partial y}\right)^2 - \frac{\partial^2 W}{\partial x^2}\frac{\partial^2 W}{\partial y^2}\right]$$

where $W(x,y,t)$ and ϕ show the plate displacement and the airy stress function, respectively.

Other parameters ρ, D and E represent the respective plate density, bending stiffness, and the elasticity modulus of plate. $P_E(x,y,t)$ and $P_i(x,y,t)$ are the respective external excitation and acoustic pressure of the air cavity at $z=-c$.

P.6.4.a. Assume the plate boundary conditions as SSSS (simply-supported on all four edges), and use Galerkin procedure to find the first two modes of vibration of the system.
P.6.4.b. Implement HAM to find the analytical solution for the differential equation obtained, and compare the results with those obtained from numerical simulation.
P.6.4.c. Determine the natural frequency and the mode shape of the system.
P.6.4.d. Assume that the plate boundary conditions are SSCC (simply-supported on two edges and clamped-clamped on the other two). Use the Galerkin procedure to find the differential equations and apply HAM to find the natural frequency and the approximate solutions of the system.

Project:

- Use Java program to write a code to obtain the 10th iteration of solutions from the HAM for the two different boundary conditions SSSS and SSCC.
- Show the influence of the boundary conditions on the natural frequency of the system.

P.6.5. Internet is quickly popularized, and accordingly, computers are rapidly connecting to the Internet. Thus, it is so crucial to provide an anti-virus to protect the computers, but in fact, no anti-virus software can be found, which has the capability of detecting, fighting and finally removing all sorts of computer viruses. For a better understanding of virus propagation in computers, few mathematical models have been developed. As one example, Yang et al. developed the following model for the epidemic model of virus propagation in computers [142].

$$\begin{aligned}
\dot{S} &= b_1 + \gamma_2 L + \gamma_2 B + \eta R - \gamma_1 S - \mu S - \beta(L+B)S \\
\dot{L} &= b_2 + \beta(L+B)S - \gamma_1 L - \gamma_2 L - \mu L - \alpha L \\
\dot{B} &= b_3 + \alpha L - \gamma_1 B - \gamma_2 B - \mu B \\
\dot{R} &= b_4 + \gamma_1 L + \gamma_1 B - \mu R - \eta R
\end{aligned}$$

where S, L, B and R represent the average numbers of susceptible, latent, breaking and recovered computers, respectively, and their respective entering rates are b_1, b_2, b_3 and b_4. Coefficients $\alpha, \beta, \gamma_1, \mu, \eta$ and γ_2 are the breaking rate, incidence rate, recovered rate, rate at which a computer leaves the internet, rate of a recovered computer losing immunity, and the rate of an infected computer reinstalling the operating system, respectively, all been adopted from Ref. [142].

P.6.5.a. Employ VIM, HAM and ADM to find the approximate solutions for the objective system (Third approximations).
P.6.5.b. Compare the results obtained with those obtained from numerical simulation.

P.6.6. Very recently, R.A. van Gorder demonstrated that the VIM (and all its optimal analogous) are specific cases of the more general form of HAM. Confirm that VIM is a specific case of HAM. Provide a real-world example and then show that HAM and VIM reach the analogous result for that, adopted from the article [78].

P.6.6.a. The unsteady flow and the heat transfer in a two-dimensional squeezing Nano fluid between the two s by the following set of equations:

$$\frac{\partial u}{\partial x} + \frac{\partial v}{\partial y} = 0$$

$$\rho_{nf}\left(\frac{\partial u}{\partial t} + u\frac{\partial u}{\partial x} + v\frac{\partial u}{\partial y}\right) = -\frac{\partial p}{\partial x} + \mu_{nf}\left(\frac{\partial^2 u}{\partial x^2} + \frac{\partial^2 u}{\partial y^2}\right)$$

$$\rho_{nf}\left(\frac{\partial u}{\partial t} + u\frac{\partial u}{\partial x} + v\frac{\partial u}{\partial y}\right) = -\frac{\partial p}{\partial y} + \mu_{nf}\left(\frac{\partial^2 u}{\partial x^2} + \frac{\partial^2 u}{\partial y^2}\right)$$

$$\left(\frac{\partial T}{\partial t} + u\frac{\partial T}{\partial x} + v\frac{\partial T}{\partial y}\right) = -\frac{k_{nf}}{(\rho C_p)_{nf}}\left(\frac{\partial^2 T}{\partial x^2} + \frac{\partial^2 T}{\partial y^2}\right) + \frac{\mu_{nf}}{(\rho C_p)_{nf}}\left(4\left(\frac{\partial u}{\partial x}\right)^2 + \left(\frac{\partial u}{\partial x} + \frac{\partial u}{\partial y}\right)^2\right)$$

The parameters considered in the set of partial differential equation are found in Ref. [143].

P.6.6.b. Employ the HAM to find the approximate solution for that.
P.6.6.c. Compare the result obtained from HAM with those obtained from ADM and VIM.

P.6.7. Use VIM, ADM and HAM to solve the following nonlinear ordinary differential equations and compare the results obtained with those found from the numerical simulation.

P.6.7.a. $\ddot{x} + \alpha x + \beta x^3 + \lambda x^5 = 0$
P.6.7.b. $\ddot{x} + x^{\frac{1}{3}} = 0$
P.6.7.c. $\ddot{x} + \frac{x}{x^2+1} = 0$
P.6.7.d. $\ddot{x} + \frac{x^3}{x^2+1} = 0$
P.6.7.e. $\ddot{x} + \frac{1}{x} = 0$
P.6.7.f. $\ddot{x} + \alpha x + \beta x^2 + \lambda x^3 = 0$
P.6.7.g. $\ddot{x} + x|x| = 0$

P.6.8. Employ the HAM for solving the Hirota equation as given below:

$$u_t + iau + ib(u_{xx} - 2\eta|u^2|u) + cu_x + d(u_{xxx} - 6\eta|u|^2) = 0$$

P.6.9. Consider a tapered beam which has the following governing equation:

$$\ddot{u} + \alpha u + \alpha_1 u^3 + \beta[u^2\ddot{u} + u\dot{u}^2] = 0$$

P.6.9.a. Employ VIM and HAM to find the third approximate solutions for the above equation.
P.6.9.b. Apply the OVIM and OHAM to find the third approximate solutions for the above equation.
P.6.9.c. Compare the results obtained from part (a) and part (b) and present the most powerful method among them for solving the equation.

P.6.10. Consider the following Mathieu equation:

$$\ddot{u} + (\omega^2 + 2\varepsilon\cos(2\tau))u = 0$$

P.6.10.a. Combine the VIM with the 'strained parameters expansion' method to identify the stable region from the unstable region.
P.6.10.b. Obtain the analytical solution for the Mathieu equation by employing the VIM.
P.6.10.c. Compare the results obtained from VIM with those obtained from numerical simulation.

P.6.11. Consider the following coupled nonlinear ordinary differential equations:

$$\ddot{x} + \omega_1^2 x = -\alpha_1 x^3 - \alpha_3 x y^2$$
$$\ddot{y} + \omega_2^2 y = -\alpha_2 y^3 - \alpha_4 x^2 y$$

in which $\alpha_1, \alpha_2, \alpha_3, \alpha_4$ and ω_1, ω_2 are constant parameters, and assume that $\alpha_1 = \alpha_2, \alpha_3 = \alpha_4$.

P.6.11.a. Utilize VIM to solve the above equation. (third approximation)
P.6.11.b. Employ ADM for finding the third approximation.
P.6.11.c. Compare results obtained from VIM and ADM with those found from numerical simulation.

P.6.12. The autocatalytic hydrolysis of cellulose is mathematically formulated by the following set of differential equations, adopted from the article referenced [144]:

$$\frac{\mathrm{d}C}{\mathrm{d}t} = kHC$$
$$\frac{\mathrm{d}P}{\mathrm{d}t} = kHC - k_{\alpha}P$$
$$\frac{\mathrm{d}H}{\mathrm{d}t} - \alpha k_{\alpha}P$$

with the initial conditions of $C(0) = C_0; P(0) = 0; \text{and}\, H(0) = H_0$.

Parameters C, H, P, k, k_{α} and α represent the starting cellulose chains, acid catalyst agent, degraded cellulose chains, second-order rate constant of the main reaction step, first-order rate constant of the reaction step, and the excess produced acidity, respectively.

Determine a series solution for the autocatalytic hydrolysis of cellulose using the VIM, HAM and DTM [144].

P.6.13. Many scientists worked on developing a mathematical model for the 'Love and Hate'. The following differential equation represent a mathematical model for the 'love and hate' of Layla and Majnun Model [145]:

$$\frac{\mathrm{d}M}{\mathrm{d}t} = a + L^2 + cM$$
$$\frac{\mathrm{d}L}{\mathrm{d}t} = b + M^2 + dL$$

where M is the love of Majnun for Layla, and L is the love of Layla for Majnun (or hate if negative). Other parameters a, b, c and d are the constants that determine the romantic styles.

P.6.13.a. Find an analytical solution for the above model using the VIM and HAM.
P.6.13.b. Consider that Majnun has another mistress, accordingly, find the mathematical model of the "love triangle".
P.6.13.c. Develop an analytical solution for the "love triangle", and then compare the behavior of Layla in the case of "love triangle" with those of the first case.

References

1. Adomian G (1988) A review of the decomposition method in applied mathematics. J Math Anal Appl 135(2):501–544
2. He JH (1999) Variational iteration method—a kind of non-linear analytical technique: some examples. Int J Non-Linear Mech 34(4):699–708
3. Liao SJ (1992) The proposed homotopy analysis technique for the solution of nonlinear problems. Shanghai Jiao Tong University
4. Adomian G (1970) Random operator equations in mathematical physics. I. J Math Phys 11 (3):1069–1084

5. Abbaoui K, Cherruault Y (1995) New ideas for proving convergence of decomposition methods. Comput Math Appl 29(7):103–108
6. Abbaoui K, Cherruault Y (1994) Convergence of Adomian's method applied to nonlinear equations. Math Comput Model 20(9):69–73
7. Abbasbandy S (2003) Improving Newton-Raphson method for nonlinear equations by modified Adomian decomposition method. Appl Math Comput 145(2–3):887–893
8. Dehghan M, Hamidi A, Shakourifar M (2007) The solution of coupled Burgers' equations using Adomian-Pade technique. Appl Math Comput 189(2):1034–1047
9. Bellomo N, Sarafyan D (1987) On Adomian's decomposition method and some comparisons with Picard's iterative scheme. J Math Anal Appl 123(2):389–400
10. Babolian E, Biazar J (2002) On the order of convergence of Adomian method. Appl Math Comput 130(2–3):383–387
11. Ji J, Zhang J, Dong Y (2012) The fractional variational iteration method improved with the Adomian series. Appl Math Lett 25(12):2223–2226
12. Wazwaz A-M, Gorguis A (2004) An analytic study of Fisher's equation by using Adomian decomposition method. Appl Math Comput 154(3):609–620
13. Wazwaz AM (2000) A new algorithm for calculating Adomian polynomials for nonlinear operators. Appl Math Comput 111(1):53–69
14. Rach R (1987) On the Adomian (decomposition) method and comparisons with Picard's method. J Math Anal Appl 128(2):480–483
15. He JH (2000) A review on some new recently developed nonlinear analytical techniques. Int J Nonlinear Sci Numer Simul 1(1):51–70
16. He JH, Wu XH (2007) Variational iteration method: new development and applications. Comput Math Appl 54(7–8):881–894
17. He JH, Austin GCWF (2010) The variational iteration method which should be followed. Nonlinear Sci Lett A 1(1):1–30
18. Rafei M, Daniali H, Ganji DD (2007) Variational iteration method for solving the epidemic model and the prey and predator problem. Appl Math Comput 186(2):1701–1709
19. He J-H (2011) A short remark on fractional variational iteration method. Phys Lett A 375 (38):3362–3364
20. Batiha B, Noorani MSM, Hashim I (2007) Application of variational iteration method to heat- and wave-like equations. Phys Lett A 369(1–2):55–61
21. Assas LMB (2007) Approximate solutions for the generalized KdV-Burgers' equation by He's variational iteration method. Phys Scr 76(2):161–164
22. Liao S (2004) On the homotopy analysis method for nonlinear problems. Appl Math Comput 147(2):499–513
23. Liao S-J (1995) An approximate solution technique not depending on small parameters: A special example. Int J Non-Linear Mech 30(3):371–380
24. Liao SJ (1999) A uniformly valid analytic solution of two-dimensional viscous flow over a semi-infinite flat plate. J Fluid Mech 385:101–128
25. Liao SJ, Chwang AT (1998) Application of homotopy analysis method in nonlinear oscillations. J Appl Mech Trans ASME 65(4):914–922
26. Abbasbandy S (2006) The application of homotopy analysis method to nonlinear equations arising in heat transfer. Phys Lett Sect A: Gen At Solid State Phys 360(1):109–113
27. Abbasbandy S (2008) Solitary wave solutions to the Kuramoto-Sivashinsky equation by means of the homotopy analysis method. Nonlinear Dyn 52(1–2):35–40
28. Sajid M, Hayat T (2008) The application of homotopy analysis method to thin film flows of a third order fluid. Chaos, Solitons Fractals 38(2):506–515
29. Xu H, Liao SJ, You XC (2009) Analysis of nonlinear fractional partial differential equations with the homotopy analysis method. Commun Nonlinear Sci Numer Simul 14(4):1152–1156
30. Motsa SS, Sibanda P, Shateyi S (2010) A new spectral-homotopy analysis method for solving a nonlinear second order BVP. Commun Nonlinear Sci Numer Simul 15(9):2293–2302

31. Bataineh AS, Noorani MSM, Hashim I (2008) The homotopy analysis method for Cauchy reaction–diffusion problems. Phys Lett A 372(5):613–618
32. Wazwaz AM (1999) A reliable modification of Adomian decomposition method. Appl Math Comput 102(1):77–86
33. Rèpaci A (1990) Nonlinear dynamical systems: On the accuracy of adomian's decomposition method. Appl Math Lett 3(4):35–39
34. Wazwaz AM (2001) A new algorithm for solving differential equations of Lane-Emden type. Appl Math Comput 118(2–3):287–310
35. Wazwaz AM (1999) Analytical approximations and padé approximants for volterra's population model. Appl Math Comput 100(1):13–25
36. Shawagfeh NT (2002) Analytical approximate solutions for nonlinear fractional differential equations. Appl Math Comput 131(2–3):517–529
37. Wazwaz AM (1999) The modified decomposition method and Padé approximants for solving the Thomas-Fermi equation. Appl Math Comput 105(1):11–19
38. Kaya D, El-Sayed SM (2003) An application of the decomposition method for the generalized KdV and RLW equations. Chaos, Solitons Fractals 17(5):869–877
39. Abbasbandy S (2007) A new application of He's variational iteration method for quadratic Riccati differential equation by using Adomian's polynomials. J Comput Appl Math 207 (1):59–63
40. Wazwaz A-M (2005) Adomian decomposition method for a reliable treatment of the Bratu-type equations. Appl Math Comput 166(3):652–663
41. El-Sayed SM (2003) The decomposition method for studying the Klein-Gordon equation. Chaos, Solitons Fractals 18(5):1025–1030
42. Rida SZ, El-Sherbiny HM, Arafa AAM (2008) On the solution of the fractional nonlinear Schrödinger equation. Phys Lett A 372(5):553–558
43. Babolian E, Biazar J, Vahidi AR (2004) The decomposition method applied to systems of Fredholm integral equations of the second kind. Appl Math Comput 148(2):443–452
44. Cherruault Y, Adomian G (1993) Decomposition methods: a new proof of convergence. Math Comput Model 18(12):103–106
45. Wazwaz AM (2007) A comparison between the variational iteration method and Adomian decomposition method. J Comput Appl Math 207(1):129–136
46. Kohn W (1949) A variational iteration method for solving secular equations. J Chem Phys 17(7):670
47. He J (1997) A new approach to nonlinear partial differential equations. Commun Nonlinear Sci Numer Simul 2(4):230–235
48. He J (1998) Approximate analytical solution of Blasius' equation. Commun Nonlinear Sci Numer Simul 3(4):260–263
49. Dehghan M, Tatari M (2006) The use of He's variational iteration method for solving a Fokker-Planck equation. Phys Scr 74(3):310–316
50. Abdou MA, Soliman AA (2005) Variational iteration method for solving Burger's and coupled Burger's equations. J Comput Appl Math 181(2):245–251
51. Sweilam NH (2007) Variational iteration method for solving cubic nonlinear Schrödinger equation. J Comput Appl Math 207(1):155–163
52. Wazwaz AM (2007) The variational iteration method for exact solutions of Laplace equation. Phys Lett Sect A: Gen At Solid State Phys 363(4):260–262
53. Abassy TA, El-Tawil MA, El-Zoheiry H (2007) Modified variational iteration method for Boussinesq equation. Comput Math Appl 54(7–8):955–965
54. He J-H (1998) Approximate analytical solution for seepage flow with fractional derivatives in porous media. Comput Methods Appl Mech Eng 167(1–2):57–68
55. He J-H (2000) Exact resonances of nonlinear vibration of rotor-bearings system without small parameter. Mech Res Commun 27(4):451–456
56. Momani S, Odibat Z (2006) Analytical approach to linear fractional partial differential equations arising in fluid mechanics. Phys Lett A 355(4–5):271–279

57. Drăgănescu GE (2006) Application of a variational iteration method to linear and nonlinear viscoelastic models with fractional derivatives. J Math Phys 47(8):082902
58. Askari H, Esmailzadeh E (2014) Chaotic and periodic vibration of a carbon nanotube supported by nonlinear foundation. In: 14th IEEE International conference on nanotechnology, IEEE-NANO 2014
59. Sadri M, Younesian D (2014) Nonlinear free vibration analysis of a plate-cavity system. Thin-Walled Struct 74:191–200
60. Younesian D, Saadatnia Z, Askari H (2012) Analytical solutions for free oscillations of beams on nonlinear elastic foundations using the variational iteration method. J Theor Appl Mech 50(2):639–652
61. Younesian D et al (2012) Analytical solution for nonlinear wave propagation in shallow media using the variational iteration method. Waves Random Complex Media 22(2):133–142
62. Ramos JI (2008) On the variational iteration method and other iterative techniques for nonlinear differential equations. Appl Math Comput 199(1):39–69
63. Ramos JI (2008) On the Picard-Lindelof method for nonlinear second-order differential equations. Appl Math Comput 203(1):238–242
64. Odibat ZM, Momani S (2006) Application of variational iteration method to nonlinear differential equations of fractional order. Int J Nonlinear Sci Numer Simul 7(1):27–34
65. Hosseini MM et al (2010) Auxiliary parameter in the variational iteration method and its optimal determination. Int J Nonlinear Sci Numer Simul 11(7):495–502
66. He J-H (2012) Notes on the optimal variational iteration method. Appl Math Lett 25 (10):1579–1581
67. Turkyilmazoglu M (2011) An optimal variational iteration method. Appl Math Lett 24 (5):762–765
68. Wu GC, Lee EWM (2010) Fractional variational iteration method and its application. Phys Lett Sect A: Gen At Solid State Phys 374(25):2506–2509
69. Hilal EMA, Elzaki TM (2014) Solution of nonlinear partial differential equations by new laplace variational iteration method. J Func Spaces
70. He J-H (1999) Homotopy perturbation technique. Comput Methods Appl Mech Eng 178(3–4):257–262
71. Noor MA, Mohyud-Din ST (2008) Variational homotopy perturbation method for solving higher dimensional initial boundary value problems. Math Prob Eng
72. Liao S-J (1997) A kind of approximate solution technique which does not depend upon small parameters—II. An application in fluid mechanics. Int J Non-Linear Mech 32(5):815–822
73. Liao SJ (1999) An explicit, totally analytic approximate solution for Blasius' viscous flow problems. Int J Non-Linear Mech 34(4):759–778
74. Wang C, Pop I (2006) Analysis of the flow of a power-law fluid film on an unsteady stretching surface by means of homotopy analysis method. J Nonnewton Fluid Mech 138(2–3):161–172
75. Liao SJ, Cheung KF (2003) Homotopy analysis of nonlinear progressive waves in deep water. J Eng Math 45(2):105–116
76. Abbasbandy S, Shirzadi A (2011) A new application of the homotopy analysis method: Solving the Sturm-Liouville problems. Commun Nonlinear Sci Numer Simul 16(1):112–126
77. Sami Bataineh A, Noorani MSM, Hashim I (2007) Solutions of time-dependent Emden-fowler type equations by homotopy analysis method. Phys Lett Sect A: Gen At Solid State Phys 371(1–2):72–82
78. Van Gorder RA, Vajravelu K (2008) Analytic and numerical solutions to the Lane-Emden equation. Phys Lett Sect A: Gen At Solid State Phys 372(39):6060–6065
79. Tan Y, Xu H, Liao SJ (2007) Explicit series solution of travelling waves with a front of Fisher equation. Chaos, Solitons Fractals 31(2):462–472

80. Abbasbandy S, Babolian E, Ashtiani M (2009) Numerical solution of the generalized Zakharov equation by homotopy analysis method. Commun Nonlinear Sci Numer Simul 14 (12):4114–4121
81. Hayat T, Sajid M (2007) On analytic solution for thin film flow of a fourth grade fluid down a vertical cylinder. Phys Lett Sect A: Gen At Solid State Phys 361(4–5):316–322
82. Hayat T, Abbas Z, Sajid M (2007) On the analytic solution of magnetohydrodynamic flow of a second grade fluid over a shrinking sheet. J Appl Mech Trans ASME 74(6):1165–1171
83. Nadeem S, Hussain A, Khan M (2010) HAM solutions for boundary layer flow in the region of the stagnation point towards a stretching sheet. Commun Nonlinear Sci Numer Simul 15 (3):475–481
84. Mustafa M et al (2012) Influence of wall properties on the peristaltic flow of a nanofluid: Analytic and numerical solutions. Int J Heat Mass Transf 55(17–18):4871–4877
85. Marinca V et al (2009) An optimal homotopy asymptotic method applied to the steady flow of a fourth-grade fluid past a porous plate. Appl Math Lett 22(2):245–251
86. Wazwaz AM (2011) Linear and nonlinear integral equations: methods and applications. Higher Education Press
87. Volterra V (1912) Sur les équations intégro-différentielles et leurs applications. Acta Math 35(1):295–356
88. Brunner H (1990) On the numerical solution of nonlinear Volterra-Fredholm integral equations by collocation methods. SIAM J Numer Anal 27(4):987–1000
89. Brunner H, Hu Q, Lin Q (2001) Geometric meshes in collocation methods for Volterra integral equations with proportional delays. IMA J Numer Anal 21(4):783–798
90. Maleknejad K, Hadizadeh M (1999) New computational method for Volterra-Fredholm integral equations. Comput Math Appl 37(9):1–8
91. Maleknejad K, Mahmoudi Y (2003) Taylor polynomial solution of high-order nonlinear Volterra-Fredholm integro-differential equations. Appl Math Comput 145(2–3):641–653
92. Crisci MR et al (1995) Stability of continuous and discrete volterra integro-differential equations by liapunov approach. J Integr Equat Appl 7(4):393–411
93. Wazwaz A-M (1999) Analytical approximations and Padé approximants for Volterra's population model. Appl Math Comput 100(1):13–25
94. Dafermos CM (1970) An abstract Volterra equation with applications to linear viscoelasticity. J Diff Equat 7(3):554–569
95. Song Y, Baker CTH (2004) Perturbation of Volterra difference equations. J Diff Equat Appl 10(4):379–397
96. Liu B, Zhang Y, Chen L (2005) The dynamical behaviors of a Lotka-Volterra predator-prey model concerning integrated pest management. Nonlinear Anal Real World Appl 6(2):227–243
97. Wazwaz AM (2010) The combined Laplace transform-Adomian decomposition method for handling nonlinear Volterra integro-differential equations. Appl Math Comput 216(4):1304–1309
98. Yousefi SA, Lotfi A, Dehghan M (2009) He's variational iteration method for solving nonlinear mixed Volterra-Fredholm integral equations. Comput Math Appl 58(11–12):2172–2176
99. Simos TE, Williams PS (1999) On finite difference methods for the solution of the Schrödinger equation. Comput Chem 23(6):513–554
100. Biswas A, Khalique CM (2011) Stationary solutions for nonlinear dispersive Schrödinger's equation. Nonlinear Dyn 63(4):623–626
101. Biswas A et al (2010) An exact solution for the modified nonlinear Schrödinger's equation for Davydov Solitons in α-helix proteins. Math Biosci 227(1):68–71
102. Feit MD, Fleck JA Jr, Steiger A (1982) Solution of the Schrödinger equation by a spectral method. J Comput Phys 47(3):412–433
103. Kosloff D, Kosloff R (1983) A fourier method solution for the time dependent Schrödinger equation as a tool in molecular dynamics. J Comput Phys 52(1):35–53

104. Kevrekidis PG, Rasmussen KØ, Bishop AR (2001) The discrete nonlinear schrödinger equation: A survey of recent results. Int J Mod Phys B 15(21):2833–2900
105. Tian B et al (2005) Transformations for a generalized variable-coefficient nonlinear Schrödinger model from plasma physics, arterial mechanics and optical fibers with symbolic computation. Europ Phys J B 47(3):329–332
106. Hayashi N, Naumkin PI (1998) Asymptotics for large time of solutions to the nonlinear Schrödinger and Hartree equations. American Journal of Mathematics 120(2):369–389
107. Datta BK (1990) A technique for approximate solutions to schrödinger-like equations. Comput Math Appl 20(1):61–65
108. Adomian G, Rach R (1991) Linear and nonlinear Schrödinger equations. Found Phys 21 (8):983–991
109. Alomari AK, Noorani MSM, Nazar R (2009) Explicit series solutions of some linear and nonlinear Schrodinger equations via the homotopy analysis method. Commun Nonlinear Sci Numer Simul 14(4):1196–1207
110. Adomian G, Meyers RE (1995) Generalized nonlinear Schrödinger equation with time-dependent dissipation. Appl Math Lett 8(6):7–8
111. Ginoux J, Letellier C (2012) Van der Pol and the history of relaxation oscillations: toward the emergence of a concept. Chaos 22(2)
112. Cartwright ML, Littlewood JE (1945) On non-linear differential equations of the second order: I. The Equation $-k(1-y2)+y=b\lambda k\cos(\lambda l+\alpha)$, *k Large*. J Lond Math Soc s1–20 (3):180–189
113. Mahboob I et al (2015) Dispersive and dissipative coupling in a micromechanical resonator embedded with a nanomechanical resonator. Nano Lett 15(4):2312–2317
114. Foutse M et al (2015) Edge-emitting semiconductor laser driven by a van der Pol oscillator: analytical and numerical analysis. Opt Quant Electron 47(3):705–720
115. Le T-H, Caracoglia L (2015) Reduced-order wavelet-Galerkin solution for the coupled, nonlinear stochastic response of slender buildings in transient winds. J Sound Vib 344:179–208
116. Zhang J, Tang Y (2015) Fatigue analysis of deep-water risers under vortex-induced vibration considering parametric excitations. J Coast Res 652–659
117. Suchorsky MK, Rand RH (2012) A pair of van der Pol oscillators coupled by fractional derivatives. Nonlinear Dyn 69(1–2):313–324
118. Suchorsky MK, Sah SM, Rand RH (2010) Using delay to quench undesirable vibrations. Nonlinear Dyn 62(1–2):407–416
119. Grudziński K, Żebrowski JJ (2004) Modeling cardiac pacemakers with relaxation oscillators. Phys A 336(1–2):153–162
120. Lopes I, Passos D (2009) Evidence for a long-term variation of the dynamo action responsible for the solar magnetic cycle. Mon Not R Astron Soc 397(1):320–324
121. Koštrun M, Smith WW, Javanainen J (1998) Theory of the $$\pi${}$-periodic motion of two ions in a Paul trap. Phys Rev A 57(4):2895–2904
122. Huang YJ, Liu HK (2013) A new modification of the variational iteration method for van der Pol equations. Appl Math Model 37(16–17):8118–8130
123. Ramana PV, Raghu BK (2014) Prasad, modified Adomian decomposition method for Van der Pol equations. Int J Non-Linear Mech 65:121–132
124. Zerroukat M, Allen T (2015) A moist Boussinesq shallow water equations set for testing atmospheric models. J Comput Phys 290:55–72
125. Ha T, Cho YS (2015) Tsunami propagation over varying water depths. Ocean Eng 101:67–77
126. Chang C-H, Wang K-H (2015) Numerical study on three-dimensional waves produced by a bottom jet. Appl Ocean Res 50:141–154
127. Mehrvarzi CO, Paul MR (2014) Front propagation in a chaotic flow field. Phys Rev E—Stat Nonlinear Soft Matt Phys 90(1)
128. Chen F, Liu QQ (2014) Modified asymptotic Adomian decomposition method for solving Boussinesq equation of groundwater flow. App Math Mech (Engl Ed) 35(4):481–488

129. Yao Q, Pan X, Zhu Q (2014) Numerical simulation of the hydrogen dispersion behavior by a parallel characteristic curve method. Abst Appl Anal
130. Korteweg DJ, de Vries G (1895) XLI. On the change of form of long waves advancing in a rectangular canal, and on a new type of long stationary waves. Philos Mag Ser 5 39 (240):422–443
131. Carstea AS, Tokihiro T (2015) Coupled discrete KdV equations and modular genetic networks. J Phys A: Math Theor 48(5):055205
132. Zhuang K, Du Z, Lin X (2015) Solitary waves solutions of singularly perturbed higher-order KdV equation via geometric singular perturbation method. Nonlinear Dyn 80(1–2):629–635
133. Colorado E (2014) Existence of bound and ground states for a system of coupled nonlinear Schrödinger–KdV equations. Comptes Rendus Mathematique
134. El-Ajou A, Arqub OA, Momani S (2015) Approximate analytical solution of the nonlinear fractional KdV–Burgers equation: a new iterative algorithm. J Comput Phys 293:81–95
135. Razborova P, Kara A, Biswas A (2015) Additional conservation laws for Rosenau–KdV–RLW equation with power law nonlinearity by Lie symmetry. Nonlinear Dyn 79(1):743–748
136. Biswas A et al (2013) Topological soliton and other exact solutions to KdV–Caudrey–Dodd–Gibbon equation. RM 63(1–2):687–703
137. Assas LMB (2008) Variational iteration method for solving coupled-KdV equations. Chaos, Solitons Fractals 38(4):1225–1228
138. Abdou MA, Soliman AA (2005) New applications of variational iteration method. Physica D 211(1–2):1–8
139. Wang Q (2008) Homotopy perturbation method for fractional KdV-Burgers equation. Chaos, Solitons Fractals 35(5):843–850
140. Zhao L et al (2013) Rumor spreading model considering forgetting and remembering mechanisms in inhomogeneous networks. Phys A 392(4):987–994
141. De Rossi A et al (2015) A food chain ecoepidemic model: infection at the bottom trophic level. Ecol Complex 21:233–245
142. Yang LX, Yang X (2015) The impact of nonlinear infection rate on the spread of computer virus. Nonlinear Dyn 82(1–2):85–95
143. Pourmehran O et al (2015) Analytical investigation of squeezing unsteady nanofluid flow between parallel plates by LSM and CM. Alexandria Eng J 54(1):17–26
144. Chen RX, Liu Z, He JH (2015) Series solution of the autocatalytic hydrolysis of cellulose. Cellulose 22(5):3099–3104
145. Jafari S, Sprott JC, Golpayegani SMRH (2016) Layla and Majnun: a complex love story. Nonlinear Dyn 83(1–2):615–622

Chapter 7
Nonlinearities in Nano- and Microsystems

Abstract Existence of nonlinearities in very small-sized structures is scrutinized in this chapter and shown how nonlinear behavior will occur in nano- and microsystems. The mathematical model of micro–nanosystems, based on Duffing equation, is presented and several examples on the modeling of nanoresonators for mass detection, vibration modeling of protein microtubules, and the nonlinear vibration of carbon nanotube resonators are analyzed. The focus of the next section is on systems having their governing equations of motion comprised of either self-excitation terms or parametrically excited ones. The nonlinearities that appeared in the coupled nonlinear micro- and nanoscaled systems are fully discussed and possibilities of quadratic nonlinearities, reported in nanoscale systems, are illustrated in the last section of this chapter.

Keywords Nano/microsystems · Duffing equation · Resonator
Protein microtubule · Carbon nanotubes · Self-excitation · Quadratic nonlinearity

Introduction

Existence of nonlinearities in very small structures is identified in this chapter. It is shown how the nonlinear behavior emerges in the nano- and microsystems. Nonlinear phenomena are easily detected in micro and nanosystems. Accordingly, it is quite important to understand the nonlinear behavior of these systems. It should be highlighted that the typical nonlinear MEMS and NEMS resonators are characterized by extremely high frequencies, beyond 1 GHz. This is an excellent indication that there are many types of energy present in these systems, and probability for delineating nonlinear behavior. Consequently, the present chapter is categorized in this direction.

Three different miniature-scale systems are studied in the first section and their dynamical behavior is mathematically modeled with the well-known Duffing equation. Dynamical model of nanosystems, which have a wide range of applications in mass detection, is discussed [1–5]. As a second model, the nonlinear vibration of carbon nanotubes having various parameters, such as surface elasticity, temperature, curvature, and the fluid passage is studied. The nanoscale vibration of

E. Esmailzadeh et al., *Analytical Methods in Nonlinear Oscillations*, Solid Mechanics and Its Applications 252, https://doi.org/10.1007/978-94-024-1542-1_7

the protein microtubules, based on the modified strain gradient theory, has also been reviewed [6].

The focus of the second section is on those systems having their governing equations of motion comprised of either self-excitation terms or parametrically excited ones. Accordingly, the vibrations of microbeam actuators and micro-gyroscopes are analyzed in detail. The nonlinearities appeared in the coupled nonlinear micro- and nanoscaled systems are fully discussed in the third section. Different types of nonlinearities are modeled in the governing equations of double-walled carbon nanotubes as the first example and the effects of higher modes of vibration in single-walled carbon nanotubes are discussed in the next one. Finally, brief reviews on the vibration of nanowires and dynamic behavior of micro- and nanoscale systems, using the Timoshenko beam theory, are presented.

The possibility of quadratic nonlinearities present in the nanoscale systems is illustrated in the last section of this chapter. Furthermore, it is shown that the non-natural oscillation phenomenon could be present in a nano–microsystem. Finally, the influence of the Rydberg–Varshni potentials and Casimir force on the governing equation of vibration of a nanoscale system has been investigated.

7.1 Duffing Equation in NEMS and MEMS

7.1.1 Mass Sensors

Nano- and microsystems have many applications in sensing devices [7, 8]. The idea of NEMS and MEMS resonators has been extensively implemented for the biological detection, such as cell [9], virus, protein [10], DNA and enzymatic activity [11]. Indeed, it is vital for medical diagnostics and food supply monitoring to have the capability of detection even small amount of materials, especially pathogenic bacteria [9]. The idea of atomic force microscopy (AFM) has been utilized so far to detect and quantify various viruses such as, Dengue, HIV, Influenza and the Orthopoxvirus [12]. The basic concept behind all those sensing devices is either to have a linear or nonlinear resonator.

As one of the earliest attempts for cell detection is an article published by Ilic et al. [9] in which, a relationship between the number of cells and the resonant frequency shift was presented. Based on the above example, specific binding between the *Escherichia coli* antibody monolayers and the single *E. coli* cells was detected [9] as can be seen in Fig. 7.1.

The properties of the nanocrystalline materials, and also diagnosing diseases, such as the human immunodeficiency viruses (HIVs) were studied by Shaat and Abdelkefi [13] by using nanoresonators to detect a tiny mass. A new distributed parametric model, using the classical beam theory, was developed to simulate the behavior of the considered system. Their proposed model is capable of detecting the HIV particles in a liquid sample [13] as shown in Fig. 7.2.

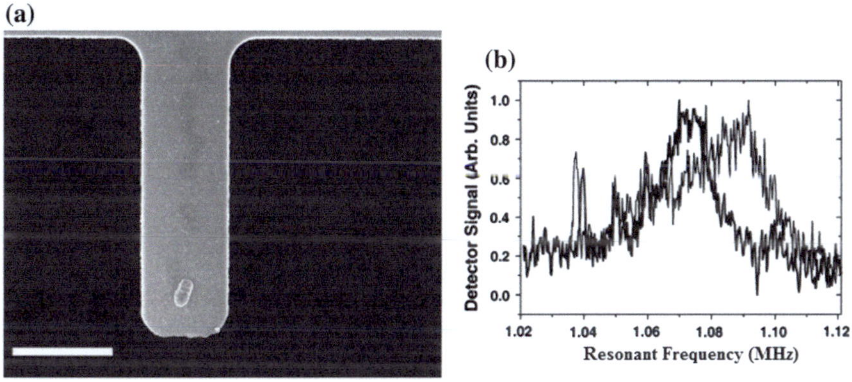

Fig. 7.1 **a** Scanning electron micrograph (SEM) of a single *E. coli* O157: H7 cell bound to the immobilized antibody layer on top of the oscillator. **b** The corresponding transverse vibration spectra of the cantilever due to the thermal effect and ambient noise. *Source* Figures with their captions are adopted with permission from Ref. [9]

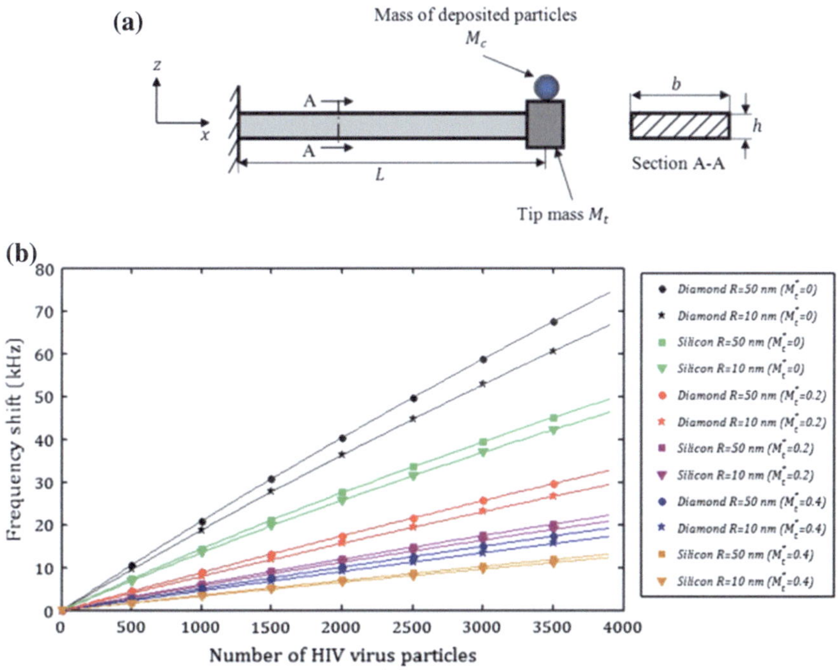

Fig. 7.2 **a** Schematic diagram of a mechanical cantilever resonator used for mass detection, disease diagnosis, and material characterization. **b** Resonant frequency shift as a function of the number of HIV virus particles. *Source* Figures with their captions are adopted with permission from Ref. [13]

A resonating mechanical cantilever was considered by Ilic et al. [9] to identify the immune-specific binding of viruses, captured from a specific liquid. The mass of single-virus particles precipitated on the cantilever can be detected by using this resonator. It was shown that the resonant frequency shift, due to the mass of the virus particles, identify the solutions of virus concentration to be between 105 and 107 pfu/ml. It has been confirmed that the functionalized NEMS are a practical approach to detect any tiny mass [14].

A micro-cantilever resonator is immersed in a liquid and used for the in situ monitoring of proteolysis of peptide chain attributed to protease (Cathepsin B). Resonant frequency measurement for the detection of objective particle provides the required information. The resonant frequency shift was obtained from the proteolysis of peptides, which indeed is the result of the overall mass decrement of the cantilever as illustrated in Fig. 7.3.

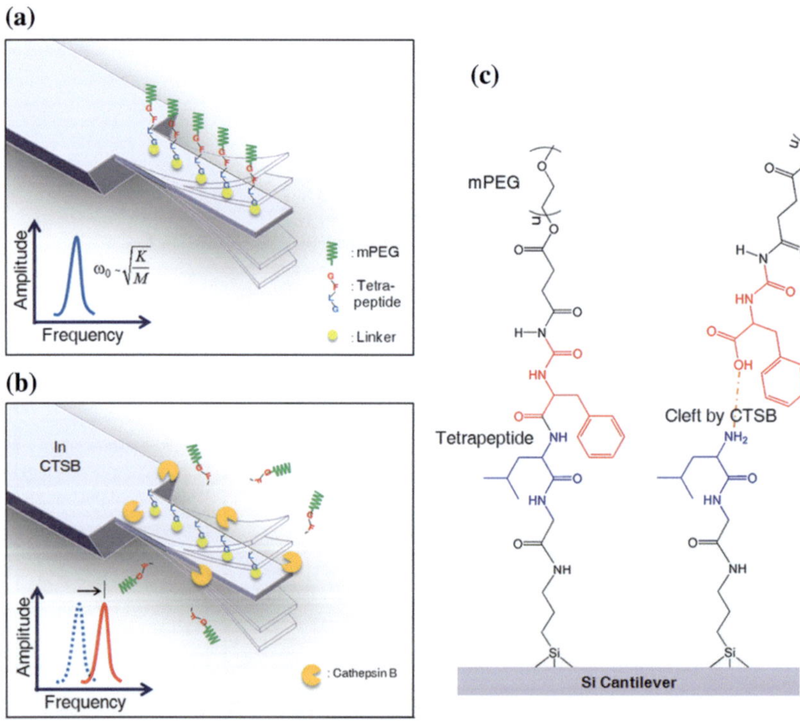

Fig. 7.3 Illustration of nanomechanical detection of enzymatic cleavage of peptide chain using resonant micro-cantilevers immersed in buffer solution. **a** The resonant frequency of a micro-cantilever functionalized by peptide chains is inversely proportional to the square root of the effective mass of a functionalized cantilever. **b** Enzymatic cleavage of tetrapeptide due to Cathepsin B reduces the effective mass of a functionalized cantilever, which eventually increases the resonant frequency of such cantilever. **c** Chemical structure of tetrapeptide that can be cleft by an enzyme (i.e., Cathepsin B). *Source* Figures are adopted with permission from Ref. [11]

The Euler–Bernoulli beam theory is repeatedly implemented to mathematically probe the nano–micromechanical mass sensors and developed its governing equation. As a well-known case, one could refer to an innovative article by Cho et al. [1] in which, a nanomechanical resonator with the geometric nonlinearity, shown in Fig. 7.4, was modeled.

The following partial differential equation, using Euler–Bernoulli beam theory, for their system considering foreign mass m_c was developed as

$$(\rho A + m_c\delta(x-L))w_{tt} + \left(\frac{m\omega_0}{Q}\right)w_t + EIw_{xxxx} - \left(\frac{EA}{4L}\right)w_{xx}\int_0^{2L} w_x^2 \mathrm{d}x = F\cos\omega t\delta(x-L) \tag{7.1.1}$$

in which $w(x,t)$, E, ρ, A, L, I, F, ω, ω_0, and Q are the transverse beam displacement, elasticity modulus, mass density, cross-sectional area, length, area moment of inertia, applied force to the middle, driving frequency, linearized natural frequency, and the resonator quality factor in the linear dynamic region, respectively. Applying the Galerkin procedure and considering the principal vibration mode of the system, the following nonlinear ordinary differential equation is obtained, which is of the type of the Duffing equation.

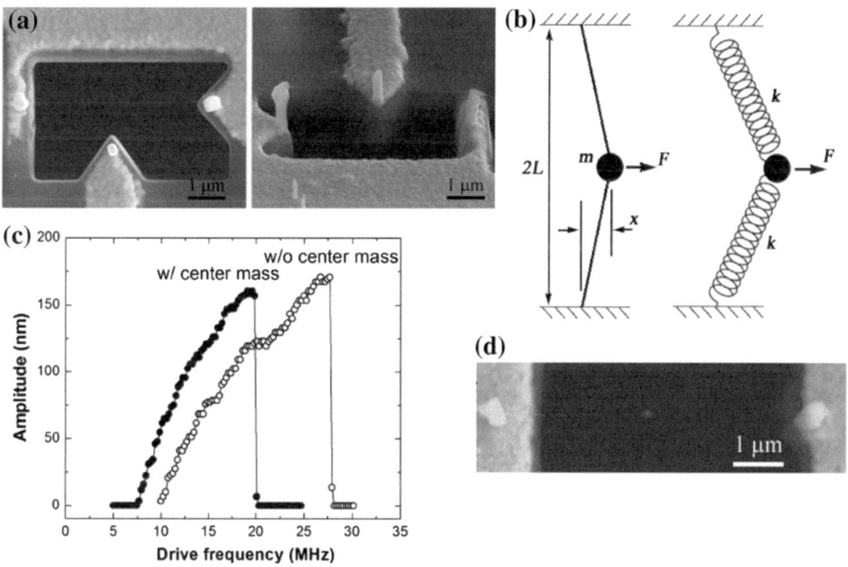

Fig. 7.4 **a** SEM images in the top and tilted views of a nanoresonator employing a CNT suspended in between and fixed at both ends on the fabricated platinum electrode posts. **b** Schematic diagram of a simple doubly clamped mechanical beam and its equivalent spring model having an intrinsic geometric nonlinearity. **c** The acquired response spectrum of CNT nonlinear nanoresonator: Sign O indicates before and • after depositing a center mass with electron-beam-induced deposition. **d** SEM image showing the Pt deposit in the middle of a suspended CNT. *Source* Figures with captions are adopted with permission from Ref. [1]

$$(1+M)\ddot{u}+\frac{\omega_0}{Q}u+\omega_0^2u+\alpha u^3=q\cos(\omega t) \tag{7.1.2}$$

where $M=\left(\frac{m_c}{2\rho AL}\right)W_1^2(L)=\left(\frac{m_c}{m_0}\right)W_1^2(L)$ and $q=\frac{W(L)F}{m_0}$.

The nonlinear coefficient appeared in Eq. (7.1.2) is obtained by

$$\alpha=-\frac{EA}{32\rho L^4}\int_0^{2L}WW''\mathrm{d}x\int_0^{2L}(W')^2\mathrm{d}x \tag{7.1.3}$$

The typical form of Eq. (7.1.2) is emerged in different types of mass sensors. As an example, the nonlinear graphene resonators were implemented for the mass sensing. The following partial differential equation is used for their system, using the plate model, shown in Fig. 7.5 [15]:

$$\rho_0\partial_t^2w(x,y,t)+k\nabla^4w(x,y,t)-\frac{E_s}{2(1-v)}\{(\partial_xw(x,y,t))^2+(\partial_yw(x,y,t))^2\}\nabla^2w(x,y,t)$$
$$-\frac{E_s}{1+\upsilon}\partial_xw(x,y,t)\partial_yw(x,y,t)\partial_{xy}w(x,y,t)-N_0\partial^2w(x,y,t)=f \tag{7.1.4}$$

where $k, E_s, h,$ $w(x,y,t)$, N_0, f and ρ_0 correspond to the bending rigidity, axial stretching modulus, thickness, out-of-plane deflection of graphene, axial tension constant, applied force magnitude, and the mass density of graphene including the added mass, respectively.

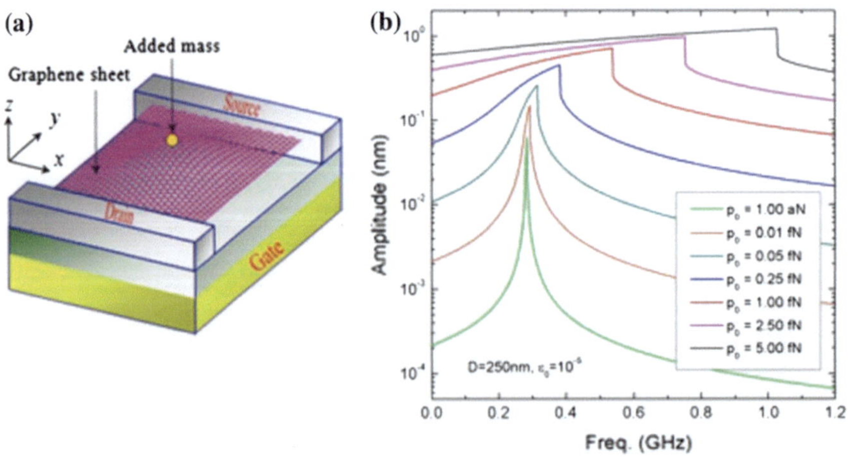

Fig. 7.5 **a** Schematic illustration of a suspended graphene resonator onto whose surface atomic mass is adsorbed. **b** Resonance curves of a graphene resonator (with its size of D = 250 nm) as a function of actuation amplitude. *Source* Figures are adopted with permission from Ref. [15]

The nonlinear Duffing equation is developed for the vibration of graphene in the same manner as the previous system.

7.1.2 Vibration of Carbon Nanotubes

The vibration analysis of carbon nanotubes (CNTs) is an essential area of concern owing to their many promising applications in tiny instruments. The unique and interesting properties of CNTs, particularly, their mechanical and electrical features, have drawn interest from scientists and industries to implement CNTs for production of different electromechanical devices [16]. The nonlinearities are emerged in CNTs due to their small-scale, geometry, and material properties.

This part of the chapter is aimed to reveal the dynamical behavior of CNTs, which is mathematically modeled by Duffing equation. The following partial differential equation is considered for the vibration of fluid conveying CNTs, and according to the article given in Ref. [16] both the electromagnetic and thermal effects are considered:

$$
\begin{aligned}
&EI\frac{\partial^4 w}{\partial x^4} + \rho A\frac{\partial^2 w(x,t)}{\partial t^2} + \mu\frac{\partial w(x,t)}{\partial t} + k_1 w(x,t) + k_3 w^3(x,t) - k_p\frac{\partial^2 w(x,t)}{\partial x^2} - f_{\mathrm{em}}(x,t) - F(x,t) \\
&+ F_{\mathrm{ff}} - \left(\frac{EA}{2L}\int_0^L \left(\frac{\partial w}{\partial x}\right)^2 \mathrm{d}x + N_{\mathrm{tt}}\right)\frac{\partial^2 w(x,t)}{\partial x^2} - (e_0^2 l_i^2)\left(\rho A\frac{\partial^4 w(x,t)}{\partial x^2 \partial t^2} + \mu\frac{\partial^3 w(x,t)}{\partial x^2 \partial t} + k_1\frac{\partial^2 w(x,t)}{\partial x^2}\right. \\
&+ k_3\left(6w(x,t)\left(\frac{\partial w}{\partial x}\right)^2 + 3w^2\left(\frac{\partial w}{\partial x}\right)\right) + \frac{\partial^2}{\partial x^2}F_{\mathrm{ff}} - \frac{\partial^2}{\partial x^2}f_{\mathrm{em}}(x,t) - k_p\frac{\partial^4 w(x,t)}{4} - \frac{\partial^2 F(x,t)}{\partial x^2} \\
&\left.- \left(\frac{EA}{2L}\int_0^L \left(\frac{\partial w}{\partial x}\right)^2 \mathrm{d}x + N_{\mathrm{tt}}\right)\frac{\partial^4 w}{\partial x^4}\right) = 0
\end{aligned}
\tag{7.1.5}
$$

The force per unit length induced by the fluid flow is

$$
F_{\mathrm{ff}} = m_{\mathrm{f}}\left(2v\frac{\partial^2 w(x,t)}{\partial x \partial t} + v^2\frac{\partial^2 w(x,t)}{\partial x^2} + \frac{\partial^2 w(x,t)}{\partial t^2}\right) \tag{7.1.6}
$$

and the coefficient of the thermal effect is given by

$$
N_{\mathrm{tt}} = -\frac{EA}{1-2v}\alpha_x \theta \tag{7.1.7}
$$

The applied force per unit length due to the longitudinal electromagnetic field is formulated as

$$f_{\mathrm{em}}(x,t) = \xi A H_x^2 w_{xx}(x,t) \tag{7.1.8}$$

Applying the Galerkin procedure and considering the primary mode of vibration, the following equation can be written as

$$w(x,t) = \phi_1(x) u(t) \tag{7.1.9}$$

Finally, by substituting Eq. (7.1.9) in Eq. (7.1.5), the following equation is obtained.

$$\ddot{u} + \omega_1^2 u + \hat{\mu}\dot{u} + \alpha_3 u^3 = f_1 \cos \Omega t \tag{7.1.10}$$

in which

$$\omega_1^2 = \frac{\begin{array}{r}\left(m_f v^2 \int_0^L \phi'' \phi \mathrm{d}x - m_f v^2 (e_0 l_i)^2 \int_0^L \phi''''(x)\phi(x)\mathrm{d}x + EI \int_0^L \phi''''(x)\phi(x)\,\mathrm{d}x\right.\\ + k_1 \int_0^L \phi^2(x)\mathrm{d}x - (e_0 l_i)^2 \xi A H_x^2 \int_0^L \phi''''''(x)\phi(x) - N_{tt}(e_0 l_i)^2 \int_0^L \phi''''(x)\phi(x)\mathrm{d}x\\ - k_p \int_0^L \phi''(x)\phi(x)\mathrm{d}x - k_1 (e_0 l_i)^2 \int_0^L \phi''(x)\phi(x)\mathrm{d}x + (e_0 l_i)^2 k_p \int_0^L \phi''''(x)\phi(x)\mathrm{d}x\\ - \xi A H_x^2 \int_0^L \phi''''(x)\phi(x) - N_{tt} \int_0^L \phi''(x)\phi(x)\mathrm{d}x\end{array}}{\rho A\left(\int_0^L \phi^2(x)\mathrm{d}x - (e_0 l_i)^2 \int_0^L \phi''(x)\phi(x)\mathrm{d}x\right) + m_f v^2 \left(\int_0^L \phi^2(x)\mathrm{d}x - (e_0 l_i)^2 \int_0^L \phi''(x)\phi(x)\mathrm{d}x\right)} \tag{7.1.11}$$

$$\hat{\mu} = \frac{\begin{array}{c}\mu\left(\int_0^L \phi^2(x)\mathrm{d}x - (e_0 l_i)^2 \int_0^L \phi''\phi(x)\mathrm{d}x\right)\\ + 2 v m_{\mathrm{f}}\left(\int_0^L \phi(x)\phi'(x)\mathrm{d}x - (e_0 l_i)^2 \int_0^L \phi'''(x)\phi(x)\mathrm{d}x\right)\end{array}}{\begin{array}{c}\rho A\left(\int_0^L \phi^2(x)\mathrm{d}x - (e_0 l_i)^2 \int_0^L \phi''(x)\phi(x)\mathrm{d}x\right)\\ + m_{\mathrm{f}} v^2\left(\int_0^L \phi^2(x)\mathrm{d}x - (e_0 l_i)^2 \int_0^L \phi''(x)\phi(x)\mathrm{d}x\right)\end{array}} \tag{7.1.12}$$

$$\alpha_3 = \frac{k_3\int_0^L \phi^4(x)\mathrm{d}x - \frac{EA}{L}\int_0^L \phi(x)\phi''(x)\mathrm{d}x\int_0^L \frac{1}{2}\phi'^2(x)\mathrm{d}x - 6k_3(e_0l_i)^2\int_0^L \phi^2(x)\phi'^2(x)\mathrm{d}x + 3k_3(e_0l_i)^2\int_0^L \phi^3\phi'\mathrm{d}x + \frac{EA}{2L}(e_0l_i)^2\int_0^L \phi''''(x)\phi(x)\mathrm{d}x\int_0^L \phi'^2(x)\mathrm{d}x}{\rho A\left(\int_0^L \phi^2(x)\mathrm{d}x - (e_0l_i)^2\int_0^L \phi''(x)\phi(x)\mathrm{d}x\right) + m_f v^2\left(\int_0^L \phi^2(x)\mathrm{d}x - (e_0l_i)^2\int_0^L \phi''(x)\phi(x)\mathrm{d}x\right)} \tag{7.1.13}$$

and

$$f_1 = \frac{F}{\rho A\left(\int_0^L \phi^2(x)\mathrm{d}x - (e_0l_i)^2\int_0^L \phi''(x)\phi(x)\mathrm{d}x\right) + m_f v^2\left(\int_0^L \phi^2(x)\mathrm{d}x - (e_0l_i)^2\int_0^L \phi''(x)\phi(x)\mathrm{d}x\right)}\int_0^L \phi(x)\delta(x-x_0)\,\mathrm{d}x \tag{7.1.14}$$

7.1.3 Vibration of Microtubules

Cytoplasm is the main part of eukaryotic cells, which consists of three types of fibers, called the protein microtubules, actin filaments, and the intermediate filaments. It has been experimentally confirmed that microtubules (MTs) are approximately 100 times stronger than the acting and intermediate filaments. It should be mentioned that MTs play the essential role in biological purposes, namely the intracellular transport, separation of chromosomes during mitosis, and the cell division. It is quite interesting to note that the microtubule structure is both polar and nonlinear [17]. In order to investigate the dynamical behavior of cells, it is important to identify the mechanical properties of MTs [18]. Accordingly, few scientists have focused their attention on the mechanical behavior of this structure.

As one of the earlier studies on the vibration of microtubules, the article by Sirenko et al. may be examined in which, they showed that MT-water system supports the interface elastic waves with maximal frequencies in a gigahertz range [19]. In other primary works on the vibration of MTs, it was reported that the highest frequency of oscillation in microtubules is in the order of 10^8 Hz [20] while Foster and Baish [21] developed a theoretical model of MTs with the addition of viscous damping.

The vibration of MTs was further modeled by an orthotropic elastic shell, wherein one of the assumptions was based on the relation of longitudinal modulus with the shear modulus and the circumferential modulus of microtubules [22]. It has been experimentally confirmed that MTs are characterized by extremely low shear modulus and, therefore, the Timoshenko beam theory was originally implemented to investigate the effect of the transverse shearing due to the low shear modulus of MTs [23].

To demonstrate the effect of small scale on the dynamic behavior of MTs, the nonlocal model was considered by Heireche et al., and the influence of small scale on the vibration of MTs was examined [24]. Large-amplitude flexural vibration behavior of microtubules (MTs), embedded in an elastic matrix of cytoplasm, was studied by H. S. Shen and the thermal and small-scale effects were considered for the vibration of the system [25]. An orthotropic-Pasternak model for the vibration of MTs was developed to confirm that the flexural vibration of MTs is increased with stiffening of the elastic medium [26].

In this part, a nonlinear model for the vibration of microtubules, based on the modified strain gradient theory [18] is developed and the equation of vibration of MTs inside the cytoplasm is obtained as

$$V\frac{\partial^6 w}{\partial x^6} - B\frac{\partial^4 w}{\partial x^4} - \rho A\frac{\partial^2 w}{\partial t^2} + \left(-E'A\alpha\Delta T + \frac{E'A}{2L}\int_0^L \left(\frac{\partial w}{\partial x}\right)^2 \mathrm{d}x\right) \times \frac{\partial^2 w}{\partial x^2} - 2b\left(k_1 w - k_2\frac{\partial^2 w}{\partial x^2}\right) = 0 \tag{7.1.15}$$

in which

$$B = E'I + \mu A l_2^2 + 2\mu A l_0^2 + \frac{8}{15}\mu A l_1^2 \tag{7.1.16}$$

and

$$V = 2\mu I\left(l_0^2 + \frac{2}{5}l_1^2\right) \tag{7.1.17}$$

Using the Galerkin procedure and substituting $w(x,t) = \phi(x)u(t)$ in Eq. (7.1.15), a nonlinear differential equation in the form of Duffing equation is obtained.

$$\ddot{u}(t) + \alpha_1 u(t) + \alpha_2 u(t)^3 = 0 \tag{7.1.18}$$

in which

$$\alpha_1 = \beta_2 + \beta_3 + \beta_4 - \beta_1 - \beta_5 \text{ and } \alpha_2 = \beta_6 \tag{7.1.19}$$

and the six functions of β_i are obtained as

$$
\begin{aligned}
\beta_1 &= \left(V \int_0^L \phi^{(6)} \phi \, \mathrm{d}x \right) \Big/ \left(\rho A \int_0^L \phi^2 \, \mathrm{d}x \right); \\
\beta_2 &= \left(B \int_0^L \phi^{(4)} \phi \, \mathrm{d}x \right) \Big/ \left(\rho A \int_0^L \phi^2 \, \mathrm{d}x \right); \\
\beta_3 &= \left(E' A \alpha \Delta T \int_0^L \phi^{(4)} \phi \, \mathrm{d}x \right) \Big/ \left(\rho A \int_0^L \phi^2 \, \mathrm{d}x \right); \\
\beta_4 &= \left(2 k_1 b \int_0^L \phi^2 \, \mathrm{d}x \right) \Big/ \left(\rho A \int_0^L \phi^2 \, \mathrm{d}x \right) \\
\beta_5 &= \left(2 k_2 b \int_0^L \phi'' \phi \, \mathrm{d}x \right) \Big/ \left(\rho A \int_0^L \phi^2 \, \mathrm{d}x \right) \\
\beta_6 &= \left(-\frac{E' A}{2L} \int_0^L \phi'' \phi \left(\int_0^L \phi'^2 \, \mathrm{d}x \right) \right) \Big/ \left(\rho A \int_0^L \phi^2 \, \mathrm{d}x \right)
\end{aligned}
\tag{7.1.20}
$$

7.2 Parametric and Self-Excited Oscillations

7.2.1 Electrically Actuated Microbeams

There are several potential applications of microbeam vibration and in fact, they have been implemented in different sensing instruments, namely, the mass sensing gadgets [27], high-viscosity sensors [28], acceleration sensors [29], energy harvesters [30, 31], and the biosensors. Recently, a self-excited coupled micro-cantilever was developed to measure a tiny mass. The idea of this system is based on the relative changes in the amplitude ratio of the first oscillation mode using the self-excited coupled vibration of micro-cantilevers [27] as shown in Fig. 7.6.

The idea of the self-excited oscillations was also implemented in the microbeams for high-viscosity sensing system, shown in Fig. 7.7 [28].

In pursuance of developing the governing equation of microbeam vibration, the Euler–Bernoulli beam theory and Timoshenko beam theory were implemented. For the sake of analyzing a typical example of microbeam vibration, the mathematical model of an electronically actuated microbeam is presented [32]. The transverse deflection of the microbeam $w(x, t)$, in accordance with the Euler–Bernoulli beam theory, is developed by the following governing equation:

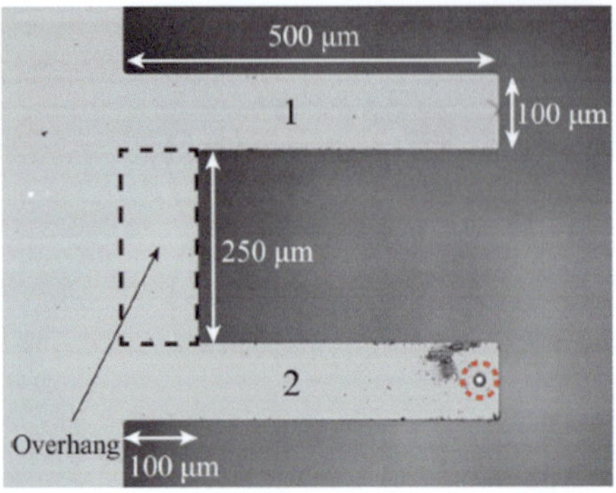

Fig. 7.6 Schematic diagram of coupled micro-cantilevers, fabricated by etching from an SOI wafer and a polystyrene microsphere, to measure the tip of cantilever 2 (surrounded by a red dashed circle) [27]. *Source* Figure is adopted with permission from Ref. [27]

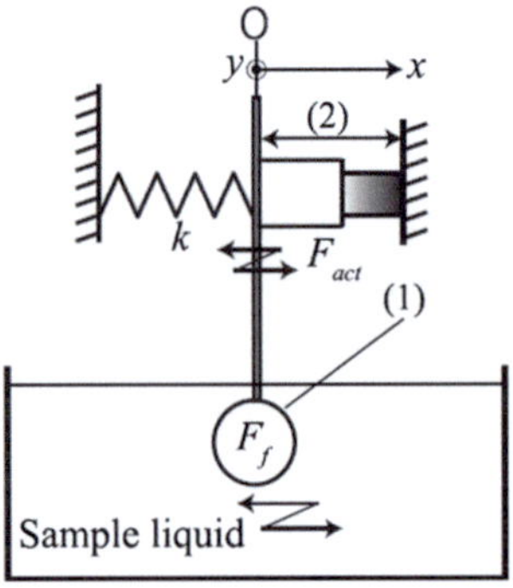

Fig. 7.7 Schematic diagram of a self-excited vibrational viscometer for high-viscosity measurement corresponding to the experimental apparatus shown in Fig. 7.7, where (1) is a rigid vibrational disk immersed in a liquid sample, and (2) is a linear motor for the positive velocity feedback control to produce the self-excited oscillation. *Source* Figure is adopted with permission from Ref. [28]

$$EI\frac{\partial^4 w(x,t)}{\partial x^4}+\rho bh\frac{\partial^2 w(x,t)}{\partial t^2}+c\frac{\partial w(x,t)}{\partial t}-\left[\tilde{N}+\frac{Ebh}{2l}\int\limits_0^l\left[\frac{\partial w(x,t)}{\partial x}\right]^2\mathrm{d}x\right]\frac{\partial^2 w(x,t)}{\partial x^2}$$
$$=\frac{1}{2}\varepsilon_0\frac{bC_{n1}[V_{ac}\cos(\Omega t)-V_{dc}]^2}{(g_a-w(x,t))^2}H_1(x)-\frac{1}{2}\varepsilon_0\frac{bC_{n2}[V_{ac}\cos(\Omega t)-V_{dc}]^2}{(g_a-w(x,t))^2}H_2(x) \tag{7.2.1}$$

where

$$H_1(x) = H\left(x - \frac{l + l_a}{2}\right) - H\left(x - \frac{l - l_a}{2}\right) \tag{7.2.2}$$

and

$$H_2(x) = H\left(x - \frac{l + l_d}{2}\right) - H\left(x - \frac{l - l_d}{2}\right) \tag{7.2.3}$$

The following boundary conditions are assumed for the microbeam system:

$$w(0,t) = w(l,t) = 0, \text{ and } \frac{\partial w}{\partial x}(0,t) = \frac{\partial w}{\partial x}(l,t) = 0 \tag{7.2.4}$$

The Galerkin procedure is implemented after normalization of Eq. (7.2.1), and accordingly, the following expression is considered to discretize Eq. (7.2.1):

$$w(x,t) = \phi(x)x(t) \tag{7.2.5}$$

Substituting Eq. (7.2.5) in the normalized form of Eq. (7.2.1), the following nonlinear ordinary differential equation is obtained:

$$\begin{aligned} &\ddot{x} + c\dot{x} + \omega_n^2 x + \mu_1 x\ddot{x} + \mu_2 x^2\ddot{x} + \mu_3 x^3\ddot{x} + \mu_4 x^4\ddot{x} + c\mu_1 x\dot{x} + c\mu_2 x^2\dot{x} \\ &+ c\mu_3 x^3\dot{x} + c\mu_4 x^4\dot{x} + \lambda_2 x^2 + \lambda_3 x^3 + \lambda_4 x^4 + \lambda_5 x^5 + \lambda_6 x^6 + \lambda_7 x^7 + v \\ &+ \xi_0\cos(\tilde{\Omega}t) + \xi_1 x\cos(\tilde{\Omega}t) + \xi_2 x^2\cos(\tilde{\Omega}t) \\ &+ \xi_3\cos(2\tilde{\Omega}t) + \xi_4 x\cos(2\tilde{\Omega}t) + \xi_5 x^2\cos(2\tilde{\Omega}t) = 0 \end{aligned} \tag{7.2.6}$$

The corresponding parameters appeared in Eq. (7.2.6) are explained in the referenced article [32]. Both types of the parametric and self-excited oscillations can be closely observed in Eq. (7.2.6). Furthermore, higher order nonlinearities exist in the governing equation of an electronically actuated microbeam.

7.2.2 Micro-Gyroscope

There is a big interest among different industries to implement micro-electromechanical gyroscopes due to their small size, batch fabrication, light-weight, low power consumption, low cost, suitable and reasonable performance for several applications [33]. There are widely applications reported in virtual reality, namely, automotive engineering, micro-machined accelerometers, inertial mouse for computers, ride and video camera stabilization, and robotic and military systems [33–37]. This section presents a mathematical model for the nonlinear vibration of

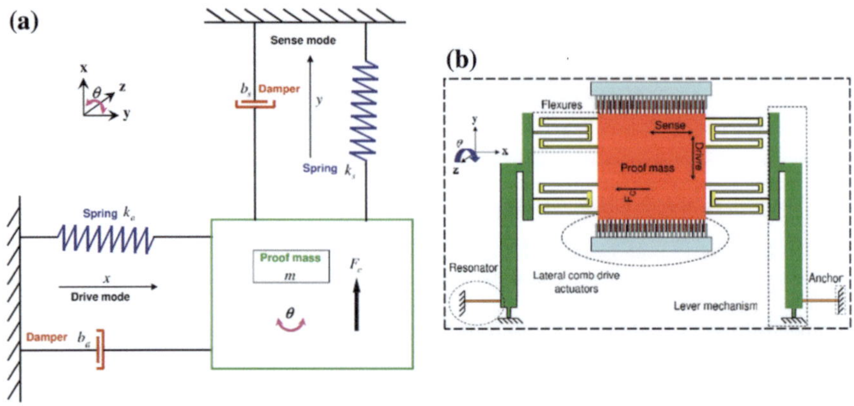

Fig. 7.8 **a** Lumped-parameter model of a vibrating micro-gyroscope. **b** A simple mass resonant gyroscope. *Source* Figures are adopted with permission from Ref. [33]

micro-gyroscope in accordance with the developed system by Kacem et al. [33]. The schematic diagram of the lumped-parameter model, shown in Fig. 7.8, is utilized to study the vibration of a micro-gyroscope.

The following series of coupled differential equations could be considered for the modeling and analysis of a micro-gyroscope being subjected to an axial time-varying Coriolis force F_c.

$$\frac{\mathrm{d}^2 Y(t)}{\mathrm{d}t^2} + \frac{\delta}{Q}\frac{\mathrm{d}Y(t)}{\mathrm{d}t} + \delta Y(t) = \frac{F_e}{M}\cos\delta t \tag{7.2.7}$$

$$EI\frac{\partial^4 w(x,t)}{\partial x^4} + \rho bh\frac{\partial^2 w(x,t)}{\partial x^2} + c\frac{\partial w(x,t)}{\partial t} - \frac{Ebh}{2l}\int_0^l \left[\frac{\partial w(x,t)}{\partial x}\right]^2 \mathrm{d}x\frac{\partial^2 w(x,t)}{\partial x^2}$$
$$- [N + AF_c\cos\delta t]\frac{\partial^2 w(x,t)}{\partial x^2} = \frac{1}{2}\varepsilon_0\frac{bC_n[V_{\mathrm{dc}} + V_{\mathrm{ac}}\cos(\Omega t)]^2}{(g - w(x,t))^2} \tag{7.2.8}$$

and

$$F_{\mathrm{c}} = 2M\theta\frac{\mathrm{d}Y(t)}{\mathrm{d}t} \tag{7.2.9}$$

where the definition of parameters $Y, M, \delta, Q, F_e, A, x, I, N, t, \rho, h, g, F_c$ and ε can be found in the published article [33]. The following boundary conditions for the model given by Eq. (7.2.8) were assumed:

$$w(0,t) = w(l,t) = 0;\ \text{and}\ \frac{\partial w}{\partial x}(0,t) = \frac{\partial w}{\partial x}(l,t) = 0 \tag{7.2.10}$$

To obtain the general solutions, it is preferred to use nondimensional variables given as

$$W = \frac{w}{g};\ \ \hat{x} = \frac{x}{l};\ \ \text{and}\ \hat{t} = \frac{t}{\tau} \tag{7.2.11}$$

in which $\tau = \left(\frac{2l^2}{h}\right)\sqrt{3\rho/E}$.

Subsequently, Eq. (7.2.11) is substituted in both Eqs. (7.2.7) and (7.2.8), and the Galerkin method is then implemented. Following equation is assumed as the general form of deflection:

$$W(x,t) = u(\hat{t})\phi(x) \tag{7.2.12}$$

where $\phi(x)$ is the linear fundamental mode shape of the straight microbeam. Accordingly, the following nonlinear differential equation is obtained:

$$\begin{aligned} &\ddot{u}(\hat{t}) + c\dot{u}(\hat{t}) + (500.564 - 2\alpha_2\xi_0 + 12.3N)u(\hat{t}) + \alpha_2\xi_0(0.831 + 4u^2(\hat{t}) \\ &+ 13.255u^4(\hat{t})) + 12.3F_c u(\hat{t})\cos\delta\hat{t} - 23.17\alpha_2\xi_0 u^5(\hat{t}) \\ &+ (151.35\alpha_1 - 7.403\alpha_2\xi_0)u^3(\hat{t}) = 0 \end{aligned} \tag{7.2.13}$$

in which $\xi_0 = V_{\text{dc}}^2 + V_{\text{ac}}\cos\hat{\Omega}\hat{t}$.

The values of the coefficients c, α_2 and $\hat{\Omega}$ are selected as those reported in the published article [33]. Equation (7.2.13) is shown to be in the form of the nonlinear Mathieu equation, which considers as the vibration-type cases of parametric and self-excited oscillations.

7.3 Nonlinear Coupled Oscillators

7.3.1 Doubled-Walled Carbon Nanotubes (DWCNTs)

Vibration analysis of double-walled carbon nanotubes (DWCNTs) is a fascinating research topic for scientists who are working in the area of nanostructures. Diverse types of experimental and theoretical approaches have been considered to explore the influences of different geometrical and physical parameters on the dynamic behavior of multi-walled carbon nanotubes (MWCNTs), specifically double-walled carbon nanotubes (DWCNTs).

Research performed by Li and Chou [38] could be cited as one of the earliest attempts for analyzing the vibration of multi-walled carbon nanotubes. They showed that vibration performances of single-walled carbon nanotubes (SWCNTs) and multi-walled carbon nanotubes (MWCNTs) have been significantly different. They also reported that fundamental frequencies of MWCNTs are about 10% less than those of SWCNTs when both carbon nanotubes have the same diameters [38].

Zhang et al. [39] studied the transverse vibration of DWCNTs using the Euler–Bernoulli beam theory when it was subjected to axial compressive loads. They reported that natural frequencies depend largely on the axial load, and decreased with an increase in the axial load [39]. Another highly cited article by Zhang et al. is the implementation of the nonlocal theory in the vibration of DWCNTs. They reported that the small-scale length is highly influential on the natural frequencies of DWCNTs [40].

In order to study the interlayer forces in DWCNTs, namely the van der Waals (vdW) forces, one should consider both the linear and nonlinear models. The vdW force per unit area is determined as [41–43]

$$P = \frac{\partial U}{\partial \delta} \tag{7.3.1}$$

in which U is the interlayer potential per unit area, which can be formulated in terms of the interlayer spacing δ as

$$U(\delta) = K\left[\left(\frac{\delta_0}{\delta}\right)^4 - 0.4\left(\frac{\delta_0}{\delta}\right)^{10}\right] \tag{7.3.2}$$

where $K = -61.655$ meV/atom, and $\delta_0 = 0.34$ nm is the equilibrium interlayer spacing.

The geometry and large-amplitude oscillations might be another reason for arising the nonlinearities in DWCNTs [44]. As an example, the large-amplitude vibration of DWCNTs could lead to a cubic nonlinearity. However, for the geometrical nonlinearity, one could state that the modeling of waviness in DWCNTs engenders the quadratic nonlinearity [45].

In this section, a model for the vibration of DWCNTs is developed, which includes the vdW interaction, geometrical nonlinearity, and the large-amplitude oscillation [45]. According to the Euler–Bernoulli beam theory, the following partial differential equation represents the nonlinear vibration of curved DWCNTs:

$$\begin{aligned} &E_1 I_1 \frac{\partial^4 w_1(x,t)}{\partial x^4} + \rho_1 A_1 \frac{\partial^2 w_1(x,t)}{\partial t^2} = \\ &\frac{E_1 A_1}{L} \int_0^L \left(\frac{\mathrm{d}Z}{\mathrm{d}x} \frac{\partial w_1(x,t)}{\partial x} + \frac{1}{2}\left(\frac{\partial w_1(x,t)}{\partial x}\right)^2 \right) \mathrm{d}x \left(\frac{\partial^2 w_1(x,t)}{\partial x} + \frac{\mathrm{d}^2 Z}{\mathrm{d}x^2} \right) \\ &+ p_v(x,t) \end{aligned} \tag{7.3.3}$$

and

$$E_2 I_2 \frac{\partial^4 w_2(x,t)}{\partial x^4} + \rho_2 A_2 \frac{\partial^2 w_2(x,t)}{\partial t^2} =$$
$$\frac{E_2 A_2}{L} \int_0^L \left(\frac{\mathrm{d}Z}{\mathrm{d}x} \frac{\partial w_2(x,t)}{\partial x} + \frac{1}{2} \left(\frac{\partial w_2(x,t)}{\partial x} \right)^2 \right) \mathrm{d}x \left(\frac{\partial^2 w_2(x,t)}{\partial x} + \frac{\mathrm{d}^2 Z}{\mathrm{d}x^2} \right) \tag{7.3.4}$$
$$- p_v(x,t) + p_m(x,t)$$

where $Z(x)$, $P_m(x,t)$ and $p_v(x,t)$ are the initial curvature (waviness) of the cylindrical tubes, the interactive force of medium and tube, and the nonlinear vdW force, respectively. The interactive forces can be presented by the following equation:

$$p_m(x,t) = -k w_2(x,t) - k_1 w_2^3(x,t) + k_p \frac{\partial^2 w}{\partial x^2} \tag{7.3.5}$$

in which k, k_1, and k_p are the coefficients of the linear, nonlinear, and Pasternak stiffness, respectively. Based on Eq. (7.3.1), the vdW force per unit area for two concentric tubes is

$$p_v(x,t) = p_1(w_2 - w_1) + p_2(w_2 - w_1)^3 \tag{7.3.6}$$

where $p_1 = 2r_1(\partial^2 U/\partial \delta^2)\big|_{\delta=\delta_0}$, and $p_2 = 2r_1\left(\frac{1}{6}\right)\left(\frac{\partial^4 U}{\partial \delta^4}\right)\Big|_{\delta=\delta_0}$.

Parameter r_1 is the inner tube radius and U is potential energy, which was defined in Eq. (7.3.2).

$$E_1 I_1 \frac{\partial^4 w_1(x,t)}{\partial x^4} + \rho_1 A_1 \frac{\partial^2 w_1(x,t)}{\partial t^2} =$$
$$\frac{E_1 A_1}{L} \int_0^L \left(\frac{\mathrm{d}Z}{\mathrm{d}x} \frac{\partial w_1(x,t)}{\partial x} + \frac{1}{2} \left(\frac{\partial w_1(x,t)}{\partial x} \right)^2 \right) \mathrm{d}x \left(\frac{\partial^2 w_1(x,t)}{\partial x} + \frac{\mathrm{d}^2 Z}{\mathrm{d}x^2} \right) \tag{7.3.7}$$
$$+ p_1(w_2 - w_1) + p_2(w_2 - w_1)^3$$

and

$$E_2 I_2 \frac{\partial^4 w_2(x,t)}{\partial x^4} + \rho_2 A_2 \frac{\partial^2 w_2(x,t)}{\partial t^2} =$$
$$\frac{E_2 A_2}{L} \int_0^L \left(\frac{\mathrm{d}Z}{\mathrm{d}x} \frac{\partial w_2(x,t)}{\partial x} + \frac{1}{2} \left(\frac{\partial w_2(x,t)}{\partial x} \right)^2 \right) \mathrm{d}x \left(\frac{\partial^2 w_2(x,t)}{\partial x} + \frac{\mathrm{d}^2 Z}{\mathrm{d}x^2} \right)$$
$$- p_1(w_2 - w_1) - p_2(w_2 - w_1)^3 \tag{7.3.8}$$
$$- k w_2(x,t) - k_1 w_2^3(x,t) + k_p \frac{\partial^2 w}{\partial x^2}$$

The nonlinear terms appeared in the above coupled partial differential equations are related to the vdW force, the large oscillation, and waviness of DWCNTs. It can be seen from Eqs. (7.3.7) to (7.3.8) that the cubic nonlinearity appears in the coupled differential equations as a result of large oscillation and the vdW interactive force. The geometrical imperfection of DWCNTs will cause quadratic nonlinearity in the vibration of the coupled system.

The Eringen theory is utilized in conjunction with the Euler–Bernoulli beam theory to develop the following governing partial differential equations of vibration of curved DWCNTs:

$$\begin{aligned}
&E_1I_1\frac{\partial^4 w_1(x,t)}{\partial x^4}+\rho_1A_1\frac{\partial^2 w_1(x,t)}{\partial t^2}=\\
&\frac{E_1A_1}{L}\int_0^L\left(\frac{\mathrm{d}Z}{\mathrm{d}x}\frac{\partial w_1(x,t)}{\partial x}+\frac{1}{2}\left(\frac{\partial w_1(x,t)}{\partial x}\right)^2\right)\mathrm{d}x\left(\frac{\partial^2 w_1(x,t)}{\partial x}+\frac{\mathrm{d}^2Z}{\mathrm{d}x^2}\right)+p_v(x,t)\\
&-(e_0a)^2\frac{\partial^2}{\partial x^2}\left(\frac{E_1A_1}{L}\int_0^L\left(\frac{\mathrm{d}Z}{\mathrm{d}x}\frac{\partial w_1(x,t)}{\partial x}+\frac{1}{2}\left(\frac{\partial w_1(x,t)}{\partial x}\right)^2\right)\left(\frac{\partial^2 w_1(x,t)}{\partial x}+\frac{\mathrm{d}^2Z}{\mathrm{d}x^2}\right)\right.\\
&\left.+p_v(x,t)-\rho_1A_1\frac{\partial^2 w_1(x,t)}{\partial t^2}\right)
\end{aligned}\tag{7.3.9}$$

and

$$\begin{aligned}
&E_2I_2\frac{\partial^4 w_2(x,t)}{\partial x^4}+\rho_2A_2\frac{\partial^2 w_2(x,t)}{\partial t^2}=\\
&\frac{E_2A_2}{L}\int_0^L\left(\frac{\mathrm{d}Z}{\mathrm{d}x}\frac{\partial w_2(x,t)}{\partial x}+\frac{1}{2}\left(\frac{\partial w_2(x,t)}{\partial x}\right)^2\right)\mathrm{d}x\left(\frac{\partial^2 w_2(x,t)}{\partial x}+\frac{\mathrm{d}^2Z}{\mathrm{d}x^2}\right)-p_v(x,t)+p_m(x,t)\\
&-(e_0a)^2\frac{\partial^2}{\partial x^2}\left(\frac{E_2A_2}{L}\int_0^L\left(\frac{\mathrm{d}Z}{\mathrm{d}x}\frac{\partial w_2(x,t)}{\partial x}+\frac{1}{2}\left(\frac{\partial w_2(x,t)}{\partial x}\right)^2\right)\left(\frac{\partial^2 w_2(x,t)}{\partial x}+\frac{\mathrm{d}^2Z}{\mathrm{d}x^2}\right)\right.\\
&\left.-p_v(x,t)+p_m(x,t)-\rho_2A_2\frac{\partial^2 w_2(x,t)}{\partial t^2}\right)
\end{aligned}\tag{7.3.10}$$

7.3.2 Higher Mode Vibration of Single-Walled Carbon Nanotubes

Vibration analysis of SWCNTs is an attractive topic in the area of nanostructures. Many experimental and theoretical approaches have been proposed to accurately

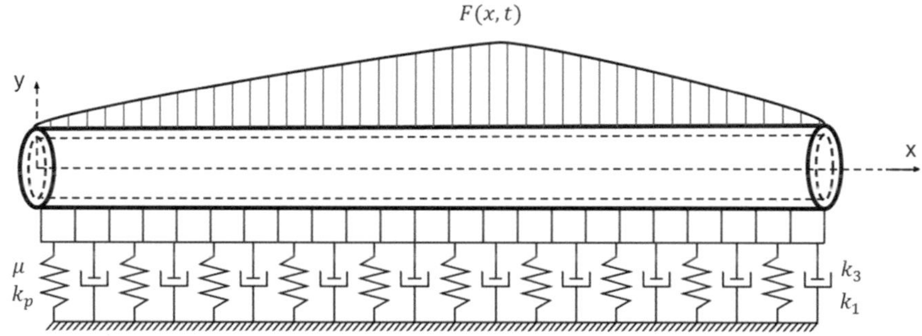

Fig. 7.9 Schematic diagram of SWCNT postulating the nonlocal parameter. *Source* Figure with caption is adopted with permission from Ref. [16]

predict the vibration behavior of SWCNTs [46]. Theories proposed by Timoshenko [47, 48], Euler–Bernoulli [49], Rayleigh [50, 51], nonlocal [52, 53], stress and strain-inertia gradient elasticity [54], and the spatial periodic strain [55] could be examined for theoretically modeling of SWCNTs, as shown in Fig. 7.9. The goal here is to develop a mathematical model of SWCNTs, based on the nonlocal and Euler–Bernoulli beam theories, when considering the first and second modes of vibration. Accordingly, the governing equation of vibration of SWCNTs is developed as

$$\begin{aligned}
&EI\frac{\partial^4 w}{\partial x^4}+\rho A\frac{\partial^2 w(x,t)}{\partial t^2}+\mu\frac{\partial w(x,t)}{\partial t}+k_1 w(x,t)+k_3 w^3(x,t)-k_p\frac{\partial^2 w(x,t)}{\partial x^2}-F(x,t)\\
&-\left(\frac{EA}{2L}\int_0^L\left(\frac{\partial w}{\partial x}\right)^2 \mathrm{d}x\right)\frac{\partial^2 w(x,t)}{\partial x^2}-(e_0^2 l_i^2)\left\{\rho A\frac{\partial^4 w(x,t)}{\partial x^2\partial t^2}+\mu\frac{\partial^3 w(x,t)}{\partial x^2\partial t}+k_1\frac{\partial^2 w(x,t)}{\partial x^2}\right.\\
&+k_3\left[6w(x,t)\left(\frac{\partial w}{\partial x}\right)^2+3w^2\left(\frac{\partial w}{\partial x}\right)\right]-k_p\frac{\partial^4 w(x,t)}{\partial x^4}-\frac{\partial^2 F(x,t)}{\partial x^2}\\
&\left.-\left[\frac{EA}{2L}\int_0^L\left(\frac{\partial w}{\partial x}\right)^2 \mathrm{d}x\right]\frac{\partial^4 w}{\partial x^4}\right\}=0
\end{aligned} \tag{7.3.11}$$

Consider the first two modes of vibration and apply the Galerkin procedure to obtain

$$\left(\rho A\frac{L}{2}+(e_0l_i)^2\rho A\frac{\pi^2}{2L}\right)\ddot{q}_1+\left(\frac{\mu L}{2}+(e_0l_i)^2\frac{\mu\pi^2}{2L}\right)\dot{q}_1+\left(\frac{EI\pi^4}{2L^3}+\frac{k_1L}{2}+\frac{k_p\pi^2}{2L}\right.$$
$$+(e_0l_i)^2\left(\frac{k_1\pi^2}{2L}+\frac{k_p\pi^4}{2L^3}\right)\Bigg)q_1+\left(\frac{3L}{8}k_3+\left(\frac{EA}{2L}\right)\left(\frac{\pi^2}{2L}\right)^2+\left((e_0l_i)^2\left(\frac{EA}{2L}\right)\left(\frac{\pi^2}{2L}\right)\left(\frac{\pi^4}{2L^3}\right)\right.\right.$$
$$\left.+3k_3(e_0l_i)^2\left(\frac{3\pi^2}{8L}\right)\right)q_1^3+q_1q_2^2\left(\frac{3L}{4}k_3+\left(\frac{EA}{2L}\right)\left(\frac{\pi^4}{L^2}\right)+(e_0l_i)^2\left(\frac{EA}{2L}\right)\left(\frac{2\pi^2}{L}\right)\left(\frac{\pi^4}{2L^3}\right)\right.$$
$$\left.+\frac{3\pi^2}{4l}(e_0l_i)^2k_3\right)=F_1 \tag{7.3.12}$$

and

$$\left(\rho A\frac{L}{2}+(e_0l_i)^2\rho A\frac{2\pi^2}{L}\right)\ddot{q}_2+\left(\frac{\mu L}{2}+(e_0l_i)^2\frac{2\mu\pi^2}{L}\right)\dot{q}_1+\left(\frac{8EI\pi^4}{L^3}+\frac{k_1L}{2}+\frac{k_p2\pi^2}{L}+(e_0l_i)^2\right.$$
$$\left(\frac{2k_1\pi^2}{L}+\frac{8k_p\pi^4}{L^3}\right)\Bigg)q_1+{q_1}^3\left(\frac{3L}{8}k_3+\left(\frac{EA}{2L}\right)\left(\frac{2\pi^2}{L}\right)^2\right)$$
$$+\left((e_0l_i)^2\left(\frac{EA}{2L}\right)\left(16\frac{\pi^2}{L}\right)\left(\frac{\pi^4}{L^3}\right)+(e_0l_i)^2\left(\frac{3\pi^2}{2L}\right)\right)q_1^3$$
$$+q_1q_2^2\left(\frac{3L}{4}k_3+\left(\frac{EA}{2L}\right)\left(\frac{\pi^4}{L^2}\right)+(e_0l_i)^2\left(\frac{EA}{2L}\right)\left(\frac{\pi^2}{2L}\right)\left(\frac{8\pi^6}{L^3}\right)+\frac{3\pi^2}{2l}(e_0l_i)^2k_3\right)=F_2 \tag{7.3.13}$$

The coupled nonlinear ordinary differential equations given above can be rewritten as [16]

$$\begin{cases}\ddot{q}_1+\omega_1^2q_1+\bar{\mu}_1\dot{q}_1+\alpha_1q_1^3+\alpha_2q_1q_2^2=f_1\cos\Omega t\\ \ddot{q}_2+\omega_2^2q_2+\bar{\mu}_2\dot{q}_2+\alpha_3q_2^3+\alpha_4q_2q_1^2=f_2\cos\Omega t\end{cases} \tag{7.3.14}$$

in which

$$\omega_1^2=\frac{1}{\left(\rho A\frac{L}{2}+(e_0l_i)^2\rho A\frac{\pi^2}{2L}\right)}\left[\frac{EI\pi^4}{2L^3}+\frac{k_1L}{2}+\frac{k_p\pi^2}{2L}+(e_0l_i)^2\left(\frac{k_1\pi^2}{2L}+\frac{k_p\pi^4}{2L^3}\right)\right] \tag{7.3.15}$$

$$\omega_2^2=\frac{1}{\left(\rho A\frac{L}{2}+(e_0l_i)^2\rho A\frac{2\pi^2}{L}\right)}\left[\frac{8EI\pi^4}{L^3}+\frac{k_1L}{2}+\frac{k_p2\pi^2}{L}+(e_0l_i)^2\left(\frac{2k_1\pi^2}{L}+\frac{8k_p\pi^4}{L^3}\right)\right] \tag{7.3.16}$$

and

$$\alpha_1 = \frac{1}{\left(\rho A \frac{L}{2} + (e_0 l_i)^2 \rho A \frac{\pi^2}{2L}\right)} \left[\frac{3L}{8} k_3 + \left(\frac{EA}{2L}\right)\left(\frac{\pi^2}{2L}\right)^2 + \left((e_0 l_i)^2 \left(\frac{EA}{2L}\right)\left(\frac{\pi^2}{2L}\right)\left(\frac{\pi^4}{2L^3}\right) + 3k_3 (e_0 l_i)^2 \left(\frac{3\pi^2}{8L}\right)\right)\right] \tag{7.3.17}$$

$$\alpha_2 = \frac{1}{\left(\rho A \frac{L}{2} + (e_0 l_i)^2 \rho A \frac{\pi^2}{2L}\right)} \left[\frac{3L}{4} k_3 + \left(\frac{EA}{2L}\right)\left(\frac{\pi^4}{L^2}\right) + (e_0 l_i)^2 \left(\frac{EA}{2L}\right)\left(\frac{2\pi^2}{L}\right)\left(\frac{\pi^4}{2L^3}\right) + \frac{3\pi^2}{4l} (e_0 l_i)^2 k_3\right] \tag{7.3.18}$$

$$\alpha_3 = \frac{1}{\left(\rho A \frac{L}{2} + (e_0 l_i)^2 \rho A \frac{2\pi^2}{L}\right)} \left[\frac{3L}{8} k_3 + \left(\frac{EA}{2L}\right)\left(\frac{\pi^2}{2L}\right)^2 + (e_0 l_i)^2 \left(\frac{EA}{2L}\right)\left(\frac{\pi^2}{2L}\right)\left(\frac{\pi^4}{2L^3}\right) + 3k_3 (e_0 l_i)^2 \left(\frac{3\pi^2}{8L}\right)\right] \tag{7.3.19}$$

$$\alpha_4 = \frac{1}{\left(\rho A \frac{L}{2} + (e_0 l_i)^2 \rho A \frac{2\pi^2}{L}\right)} \left[\frac{3L}{4} k_3 + \left(\frac{EA}{2L}\right)\left(\frac{\pi^4}{L^2}\right) + (e_0 l_i)^2 \left(\frac{EA}{2L}\right)\left(\frac{\pi^2}{2L}\right)\left(\frac{8\pi^6}{L^3}\right) + \frac{3\pi^2}{2l} (e_0 l_i)^2 k_3\right] \tag{7.3.20}$$

with

$$\bar{\mu}_1 = \frac{\left(\frac{\mu L}{2} + (e_0 l_i)^2 \frac{\mu \pi^2}{2L}\right)}{\left(\rho A \frac{L}{2} + (e_0 l_i)^2 \rho A \frac{\pi^2}{2L}\right)} \tag{7.3.21}$$

$$\bar{\mu}_2 = \frac{\left(\frac{\mu L}{2} + (e_0 l_i)^2 \frac{2\mu \pi^2}{L}\right)}{\left(\rho A \frac{L}{2} + (e_0 l_i)^2 \rho A \frac{2\pi^2}{L}\right)} \tag{7.3.22}$$

$$f_1 = \frac{F_1}{\left(\rho A \frac{L}{2} + (e_0 l_i)^2 \rho A \frac{\pi^2}{2L}\right)} \left(\int_0^L \sin\left(\frac{\pi x}{L}\right) \delta(x - x_0)\,\mathrm{d}x - (e_0 l_i)^2 \frac{\partial^2}{\partial x^2} \int_0^L \sin\left(\frac{\pi x}{L}\right) \delta(x - x_0)\,\mathrm{d}x \right) \tag{7.3.23}$$

and

$$f_2 = \frac{F_2}{\left(\rho A \frac{L}{2} + (e_0 l_i)^2 \rho A \frac{\pi^2}{2L}\right)} \left(\int_0^L \sin\left(\frac{2\pi x}{L}\right) \delta(x - x_0)\,\mathrm{d}x - (e_0 l_i)^2 \frac{\partial^2}{\partial x^2} \int_0^L \sin\left(\frac{\pi x}{L}\right) \delta(x - x_0)\,\mathrm{d}x \right) \tag{7.3.24}$$

It can be seen from Eq. (7.3.15) through Eq. (7.3.24) that all the coefficients of linear, nonlinear, and excitation force amplitude would clearly depend on the nonlocal parameter.

7.3.3 Nonlinear Oscillations of Nanowire Resonators

There is a wide range of potential applications for the nanowires [56], especially in actuators, sensors, resonators, transistors, and probes in NEMS and biotechnology. Accordingly, it is inescapable to mathematically model the physical and mechanical properties of these structures. Analogous to the SWCNTs, several theories have been proposed or implemented to investigate the mechanical behavior of nanowires. But, there are still many enigmatic phenomena that need to be fully investigated. It should be emphasized that there are few articles in which, researchers have investigated the vibration behavior of nanowires.

A thorough review is carried out to present the recent development and findings on the vibration of nanowires. As an earlier work on vibration analysis of nanowires, one could refer to an article by Wang and Feng in which, they studied the vibration and buckling of nanowires, based on the Timoshenko beam model with the surface effect [57]. In another work, the Euler–Bernoulli beam theory was utilized to investigate the effect of surface stress on the oscillation and buckling of piezoelectric nanowires [58]. It has been concluded that the resonant frequency of the piezoelectric nanowires can be tuned by adjusting the applied electric potential [58]. Oveisi [59] in his recent work showed that the surface effect is considerably effective on the free transverse vibration of piezoelectric nanowires.

Further work on the vibration of nanowires includes the nonlocal theory application in the longitudinal free vibration of tapered nanowires [60]. Vibration and instability of pre-tensioned current-carrying nanowires considering surface effect were studied by Kiani [61]. A magnetic shock was utilized for the vibrations of a conducting nanowire embedded in an elastic medium. It was shown that the general radial displacement and the nonlocal radial and hoop stresses are highly dependent on the small-scale parameter, the radial stiffness of surrounding media and the duration of the applied magnetic shock [62]. The surface energy and size effects were taken into consideration to probe the forced vibration of a current carrying nanowire in a longitudinal magnetic field [63].

The nonlocal Timoshenko beam theory based on the published work [64] is presented to mathematically model the vibration of nanowires. The following partial differential equation is considered for the vibration of nanowire:

$$\rho I(1-(ea_0)^2\nabla^2)\frac{\partial^2\psi}{\partial t^2}-(EI)^*\frac{\partial^2\psi}{\partial x^2}+k_sGA(\psi+\frac{\partial w}{\partial x})=0 \tag{7.3.25}$$

and

$$\rho A(1-(ea)^2\nabla^2)\frac{\partial^2 w}{\partial t^2}-k_sGA\left(\frac{\partial\psi}{\partial x}+\frac{\partial^2 w}{\partial x^2}\right)-N(1-(ea)^2\nabla^2)\frac{\partial^2 w}{\partial x^2}=0 \tag{7.3.26}$$

in which

$$N=\frac{1}{2}(EA)^*-l^{-1}\int_0^l w_x^2dx+\tilde{N} \tag{7.3.27}$$

where $k_s, \tilde{N}, l, e_0a, G, A, \rho, w$ and ψ are the shear stiffness, resultant stress, length, surface effect, shear modulus, cross-sectional area, density, displacement along the z-axis, and the rotational angle of the cross section, respectively. Employing the Galerkin method and assuming the simply supported boundary conditions, the following dimensionless nonlinear ordinary differential equation is obtained:

$$\begin{aligned}\left(\frac{\omega}{\omega_l}\right)\ddot{p}+\left(\chi+\frac{\gamma}{1+v^2}\right)p+\alpha p^3+\frac{\gamma\lambda}{1+v^2}\phi=0\\ \left(\frac{\omega}{\omega_l}\right)\ddot{\phi}+\zeta(1+\gamma)p+\alpha\phi+\frac{\gamma\lambda}{1+v^2}p=0\end{aligned} \tag{7.3.28}$$

and

$$p=\frac{w}{r};\ \ v=\frac{e_0a\pi}{l};\ \ \xi=\frac{\rho Al^2}{\rho I\pi^2(1+\upsilon^2)};\ \ r=\sqrt{(EI^*)/(EA)^*};\ \ \tau=\omega t;\ \ \gamma=\frac{Gk_sAl^2}{(EI)^2\pi^2};$$
$$\delta=\frac{\tilde{N}l^2}{(EI)^*\pi^2};\ \ \lambda=\frac{1}{r\pi};\ \text{ and }\ \omega_l=\pi^2l^{-2}\sqrt{(EI)^*/\rho A} \tag{7.3.29}$$

It can be seen from Eq. (7.3.28) that the resultant differential equation is a coupled nonlinear one.

7.3.4 Applications of Timoshenko Beam Theory in Nanoscale Systems

The Timoshenko beam theory was developed by the well-known Russian scientist Stephen Timoshenko (1878–1972). In his model, the effects of the shear deformation and rotational inertia have been taken into account. Accordingly, his proposed model is quite suitable for analyzing the mechanical behavior of the short, thick, and sandwiched beams, and also beams subjected to high-frequency excitation. The characteristics of the Timoshenko beam theory facilitate researchers to implement it for the modeling of micro- and nanoscale structures, such as microbeams [65], nanotubes [66], nanowires [64], and nanobeams [67, 68].

In this section, a brief review on the application of Timoshenko beam theory in the micro–nanoscale structures is presented, and finally, the Timoshenko beam theory has been employed for the modeling of multi-walled carbon nanotubes. Reddy and Pang in their earlier article combined the two theories of the Timoshenko and nonlocal beams to evaluate the vibration, buckling and static bending of carbon nanotubes [69]. The nonlocal Timoshenko beam theory was utilized to investigate the nonlinear free vibration of the single-walled carbon nanotubes [70]. Vibrations of DWCNTs were studied in the framework of nonlocal Timoshenko beam theory [44]. Thermal effect was considered for the vibration of SWCNTs, and then the nonlocal Timoshenko beam theory was employed to develop the governing differential equation of the system. It was shown that both the nonlocal parameter and temperature changes are influential on the properties of the transverse vibration of SWCNTs [48].

Considering the geometrical imperfection, Wang et al. studied the nonlinear vibration of embedded single-walled carbon nanotube subjected to harmonic loadings. It has been shown that the classical beam theory predicts a larger displacement for the nanostructures in comparison with the nonlocal beam theory. In accordance with the above result, it can be concluded that the small-scale and the nonlocal effects soften the nanostructures considerably [71]. Timoshenko beam theory combined with the von Karman geometric nonlinearity was used to analyze the nonlinear forced vibration of functionally graded carbon nanotube-reinforced composite beams. It was found that an increase in the nanotube volume fraction results in the reduction of amplitude peak, whereas an increase of the slenderness ratio leads to increment of the amplitude peak [72]. The nonlocal Timoshenko beam theory was utilized by Yang et al. for modeling of nanoparticle delivery in SWNCTs. It was shown that an increase in the confined stiffness and foundation stiffness will result in the decrease of the dynamic displacement of SWCNTs [73].

The efficiency and veracity of the Timoshenko beam theory for modeling of short and thick beams allures researchers to utilize it for modeling of nanobeams.

Recently, thermal buckling and free vibration of size-dependent FG nanobeams were fully studied. It was reported that the thermal buckling and vibration behavior of FG nanobeams is drastically influenced by the beam thickness, small-scale effect, and material distribution profile [74].

In order to demonstrate the nonlinear forced vibration characteristics of nanobeams, the Timoshenko beam theory was implemented. It was observed that the maximum amplitude occurs at the lower excitation frequencies in the case of postulating the surface stress effect [75]. The nonlinear flexural free vibration of nonuniform cross section and elastically supported nanobeams were studied with assuming both the surface and nonlocal effects [76]. Timoshenko beam theory was also utilized for the vibration modeling of microbeams and a comparison with other theories, such as the modified couple stress theory, were performed [77, 78].

In the final part of this section, a model is proposed for the vibration of multi-walled carbon nanotubes (MWNTs) based on the Timoshenko beam theory. The following governing equation of motion is developed for the MWNTs with N tubes; and hence, the 2 N coupled differential equation is obtained as

$$\begin{aligned}
&\rho A_1 \frac{\partial^2 w_1}{\partial t^2} - K_1 G A_1 \left(\frac{\partial \phi_1}{\partial x} - \frac{\partial^2 w_1}{\partial x^2}\right) = -c_1(w_2 - w_1) \\
&EI_1 \frac{\partial^2 \phi_1}{\partial x^2} - K_1 G A_1 \left(\phi_1 - \frac{\partial w_1}{\partial x}\right) + \rho I_1 \frac{\partial^2 \phi_1}{\partial t^2} = 0 \\
&\rho A_2 \frac{\partial^2 w_2}{\partial t^2} - K_2 G A_2 \left(\frac{\partial \phi_2}{\partial x} - \frac{\partial^2 w_2}{\partial x^2}\right) = -c_2(w_3 - w_2) + c_1(w_2 - w_1) \\
&EI_2 \frac{\partial^2 \phi_1}{\partial x^2} - K_2 G A_2 \left(\phi_2 - \frac{\partial w_2}{\partial x}\right) + \rho I_2 \frac{\partial^2 \phi_2}{\partial t^2} = 0 \\
&\quad\vdots \\
&\rho A_N \frac{\partial^2 w_N}{\partial t^2} - K_N G A_N \left(\frac{\partial \phi_N}{\partial x} - \frac{\partial^2 w_N}{\partial x^2}\right) = c_{N-1}(w_N - w_{N-1}) \\
&EI_N \frac{\partial^2 \phi_N}{\partial x^2} - K_N G A_N \left(\phi_N - \frac{\partial w_N}{\partial x}\right) + \rho I_N \frac{\partial^2 \phi_N}{\partial t^2} = 0
\end{aligned} \tag{7.3.30}$$

in which I_j, A_j, K_j and C_j are the second moment of area, cross-sectional area, shear correction factor, and the van der Waals interaction coefficients of the jth tube, respectively. The van der Waals interaction coefficients $C_j(j = 1, 2, \ldots, N-1)$ can be obtained by the following formula:

$$C_j = \frac{320 \times (2R_j)\,\mathrm{erg/cm^2}}{0.16\Delta^2}, \quad \Delta = 0.142\ \mathrm{nm} \tag{7.3.31}$$

where the center line radius of the jth tube is R_j and the carbon–carbon (C–C) bond length is Δ.

7.4 Other Types of Nonlinearities

7.4.1 Quadratic–Cubic Nonlinearity in Curved Nano–Micro Structures

Quadratic nonlinearities emerge in the vibration of nano- and microstructures due to their geometrical imperfections. One could refer to the carbon nanotubes (CNTs) as one of the most applicable nanostructures. For a better understanding of the dynamical behavior of these structures, it is essential to have a potent procedure for detailed modeling of their geometry [79, 80]. The fabrication process of CNTs is also convoluted as it is quite difficult to perfectly develop straight CNTs with a controlled geometry [81] as illustrated in Fig. 7.10 [82].

In pursuance of analyzing the motion of the curved miniature-scale systems, several analytical models, based on different beam and plate theories, have been developed [71, 83, 84]. In this section, the nonlocal-Euler–Bernoulli beam theory is

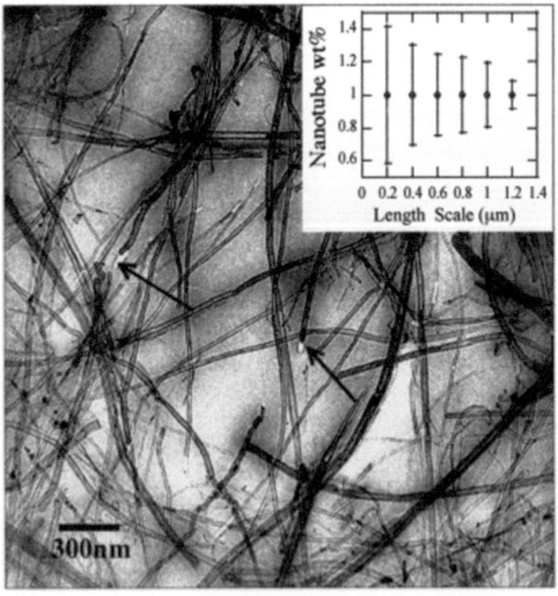

Fig. 7.10 Transmission electron microscope image showing waviness of nanotubes. *Source* Figure is adopted with permission from Ref. [82]

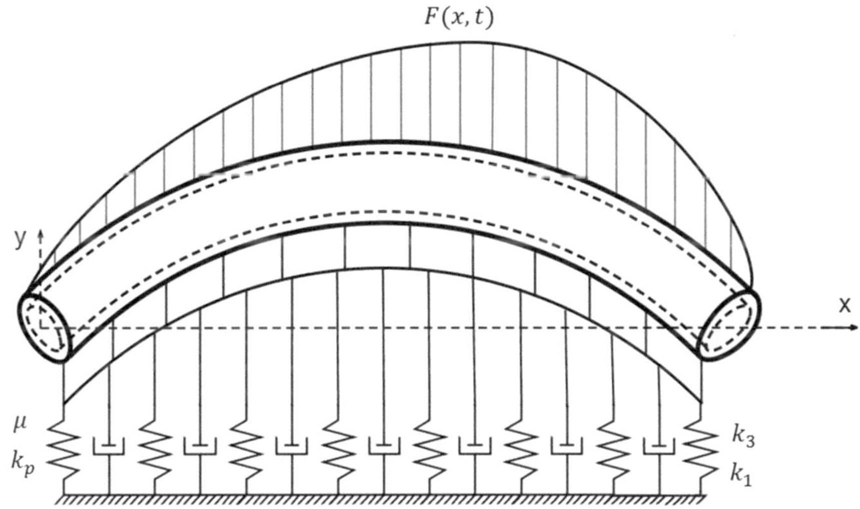

Fig. 7.11 Schematic diagram of the curved carbon nanotube. *Source* Figure is adopted with permission from Ref. [16]

utilized to exhibit the mathematical vibration model of the carbon nanotubes. It is postulated that the system is rested on the nonlinear Winkler and Pasternak foundations, as shown in Fig. 7.11.

Using the nonlocal-Euler–Bernoulli beam theory, the following integro-partial differential equation is obtained for the dynamical model of the aimed system [16]:

$$\begin{aligned}
&EI\frac{\partial^4 w}{\partial x^4} + \rho A\frac{\partial^2 w(x,t)}{\partial t^2} + \mu\frac{\partial w(x,t)}{\partial t} + k_1 w(x,t) + k_3 w^3(x,t) - k_p\frac{\partial^2 w(x,t)}{\partial x^2} - F(x,t) \\
&-\frac{EA}{L}\left(\frac{\partial^2 w}{\partial x^2} + \frac{\partial^2 w_0}{\partial x^2}\right)\int\limits_0^L\left(\frac{1}{2}\left(\frac{\partial w}{\partial x}\right)^2 + \left(\frac{\partial w}{\partial x}\right)\frac{\partial w_0}{\partial x}\right)\mathrm{d}x - (e_0^2 l_i^2)\left\{\rho A\frac{\partial^4 w(x,t)}{\partial x^2\partial t^2} + \mu\frac{\partial^3 w(x,t)}{\partial x^2\partial t}\right. \\
&+k_1\frac{\partial^2 w(x,t)}{\partial x^2} + k_3\left[6w(x,t)\left(\frac{\partial w}{\partial x}\right)^2 + 3w^2\left(\frac{\partial w}{\partial x}\right)\right] - k_p\frac{\partial^4 w(x,t)}{\partial x^4} - \frac{\partial^2 F(x,t)}{\partial x^2} \\
&\left.-\left(\frac{EA}{L}\frac{\partial^2}{\partial x^2}\left(\frac{\partial^2 w}{\partial x^2} + \frac{\partial^2 w_0}{\partial x^2}\right)\int\limits_0^L\left[\frac{1}{2}\left(\frac{\partial w}{\partial x}\right)^2 + \left(\frac{\partial w}{\partial x}\right)\frac{\partial w_0}{\partial x}\right]\mathrm{d}x\right)\right\} = 0
\end{aligned} \tag{7.4.1}$$

where the transverse vibration of the carbon nanotube is $w(x,t)$. Parameters ρ, $A, t, x, \mu, k_1, k_3, k_p, E, I, e_0 l_i, w_0$ and $F(x,t)$ are the density, area of carbon nanotube, time, axial coordinate, damping coefficient, linear Winkler foundation coefficient, nonlinear Winkler coefficient, Pasternak foundation coefficient, elasticity modulus, moment of inertia, nonlocal parameter, curvature function, and the applied force, respectively.

The following function is assumed for the vibration of the system:

$$w(x,t) = \phi_1(x)u(t) \tag{7.4.2}$$

Substituting Eq. (7.4.2) into Eq. (7.4.1) and following the steps of the Galerkin method, yields

$$\ddot{u} + \omega_1^2 u + \hat{\mu}\dot{u} + \alpha_2 u^2 + \alpha_3 u^3 = f_1 \cos\Omega t \tag{7.4.3}$$

in which

$$\omega_1^2 = \frac{\begin{array}{l} EI\int_0^L \phi''''(x)\phi(x)\,\mathrm{d}x + k_1\int_0^L \phi^2(x)\mathrm{d}x - k_p\int_0^L \phi''(x)\phi(x)\mathrm{d}x \\ \quad - k_1(e_0 l_i)^2\int_0^L \phi''(x)\phi(x)\mathrm{d}x + (e_0 l_i)^2 k_p\int_0^L \phi''''(x)\phi(x)\mathrm{d}x \\ \quad -\frac{EA}{L}\int_0^L \phi(x)\frac{\partial^2 w_0}{\partial x^2}\mathrm{d}x\int_0^L \phi'(x)\frac{\partial w_0}{\partial x}\mathrm{d}x + \frac{EA}{L}\int_0^L \phi(x)\frac{\partial^4 w_0}{\partial x^4}\mathrm{d}x\int_0^L \phi'(x)\frac{\partial w_0}{\partial x}\mathrm{d}x \end{array}}{\rho A\left(\int_0^L \phi^2(x)\mathrm{d}x - (e_0 l_i)^2\int_0^L \phi''(x)\phi(x)\mathrm{d}x\right)} \tag{7.4.4}$$

$$\hat{\mu} = \frac{\mu\left(\int_0^L \phi^2(x)\mathrm{d}x - (e_0 l_i)^2\int_0^L \phi''\phi(x)\mathrm{d}x\right)}{\rho A\left(\int_0^L \phi^2(x)\mathrm{d}x - (e_0 l_i)^2\int_0^L \phi''(x)\phi(x)\mathrm{d}x\right)} \tag{7.4.5}$$

$$\alpha_2 = \frac{\begin{array}{l} \frac{EA}{L}(e_0 l_i)^2\int_0^L \phi''''(x)\phi(x)\mathrm{d}x\int_0^L \phi'(x)\frac{\partial w_0}{\partial x} - \frac{EA}{L}\int_0^L \phi''(x)\phi(x)\mathrm{d}x\int_0^L \phi'(x)\frac{\partial w_0}{\partial x}\mathrm{d}x \\ \quad + \frac{EA}{L}(e_0 l_i)^2\frac{\partial^4 w_0}{\partial x^4}\int_0^L \frac{1}{2}(\phi')^2 - \frac{EA}{L}\frac{\partial^2 w_0}{\partial x^2}\int_0^L \frac{1}{2}(\phi')^2 \end{array}}{\rho A\left(\int_0^L \phi^2(x)\mathrm{d}x - (e_0 l_i)^2\int_0^L \phi''(x)\phi(x)\mathrm{d}x\right)} \tag{7.4.6}$$

$$\alpha_3 = \frac{\begin{array}{l} k_3\int_0^L \phi^4(x)\mathrm{d}x - \frac{EA}{L}\int_0^L \phi(x)\phi''(x)\mathrm{d}x\int_0^L \frac{1}{2}\phi'^2(x)\mathrm{d}x \\ \quad - 6k_3(e_0 l_i)^2\int_0^L \phi^2(x)\phi'^2(x)\mathrm{d}x + 3k_3(e_0 l_i)^2\int_0^L \phi^3\phi''\mathrm{d}x \\ \quad + \frac{EA}{2L}(e_0 l_i)^2\int_0^L \phi''''(x)\phi(x)\mathrm{d}x\int_0^L \phi'^2(x)\mathrm{d}x \end{array}}{\rho A\left(\int_0^L \phi^2(x)\mathrm{d}x - (e_0 l_i)^2\int_0^L \phi''(x)\phi(x)\mathrm{d}x\right)} \tag{7.4.7}$$

and

$$f_1 = \frac{F}{\rho A\left(\int_0^L \phi^2(x)\mathrm{d}x - (e_0 l_i)^2 \int_0^L \phi''(x)\phi(x)\mathrm{d}x\right)} \int_0^L \phi(x)\delta(x - x_0)\,\mathrm{d}x \qquad (7.4.8)$$

As it is evident from Eq. (7.4.3), the quadratic nonlinearity emerges in the vibration of the system due to its curvature form.

7.4.2 Rydberg–Varshni Potentials and Casimir Force

In this section, a nano-oscillator containing the Rydberg–Varshni Potentials and Casimir Force terms is considered. The Casimir force between sphere and a plate without roughness is [85]

$$F = \frac{\pi^3 \hbar c R}{360 d^3} \qquad (7.4.9)$$

in which $\hbar = \frac{h}{2\pi}$, and the parameters h, c, d and R are the Plank constant, velocity of light, distance between the sphere and the plate, and the radius of sphere, respectively.

The electrostatic force, given below, can be used for the considered system [85] as follows:

$$F = \frac{\varepsilon_0 R(2R) V^2}{2d^2} \qquad (7.4.10)$$

where ε_0 and V are the permittivity of the vacuum and the applied voltage, respectively.

The following exciting force can be assumed as a sinusoidal function [85]

$$F(t) = f_0 \sin \omega t \qquad (7.4.11)$$

where f_0 and ω are the amplitude of the applied force and the angular frequency of excitation. Based on Newton's law, the motion of a sphere is expressed by the following equation:

$$m\frac{\mathrm{d}^2 x}{\mathrm{d}t^2} + \lambda\frac{\mathrm{d}x}{\mathrm{d}t} + f_p - \frac{H}{(d_e + x)^3} - \frac{\varepsilon_0 R^2 V^2}{(d_e + x)^2} = F(t) \qquad (7.4.12)$$

Definitions of λ, f_p, d_e and x are given in the referenced article [86] and, the expression for the Rydberg potential is formulated as

$$V_R(x) = -D_e(1+bx)\exp(-bx) \tag{7.4.13}$$

where D_e and b are the dissociation energy and the anharmonicity constant, respectively.

The Varshni potential can be mathematically modeled by the following equation [87]:

$$V_v(x) = D_e\left[1 - \frac{d_e}{x+d_e}\exp(-x^2 - 2xd_e)\beta\right]^2 \tag{7.4.14}$$

in which β is a parameter that can be determined from $\beta = \frac{1}{2(\sqrt{\Delta}-1)d_e^2}$ where k_e is the force constant and $\Delta = \frac{k_e d_e^2}{2D_e}$.

Substituting Eqs. (7.4.13) and (7.4.14) into Eq. (7.4.12), the following equation of motion is obtained [85]:

$$\begin{aligned} & m\frac{\mathrm{d}^2x}{\mathrm{d}t^2} + \lambda\frac{\mathrm{d}x}{\mathrm{d}t} + 2D_e d_e \exp\left(-\beta x^2 - 2\beta x d_e\right)\left[(x+d_e)^{-2} + 2\beta\right] \\ & - 2D_e d_e^2 \exp\left(-2\beta x^2 - 4\beta x d_e\right)\left[(x+d_e)^{-3} + 2\beta(x+d_e)^{-1}\right] \\ & - \frac{H}{(d_e+x)^3} - \frac{\varepsilon_0 R^2 V^2}{(d_e+x)^2} = F(t) \end{aligned} \tag{7.4.15}$$

7.4.3 Non-natural Oscillations

Those dynamical systems in which, the kinetic energy is not a quadratic function of velocity is commonly known as the non-natural systems. Recently, different types of non-natural systems are studied and detailed information about their dynamical behavior is presented [88]. This type of vibration can become more significant when the motion of the nano–microsystems requires detailed investigation.

The vibration of a clamped–clamped microbeam, which is electrically actuated, is shown in Fig. 7.12, and the derivation of its governing equation of motion is presented in this chapter. This type of system is categorized as a non-natural system [89, 90].

The governing equation of the transverse deflection of the clamped–clamped microbeam is

$$\bar{E}I\frac{\partial^4 w}{\partial x^4} + \rho S\frac{\partial^2 w}{\partial t^2} = \left[\bar{N} + \frac{ES}{2l}\int_0^{2l}\left(\frac{\partial w}{\partial x}\right)^2\right]\frac{\partial^2 w}{\partial x^2} + q(x,t) \tag{7.4.16}$$

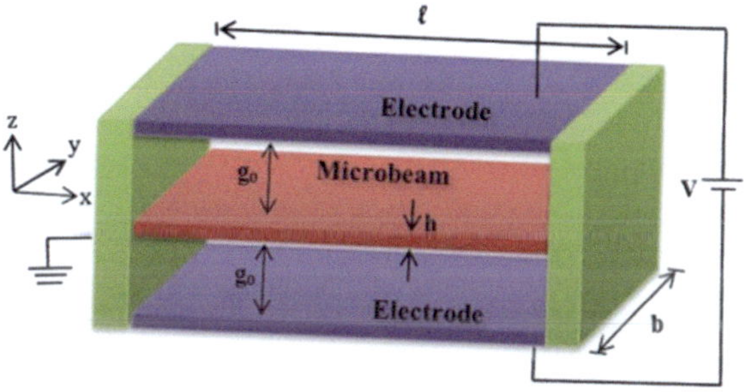

Fig. 7.12 Schematic diagram of the clamped–clamped microbeam [91]. *Source* Figure is adopted with permission from Ref. [91]

in which $\bar{E}$, E, l, h, and I are the effective elasticity modulus, Young's modulus of elasticity, beam length, width band thickness, and the moment of inertia, respectively. Parameters $\bar{N}$, S, and w are the respective tensile or compressive axial loads, area, and the transverse deflection of microbeam, and $q(x, t)$ corresponds to the force per unit length pertinent to the electrostatic excitation.

$$q(x,t) = \frac{\varepsilon_v b v^2}{2}\left[\frac{1}{(g_0 - w)^2} - \frac{1}{(g_0 + w)^2}\right] \tag{7.4.17}$$

where ε_v, v, and g_0 are the gap dielectric constant, electrostatic load, and the initial gap, respectively. Deriving the governing equation of motion into the nondimensional form and considering the clamped–clamped boundary conditions, the following form of the nonlinear ordinary differential equation is then obtained:

$$\ddot{u}(\alpha_1 u^4 + \alpha_2 u^2 + \alpha_3) + \alpha_4 u^3 + \alpha_5 u^5 + \alpha_7 u^7 = 0 \tag{7.4.18}$$

The nondimensional variables, which have been utilized, are given below:

$$\xi = x/l;\ W = \frac{w}{g_0};\ \tau = t\sqrt{\frac{\bar{E}I}{\rho b h I^4}};\ \alpha = 6\left(\frac{g_0}{h}\right)^2;\ N = \frac{\bar{N}l^2}{\bar{E}I};\quad \text{and } V^2 = \frac{24\varepsilon_v l^4 v^2}{\bar{E}h^3 g_0^3} \tag{7.4.19}$$

The Galerkin procedure is applied to develop Eq. (7.4.18), and the expression below is postulated as follows:

$$w(\xi) = u(\tau)\phi(\xi) \tag{7.4.20}$$

Accordingly, the trial function $\phi(\xi) = 16\xi^2(1-\xi)$ should be assumed.

Practice Problems

P.7.1. Develop the governing equation of vibration of a nanobeam by applying the Euler–Bernoulli and the nonlocal-Euler–Bernoulli theories. Assume that the nanobeam is rested on a nonlinear Winkler and Pasternak foundation and consider the nanobeam to undergo a periodic external force.

P.7.1.a. Use Galerkin method to find the nonlinear ordinary differential equations of the primary mode of vibration.
P.7.1.b. Determine the primary resonance of the system using the multiple time-scales method.
P.7.1.c. Examine the effect of nonlocal parameter on the primary resonance of the nanobeam.
P.7.1.d. Repeat all above for the vibration of a nanobeam using Timoshenko beam theory.

P.7.2. Nonlinear vibration of electrostatically driven nanowires can be mathematically described by the following coupled partial differential equations [92]:

$$\rho A Z_{tt} + EIZ_{xx} - \frac{EA}{2L} Z_{xx} \int_0^L \left(Y_x^2 + Z_x^2\right) \mathrm{d}x = F_f^z + F_e$$

$$\rho A Z_{tt} + EIY_{xx} - \frac{EA}{2L} Y_{xx} \int_0^L \left(Y_x^2 + Z_x^2\right) \mathrm{d}x = F_f^Y$$

where F_e is the electrostatic force on the nanowire. Also, $Y(x,t)$, $Z(x,t)$, F_t^Y, and F_t^z are the two displacements and the two viscous damping forces in the Y and Z directions, respectively [92].

P.7.2.a. Develop the nonlinear ordinary differential equations of nanowire, which is doubly clamped in the Z direction. (Hint: Apply Galerkin method based on the considered boundary conditions).
P.7.2.b. Examine the nonlinear behavior of differential equations using multiple time-scales method.
P.7.2.c. Consider a waviness for the nanowire (curved nanowire), then develop the governing equation of motion of the system using Euler–Bernoulli beam theory.
P.7.2.d. Investigate the possibility of internal resonance for the vibration of curved nanowires based on the equations obtained.
P.7.2.e. Apply a periodic force on the nanowire. Study the nonlinear behavior of the system using the multiple time-scales method.

P.7.3. Derive the equation of nonlinear vibration of electrostatically driven nanowires considering the displacements in the Y and Z direction and using nonlocal-Euler–Bernoulli beam theory.

Fig. 7.13 A suspended fabricated device [93]. Figure is adopted with permission from Ref. [93]

P.7.3.a. Consider clamped–clamped boundary conditions for the system and derive its nonlinear ordinary differential equations considering the primary mode of vibration.

P.7.4. Figure 7.13 shown is the SEM image (colorized), representing a suspended instrument fabricated in a multilayer lithographical surface nanomachining process [93].

The following differential equation of motion describes the vibration of the system:

$$m\ddot{x} + \frac{m\omega_0}{Q}\dot{x} + (k_1 + k_p(\cos(2\omega_0 t))x + k_3 x^3 = F_0\cos(\omega_0 t + \phi)$$

P.7.4.a. Study the nonlinear behavior of the system described by the nonlinear differential equation.
P.7.4.b. Investigate the possibility of chaos for this system.

P.7.5. Vibration of a nanobeam with the surface effect is described by the following equation [94]:

$$m\ddot{w} + cw + \left(\frac{Ebh^3}{12} + \frac{E^s h^3}{6} + \frac{E^s bh^2}{2}\right)(w''''(1 + w'^2) + 4w'w''w''' + w''^3)$$
$$- w'\int_s^L\int_s^L m(\dot{w}' + w'\ddot{w}'')\mathrm{d}s - H^s w''\left(1 + \frac{1}{2}w'^2\right) = 0$$

where m, c, and $w(s,t)$ are the respective mass per unit length, damping coefficient, and the lateral displacement of nanobeam.

P.7.5.a. Develop the nonlinear ordinary differential equation of the nanobeam considering principal modes.
P.7.5.b. Consider a periodic excitation force on the system as $F = D\cos\omega t$ and study the primary resonance case for the vibration of nanobeam.

P.7.5.c. In the equation of nonlinear vibration of nanobeam, $H^s = 2\tau_0 b$ in which, τ_0 being the residual surface stress and b is the width. Examine the effect of both positive and negative residual stresses on the natural frequency of the nanobeam.

P.7.6. Following differential equation represents the effect of thermal parameter on the sensitivity of carbon nanotube resonators for a mass sensing device [95]:

$$\left(\rho A + \frac{\Delta m}{L}\right)\frac{\partial^2 w}{\partial t^2} + EI\frac{\partial^4 w}{\partial x^4} - \left[\frac{EA}{2L}\int_0^L \left(\frac{\partial w}{\partial x}\right)^2 \mathrm{d}x + N_T\right]\frac{\partial^2 w}{\partial x^2} = f(x,t)$$

P.7.6.a. Develop the nondimensional equation of motion of the system considering principal modes of vibration.
P.7.6.b. Study the effect of thermal parameter on the sensitivity of carbon nanotube resonators (CNTs) for the mass detection.
P.7.6.c. Consider the first two vibration modes to study the primary and the secondary resonances of CNTs.

References

1. Cho H, Yu MF, Vakakis AF, Bergman LA, McFarland DM (2010) Tunable, broadband nonlinear nanomechanical resonator. Nano Lett 10(5):1793–1798
2. Dai MD, Eom K, Kim CW (2009) Nanomechanical mass detection using nonlinear oscillations. Appl Phys Lett 95(20):203104
3. Buks E, Yurke B (2006) Mass detection with a nonlinear nanomechanical resonator. Phys Rev E (Stat Nonlinear Soft Matter Phys) 74(4):046619
4. Chen H-J, Zhu K-D (2013) Ultrasensitive mass sensing with nonlinear optics in a doubly clamped suspended carbon nanotube resonator. J Appl Phys 114(21):213101
5. Natsuki T, Shi JX, Ni QQ (2013) Vibration analysis of nanomechanical mass sensor using double-layered graphene sheets resonators. J Appl Phys 114(9):094307
6. Eom K, Park HS, Yoon DS, Kwon T (2011) Nanomechanical resonators and their applications in biological/chemical detection: nanomechanics principles. Phys Rep 503(4–5):115–163
7. Cimalla V, Niebelschütz F, Tonisch K, Foerster Ch, Brueckner K, Cimalla I, Friedrich T, Pezoldt J, Stephan R, Hein M, Ambacher O (2007) Nanoelectromechanical devices for sensing applications. Sens Actuators B 126(1):24–34
8. Venstra WJ, Capener MJ, Elliott SR (2014) Nanomechanical gas sensing with nonlinear resonant cantilevers. Nanotechnology 25(42):425501
9. Ilic B, Czaplewski D, Zalalutdinov M, Craighead HG (2001) Single cell detection with micromechanical oscillators. J Vac Sci Technol B Nanotechnol Microelectron 19(6):2825–2828
10. Varshney M, Waggoner PS, Tan CP, Aubin K, Montagna RA, Craighead HG (2008) Prion protein detection using nanomechanical resonator arrays and secondary mass labeling. Anal Chem 80(6):2141–2148

11. Kwon T, Park J, Yang J, Yoon DS, Na S, Kim CW, Suh JS, Huh YM, Haam S, Eom K (2009) Nanomechanical in situ monitoring of proteolysis of peptide by cathepsin B. PLoS ONE 4(7): e6248
12. Caygill RL, Blair GE, Millner PA (2010) A review on viral biosensors to detect human pathogens. Anal Chim Acta 681(1–2):8–15
13. Shaat M, Abdelkefi A (2016) Modeling of mechanical resonators used for nanocrystalline materials characterization and disease diagnosis of HIVs. Microsys Technol, 2015 22(2):305–318
14. Ilic B, Yang Y, Craighead HG (2004) Virus detection using nanoelectromechanical devices. Appl Phys Lett 85(13):2604–2606
15. Dai MD, Kim CW, Eom K (2012) Nonlinear vibration behavior of graphene resonators and their applications in sensitive mass detection. Nanoscale Res Lett 7(499):1–10
16. Askari H (2014) Nonlinear vibration chaotic motion of uniform and non-uniform carbon nanotube resonators. University of Ontario Institute of Technology, Oshawa, Ontario, Canada
17. Pokorný J (1999) Conditions for coherent vibrations in the cytoskeleton. Bioelectrochem Bioenerg 48(2):267–271
18. Zeverdejani MK, Beni YT (2013) The nano scale vibration of protein microtubules based on modified strain gradient theory. Curr Appl Phys 13(8):1566–1576
19. Sirenko YM, Stroscio MA, Kim KW (1996) Elastic vibrations of microtubules in a fluid. Phys Rev E Stat Phys Plasmas Fluids Relat Interdiscip Topics 53(1 Suppl. B):1003–1010
20. Pokorný J, Jelínek F, Trkal V, Lamprecht I, Hölzel R (1997) Vibrations in microtubules. J Biol Phys 23(3):171–179
21. Foster KR, Baish JW (2000) Viscous damping of vibrations in microtubules. J Biol Phys 26 (4):255–260
22. Wang CY, Ru CQ, Mioduchowski A (2006) Vibration of microtubules as orthotropic elastic shells. Physica E 35(1):48–56
23. Shi YJ, Guo WL, Ru CQ (2008) Relevance of Timoshenko-beam model to microtubules of low shear modulus. Physica E 41(2):213–219
24. Heireche H, Tounsi A, Benhassaini H, Benzair A, Bendahmane M, Missouri M, Mokadem S (2010) Nonlocal elasticity effect on vibration characteristics of protein microtubules. Physica E 42(9):2375–2379
25. Shen HS (2011) Nonlinear vibration of microtubules in living cells. Curr Appl Phys 11 (3):812–821
26. Taj M, Zhang JQ (2012) Analysis of vibrational behaviors of microtubules embedded within elastic medium by Pasternak model. Biochem Biophys Res Commun 424(1):89–93
27. Endo D, Yabuno H, Higashino K, Yamamoto Y, Matsumoto S (2015) Self-excited coupled-microcantilevers for mass sensing. Appl Phys Lett 106(22):223105
28. Yabuno H, Higashino K, Kuroda M, Yamamoto Y (2014) Self-excited vibrational viscometer for high-viscosity sensing. J Appl Phys 116(12):124305
29. Li C, Zhao Y, Cheng R (2014) A micro resonant acceleration sensor comprising silicon support with temperature isolator and quartz doubled ended tuning fork. In: 9th IEEE international conference on Nano/Micro Engineered and Molecular Systems, IEEE-NEMS
30. Kasyap A, Phipps A, Nishida T, Sheplak M, Cattafesta L (2011) Development of MEMS-based piezoelectric vibration energy harvesters. Structural dynamics and renewable energy. In: Proceedings of the Society for Experimental Mechanics Series, vol 1, pp 1–38
31. Andò B, Baglio S, Trigona C, Dumas N, Latorre L, Nouet P (2010) Nonlinear mechanism in MEMS devices for energy harvesting applications. J Micromech Microeng 20(12):5020
32. Kacem N, Hentz S, Pinto D, Reig B, Nguyen V (2009) Nonlinear dynamics of nanomechanical beam resonators: improving the performance of NEMS-based sensors. Nanotechnology 20(27):5501
33. Kacem N, Hentz S, Baguet S, Dufour R (2011) Forced large amplitude periodic vibrations of non-linear Mathieu resonators for microgyroscope applications. Int J Non-Linear Mech 46 (10):1347–1355

34. Rasekh M, Khadem SE (2013) Design and performance analysis of a nanogyroscope based on electrostatic actuation and capacitive sensing. J Sound Vib 332(23):6155–6168
35. Ghommem M, Nayfeh AH, Choura S (2013) Model reduction and analysis of a vibrating beam microgyroscope. J Vib Control 19(8):1240–1249
36. Su J, Xiao D, Wang X, Chen Z, Wu X (2014) Vibration sensitivity analysis of the 'Butterfly-gyro' structure. Microsyst Technol 20(7):1281–1290
37. Wang H, Dong P, Xie L, Wu X (2014) Vibration characteristic analysis method for the quartz microgyroscope based on the admittance circle. AIP Adv 4(3):031318
38. Li C, Chou T-W (2004) Vibrational behaviors of multiwalled-carbon-nanotube-based nanomechanical resonators. Appl Phys Lett 84(1):121–123
39. Zhang Y, Liu G, Han X (2005) Transverse vibrations of double-walled carbon nanotubes under compressive axial load. Phys Lett A 340(1–4):258–266
40. Zhang YQ, Liu GR, Xie XY (2005) Free transverse vibrations of double-walled carbon nanotubes using a theory of nonlocal elasticity. Phys Rev B Condens Matter Mater Phys 71 (19):195404
41. Xu KY, Guo XN, Ru CQ (2006) Vibration of a double-walled carbon nanotube aroused by nonlinear intertube van der Waals forces. J Appl Phys 99(6):064303
42. Fu YM, Hong JW, Wang XQ (2006) Analysis of nonlinear vibration for embedded carbon nanotubes. J Sound Vib 296(4–5):746–756
43. He XQ, Eisenberger M, Liew KM (2006) The effect of van der Waals interaction modeling on the vibration characteristics of multiwalled carbon nanotubes. J Appl Phys 100(12):124317
44. Ke LL, Xiang Y, Yang J, Kitipornchai S (2009) Nonlinear free vibration of embedded double-walled carbon nanotubes based on nonlocal Timoshenko beam theory. Comput Mater Sci 47(2):409–417
45. Cigeroglu E, Samandari H (2014) Nonlinear free vibrations of curved double walled carbon nanotubes using differential quadrature method. Physica E 64:95–105
46. Gibson RF, Ayorinde EO, Wen Y-F (2007) Vibrations of carbon nanotubes and their composites: a review. Compos Sci Technol 67(1):1–28
47. Hsu JC, Chang RP, Chang WJ (2008) Resonance frequency of chiral single-walled carbon nanotubes using Timoshenko beam theory. Phys Lett Sect A Gen At Solid State Phys 372 (16):2757–2759
48. Benzair A, Tounsi A, Besseghier A, Heireche H, Moulay N, Boumia L (2008) The thermal effect on vibration of single-walled carbon nanotubes using nonlocal Timoshenko beam theory. J Phys D Appl Phys 41(22):225404
49. Yoon J, Ru CQ, Mioduchowski A (2005) Vibration and instability of carbon nanotubes conveying fluid. Compos Sci Technol 65(9 Spec. Iss.):1326–1336
50. Kiani K, Wang Q (2012) On the interaction of a single-walled carbon nanotube with a moving nanoparticle using nonlocal Rayleigh, Timoshenko, and higher-order beam theories. Eur J Mech A Solids 31(1):179–202
51. Kiani K (2013) Vibration behavior of simply supported inclined single-walled carbon nanotubes conveying viscous fluids flow using nonlocal Rayleigh beam model. Appl Math Model 37(4):1836–1850
52. Torkaman-Asadi MA, Rahmanian M, Firouz-Abadi RD (2015) Free vibrations and stability of high-speed rotating carbon nanotubes partially resting on Winkler foundations. Compos Struct 126:52–61
53. Lee HL, Chang WJ (2009) Vibration analysis of a viscous-fluid-conveying single-walled carbon nanotube embedded in an elastic medium. Physica E 41(4):529–532
54. Daneshmand F, Rafiei M, Mohebpour SR, Heshmati M (2013) Stress and strain-inertia gradient elasticity in free vibration analysis of single walled carbon nanotubes with first order shear deformation shell theory. Appl Math Model 37(16–17):7983–8003
55. Leung AYT, Guo X, He XQ, Kitipornchai S (2005) A continuum model for zigzag single-walled carbon nanotubes. Appl Phys Lett 86(8):083110
56. Zhang YQ, Pang M, Chen WQ (2015) Transverse vibrations of embedded nanowires under axial compression with high-order surface stress effects. Physica E 66:238–244

57. Wang GF, Feng XQ (2009) Timoshenko beam model for buckling and vibration of nanowires with surface effects. J Phys D Appl Phys 42(15):155411
58. Gang-Feng W, Xi-Qiao F (2010) Effect of surface stresses on the vibration and buckling of piezoelectric nanowires. EPL (Europhys Lett) 91(5):56007
59. Oveisi A (2014) Free vibration of piezo-nanowires using timoshenko beam theory with consideration of surface and small scale effects. Arch Mech Eng 61(1):139–152
60. Kiani K (2010) Free longitudinal vibration of tapered nanowires in the context of nonlocal continuum theory via a perturbation technique. Physica E 43(1):387–397
61. Kiani K (2015) Vibrations and instability of pretensioned current-carrying nanowires acted upon by a suddenly applied three-dimensional magnetic field. Mater Chem Phys 162(1):531–541
62. Kiani K (2012) Magneto-elasto-dynamic analysis of an elastically confined conducting nanowire due to an axial magnetic shock. Phys Lett A 376(20):1679–1685
63. Kiani K (2014) Forced vibrations of a current-carrying nanowire in a longitudinal magnetic field accounting for both surface energy and size effects. Physica E 63(1):27–35
64. Askari H, Esmailzadeh E, Zhang D (2014) Nonlinear vibration analysis of nonlocal nanowires. Compos B Eng 67:607–613
65. Asghari M, Kahrobaiyan MH, Nikfar M, Ahmadian MT (2012) A size-dependent nonlinear Timoshenko microbeam model based on the strain gradient theory. Acta Mech 223(6):1233–1249
66. Wang CM, Tan VBC, Zhang YY (2006) Timoshenko beam model for vibration analysis of multi-walled carbon nanotubes. J Sound Vib 294(4):1060–1072
67. Şimşek M, Yurtcu HH (2013) Analytical solutions for bending and buckling of functionally graded nanobeams based on the nonlocal Timoshenko beam theory. Compos Struct 97:378–386
68. Behera L, Chakraverty S (2014) Free vibration of Euler and Timoshenko nanobeams using boundary characteristic orthogonal polynomials. Appl Nanosci 4(3):347–358
69. Reddy JN, Pang SD (2008) Nonlocal continuum theories of beams for the analysis of carbon nanotubes. J Appl Phys 103(2):023511
70. Yang J, Ke LL, Kitipornchai S (2010) Nonlinear free vibration of single-walled carbon nanotubes using nonlocal Timoshenko beam theory. Physica E 42(5):1727–1735
71. Wang B, Deng ZC, Zhang K (2013) Nonlinear vibration of embedded single-walled carbon nanotube with geometrical imperfection under harmonic load based on nonlocal Timoshenko beam theory. Appl Math Mech 34(3):269–280
72. Ansari R, Shojaei MF, Mohammadi V (2014) Nonlinear forced vibration analysis of functionally graded carbon nanotube-reinforced composite Timoshenko beams. Compos Struct 113(1):316–327
73. Yang YC, Chang WJ, Lee HL (2013) Vibration behavior of nanoparticle delivery in a single-walled carbon nanotube using nonlocal Timoshenko beam theory. J Comput Theor Nanosci 10(6):1472–1476
74. Ebrahimi F, Salari E (2015) Thermal buckling and free vibration analysis of size dependent Timoshenko FG nanobeams in thermal environments. Compos Struct 128:363–380
75. Ansari R, Mohammadi V, Shojaei MF, Gholami R, Sahmani S (2014) On the forced vibration analysis of Timoshenko nanobeams based on the surface stress elasticity theory. Compos B Eng 60:158–166
76. Malekzadeh P, Shojaee M (2013) Surface and nonlocal effects on the nonlinear free vibration of non-uniform nanobeams. Compos B Eng 52:84–92
77. Farokhi H, Ghayesh MH (2015) On the nonlinear resonant dynamics of Timoshenko microbeams. Meccanica 51(1):155–169
78. Farokhi H, Ghayesh MH, Kosasih B, Akaber P (2015) Thermo-mechanical dynamics of perfect and imperfect Timoshenko microbeams. Int J Eng Sci 91:12–33
79. Askari H, Esmailzadeh E, Barari A (2015) A unified approach for nonlinear vibration analysis of curved structures using non-uniform rational B-spline representation. J Sound Vib 353:292–307

80. Askari H, Barari A, Esmailzadeh E (2013) Analysis of nonlinear oscillation of circular curved carbon nanotube. In: Proceedings of IEEE conference on nanotechnology, pp 374–378
81. Amjadipour M, Dao DV, Motta N (2015) Vibration analysis of initially curved single walled carbon nanotube with vacancy defect for ultrahigh frequency nanoresonators. Microsyst Technol 22:1115–1120
82. Qian D, Dickey EC, Andrews R, Rantell T (2000) Load transfer and deformation mechanisms in carbon nanotube-polystyrene composites. Appl Phys Lett 76(20):2868–2870
83. Yan JW, Liew KM, He LH (2013) Free vibration analysis of single-walled carbon nanotubes using a higher-order gradient theory. J Sound Vib 332(15):3740–3755
84. Ouakad HM, Younis MI (2011) Natural frequencies and mode shapes of initially curved carbon nanotube resonators under electric excitation. J Sound Vib 330(13):3182–3195
85. Kermani MM, Dehestani M (2013) Solving the nonlinear equations for one-dimensional nano-sized model including Rydberg and Varshni potentials and Casimir force using the decomposition method. Appl Math Model 37(5):3399–3406
86. Sheehan WF (1965) Rydberg potential energy function for diatomic molecules as extended to polyatomic species and activated complex. J Phys Chem 69(3):923–927
87. Bag M, Dutt R (1990) Shifted 1/N expansion for the energy levels of the Varshni potential. J Phys B At Mol Opt Phys 23(18):3075–3082
88. Askari H, Younesian D, Saadatnia Z (2015) Nonlinear oscillations analysis of the elevator cable in a drum drive elevator system. Adv Appl Math Mech 7(01):43–57
89. Qian YH, Ren DX, Lai SK, Chen SM (2012) Analytical approximations to nonlinear vibration of an electrostatically actuated microbeam. Commun Nonlinear Sci Numer Simul 17 (4):1947–1955
90. Fu Y, Zhang J, Wan L (2011) Application of the energy balance method to a nonlinear oscillator arising in the microelectromechanical system (MEMS). Curr Appl Phys 11(3):482–485
91. Younesian D, Sadri M, Esmailzadeh E (2014) Primary and secondary resonance analyses of clamped-clamped micro-beams. Nonlinear Dyn 76(4):1867–1884
92. Chen Q, Huang L, Lai YC, Grebogi C, Dietz D (2010) Extensively chaotic motion in electrostatically driven nanowires and applications. Nano Lett 10(2):406–413
93. Karabalin RB, Feng XL, Roukes ML (2009) Parametric nanomechanical amplification at very high frequency. Nano Lett 9(9):3116–3123
94. Askari H, Esmailzadeh E, Younesian D (2014) Dynamic behavior of carbon nanotubes using nonlocal Rayleigh beam. In: Proceedings of the ASME Design Engineering Technical Conference, vol 8, p V008T11A031. https://doi.org/10.1115/detc2014-35420
95. Kang D-K, Yang H-I, Kim C-W (2015) Thermal effects on mass detection sensitivity of carbon nanotube resonators in nonlinear oscillation regime. Physica E 74:39–44

MIX
Papier aus verantwortungsvollen Quellen
Paper from responsible sources
FSC® C105338

FSC
www.fsc.org

If you have any concerns about our products,
you can contact us on
ProductSafety@springernature.com

In case Publisher is established outside the EU,
the EU authorized representative is:
Springer Nature Customer Service Center GmbH
Europaplatz 3, 69115 Heidelberg, Germany

Printed by Libri Plureos GmbH
in Hamburg, Germany